De Volson Wood

The elements of coördinate geometry

De Volson Wood

The elements of coördinate geometry

ISBN/EAN: 9783337276713

Printed in Europe, USA, Canada, Australia, Japan

Cover: Foto ©berggeist007 / pixelio.de

More available books at **www.hansebooks.com**

THE

ELEMENTS

OF

COÖRDINATE GEOMETRY,

IN

THREE PARTS.

I. CARTESIAN GEOMETRY, II. QUATERNIONS,
III. MODERN GEOMETRY,

AND AN

APPENDIX

BY

DE VOLSON WOOD,
PROFESSOR MATHEMATICS AND MECHANICS IN STEVENS INSTITUTE OF
TECHNOLOGY.

SEVENTH EDITION.

THIRD THOUSAND.

NEW YORK:
JOHN WILEY AND SONS,
53 EAST TENTH STREET,
1895.

PREFACE.

This book contains some matter not heretofore found in works upon Analytical Geometry. As it is designed as a text-book, care has been taken to separate the different subjects so that they may be studied advantageously, each by itself. The Cartesian system will naturally, if not necessarily, be studied first, for it is not only the most common, but is the leading system used in the Calculus. The matter pertaining to the conic sections is considerably condensed, compared with most other works which treat of the subject. This has been accomplished by treating of the several curves under one head when discussing a property which is common to all of them. By this arrangement we trust that some time will be saved to the student in this part of the work, and thus enable him to give more time to advanced portions of the subject.

The subject of Quaternions is treated in the most elementary manner, and the examples are of the simplest kind, the object being to explain and illustrate the principles and the character of the operations without taxing the ingenuity of the student in the mere solution of problems. One cannot form a correct judgment of the power of this analysis from these examples, but to attempt to explain its higher processes would be equivalent to excluding it from our courses of study. If the presentation here made of the subject succeeds in creating an interest in it, and of establishing a correct foundation for its future study, all will be accomplished that was intended. The English works upon the subject are not numerous. The only ones known to the author are, — Hamilton's *Lectures upon Quaternions*, Hamilton's *Elements of Quaternions*, Tait's *Quaternions*, and Kelland & Tait's *Introduction to Quaternions*, the last of which is probably the best elementary work upon the subject hitherto published. There is, however, considerable literature upon the subject in English and other foreign journals, some of which is contained under the head of discussions upon more general systems of algebra. It is with pleasure that I acknowledge my indebtedness to my friend and former pupil, Mr. Henry A. Beckmeyer, for valuable assistance in the preparation of this part of the work.

The third part of the work, on Modern Geometry, is very brief, and is intended chiefly to expand still further the ideas of the student, by showing that a great variety of systems may be used. Still, scarcely enough is given to enable him to raise the veil and witness the scenes beyond. For this he must consult other works. The amount of literature upon this subject, mostly stored in foreign mathematical journals, is immense; and those only who give their time to its study can, in any sense, become masters of it. Of the English works upon the subject, we notice *Trilinear Coördinates and Methods of Modern Analytical Geometry*, by W. A. Whitworth, Salmon's *Conic Sections*, *Trilinear Coördinates* by W. J. Wright (being No. 2 of his Mathematical Tracts), *Trilinear Coördinates*, by N. M. Ferrers. There are other English, and also several French and German works which we have not been able to consult, which may be as meritorious and possibly more extensive than those above mentioned.

Several of the cuts of Higher Loci were taken—by permission of the authors—from Rice and Johnson's Elements of the Differential Calculus.   De V. W.

Hoboken, *March*, 1879.

CONTENTS.

[The numbers of the Articles are placed at the head of the page.]

PART I.—CARTESIAN GEOMETRY.

CHAPTER I.
DEFINITIONS—EQUATIONS TO A POINT.

Definitions : — Lines, when known (1); System of coördinates (2); Known point (3); Right line (4); Illustrations (5); Coördinate axes (6); Kinds of coördinates (7); The ordinate (8); Abscissa (9); Coördinates to a point (10); Equations to a point (11); EXAMPLES. *Polar System :*—Initial line (12); The pole (13); Radius vector (14); Variable angle (15); Signs of the coördinates (16); Polar coördinates of a point (17); Polar equations of a point (18); EXAMPLES. *Distance between two Points :*—Bilinear coördinates (19); Rectangular coördinates (20); Polar coördinates (21); EXAMPLES .... Pages 1-26

CHAPTER II.
THE RIGHT LINE.

Definitions :—Locus (22); Consecutive points (23); Equation of a locus (24); Analytical Geometry (25); Coördinate Geometry (26). *Equations to a Right Line :*—Bilinear equation (27); Rectangular equation (28); Intercepts (29); In terms of perpendicular (30); Constants and variables (31); Absolute term (32); Discussion (33-35); Polar equation (36); EXAMPLES. *Point and Line* (38-40); EXAMPLES. *Of two Lines :*—Intersection of (41); Parallel to (42); Line through a point (43); Angle between (44); Perpendicular to given line (45, 46); Line cutting another at a given angle (47); EXAMPLES.

CHAPTER III.
TRANSFORMATION OF COÖRDINATES.

Formulas for passing from one bilinear system to another (49-51); Rectangular to oblique coördinates (52); Oblique to rectangular (53); Rectangular to rectangular (54); Rectangular to polar (55); EXAMPLES...Pages 27-33

CHAPTER IV.

CONIC SECTIONS.

The Circle :—Definition (57-59); Discussion (60); Construction of the locus (61); Discussion (62); Definition of centre (63); Diameter and axis (64); EXAMPLES. *The Ellipse :*—Definition (65); To trace the curve (66); Definitions (67); Equations of the curve (68-72); EXAMPLES. *The Hyperbola :*—Definition (73); Construction (74, 75); Definitions (76); A branch (77); Equations of the curve (78-81); EXAMPLES. *The Parabola :*—Definition (82); To construct the curve (83), (85); Definitions (84); Equations of (86-88); EXAMPLES. *Equations to the Conic compared* (90, 91); General equation of the second degree (92); EXAMPLES; Eccentricity (93); Eccentric angle (93a); Latus rectum (94, 95); Remark (96); EXAMPLES. *Of Ordinates* (97): Intersections of a right line with conics and conics with each other (98); EXAMPLES. *Of Tangents*: —To the ellipse (100); Circle (101); Hyperbola (102); Parabola (103); Intercepts of (104); EXAMPLES; Eccentric angle (104a); Subtangents (105); Length of tangent (106); EXAMPLES; Angles between focal radii and the tangent (107); Construction of a tangent to a conic (108); *Normal and Subnormal to Conics* (109-113); In terms of eccentric angle (113a); Intercepts of the normal (114); Length of subnormal (114a); Length of normal (115); Distance of foot of normal from either focus (116); The normal bisects the angle between the focal radii (117); Construction of the normal (118); EXAMPLES; Linear equation of the conic (119); *Boscovich's Definition* (120); To construct the hyperbola (121); Supplementary chords (122, 123); Conjugate diameters (124, 125); To construct a tangent to a conic by means of conjugate diameters (126); *Oblique axes* (127); Conjugate diameters (128, 129); Equation of the tangent referred to conjugate diameters (130); Properties of conjugate diameters (131); Parabola referred to oblique axes (132, 133); Interpretation of $\frac{1}{2}\frac{p}{\sin^2\beta}$ (134, 135); Double focal ordinate of the parabola (136); *Construction of the Parabola* (137-142); To construct a tangent to a parabola parallel to a line (143); To find the axis of a parabola (144); *Parameters* (145, 146); *Pole and Polar :* — Definitions (147-149); Equations of (150-153); Polars of special points (154, 155); EXAMPLES. *The Hyperbola and its Asymptotes :*—Definition (156); Equations of the asymptotes (157); Equations of conditions for asymptotes (158); Equation of hyperbola referred to its asymptotes (159); $\varphi = 2 \sec^{-1} e$ (160); Problem (161); Equation to any chord (162); Equation to the tangent (163); Intercepts of tangent (164); Intercepts equal (165); Tangents at the extremity of conjugate diameters meet on the asymptotes (166); *Polar Equations of Conics :*—General equations (167); Of parabola (168); Of ellipse (169), (174); Of hyperbola (170), (175); Discussion (171-173); EXAMPLES................ Pages 34-118

CHAPTER V.

GENERAL DISCUSSION OF THE EQUATION OF THE SECOND DEGREE.

Rectangular equation to a conic having any position in the plane (176); Every equation of the second degree may represent a conic (177); General test (178); To make xy disappear (179); To cause the coefficient of y to disappear (180); Remark (181); Varieties of the ellipse (182); Hyperbola (183); Parabola (184); Illustrations (185); To pass a conic section through five points (186); EXAMPLES. Another method of discussing the equation of the second degree (187); To remove the term containing xy from the general equation (187a); To remove the terms containing the first powers x and y from the equation (187b); EXAMPLES............Pages 119-147

CHAPTER VI.

LOCI IN SPACE.

Of the Point, Right Line, and Plane.

Definitions (189); Axis of a plane (190); Order of the angles (191); Coördinates of a point (192); Equations of the axes (192b); Distance between two points (193); Projections (194). *The Right Line:*— Equations of (195); Intersections of (196); Passing through a point (197); Inclination of to the axes (198); Angle between two lines (199). *The Plane:*—Equation of (201); Equation of first degree (202); Discussion (203); In terms of perpendicular (204, 205); Intersection of (206); Passing through two points (207); Inclination to the axes (208); Angle between planes (209); Parallel planes (210); Perpendicular planes (211). *Line and Plane:*—Line pierces a plane (212); Line perpendicular to a plane (213-216); Angle between line and plane (217). *Transformation of Coördinates:*—To transform from rectangular to oblique axes (219); Planar to polar (220); Polar to planar (221); EXAMPLES.
Pages 148-170

CHAPTER VII.

CURVED SURFACES.

Definitions (222). *Of Cylindrical Surfaces* (223-225). *Conical Surfaces* (226-228). *Sphere* (229); Surfaces of revolution (230). *Ellipsoids* (231-233). *Paraboloids* (234-237). *Hyperboloids* (238-241). *Of Intersections* (242-252). *Discussion of Quadrics* (253-260); Tangent planes (261, 262); Of normals (263); EXAMPLES.
Pages 171-201

CHAPTER VIII.

LOCI OF HIGHER ORDERS.

Definitions (264). *Spirals:*—Spiral of Archimedes (266); Hyperbolic spiral (267); Logarithmic (268); Involute (269); Lituüs (270);

viii CONTENTS.

Parabolic spiral (271). *Trigonometrical Curves* (272-277); Logarithmic curve (278); Parabolas (279-283); Trochoids (284-294); Conchoid of Nicomedes (295, 296); Trisection of an angle (297); Remark (298); The cissoid (299-301); Two mean proportionals (302); Duplication of a cube (303); Quadratrix of Dionstratus (304, 305); Witch of Agnisi (306); Ovals (307-311); Curves of pursuit (312); Folium (313); Discontinuous curves (314-317); Loxodromic (318); Logocyclic (319); Helix (320); Conoid (321, 322).
Pages 202-228

PART II.—QUATERNIONS.

CHAPTER I.

ADDITION AND SUBTRACTION OF VECTORS.

Definitions (323); Vectors (324-328); Tensor (329). *Vector Equations :*—Equations and law of signs (330-333); EXAMPLES. Sum of independent vectors separately equal zero (334); APPLICATIONS; EXAMPLES. Medial vectors (336, 337); APPLICATIONS; EXAMPLES. Co-planar vectors (338-340); APPLICATIONS; EXAMPLES. Angle-bisectors (341, 342); APPLICATIONS; EXAMPLES. Transversals (343); APPLICATIONS, (344-349); Complete quadrilateral (350)..............................Pages 229-263

CHAPTER II.

MULTIPLICATION AND DIVISION OF VECTORS.

Notation (351); Operation (352-354); Symbol $\sqrt{-1}$ (355); Non-commutative principle (356); Associative principle (357); EXAMPLES. Square of a vector (359); Division of rectangular vectors (360); EXAMPLES. *Oblique Vectors :*—Multiplication of (361); Division of (362); EXAMPLES; Comparing results (363); Scalars (364); Versor (365); A QUATERNION (366); General expressions (367); Discussion (368); EXAMPLES. $\epsilon^{\frac{2\theta}{\pi}}$ (369); Proposition (370); EXAMPLE; Plane triangles (371-378); Three co-initial co-planar vectors (379); Triedrals (380); $\rho = ix + jy + kz$ (381); Conjugate quaternions (382)..............................Pages 264-289

CHAPTER III.

LINE, PLANE, SPHERE, AND CONICS.

Right line (383); EXAMPLES. Plane (384); Circle (385); Equation of tangent line (386); EXAMPLES. Sphere (387); Ellipse (388); Hyperbola (389); Parabola (390)....................Pages 290-300

CONTENTS. ix

PART III.—MODERN GEOMETRY.

CHAPTER I.

Modern Geometry, definition (391); Definition of locus (392). *Tangential System* :—Definition (393); Equation of four cusp hypocycloid (394); Equation of the ellipse (395); Equation of a point (396); Equation of a line (397); EXAMPLES; Equation of a line in terms of three perpendiculars from three fixed points (398). *Trilinear System* :—Notation (399); Signs of the perpendiculars (400); Relation between the coördinates (401); Bisectors (402); Equation of a line (403); Abridged notation (404). *Other Systems* :—List of twenty-two systems by Rev. Thomas Hill (405); Other systems (406, 407)..Pages 303–316

APPENDIX I.

Brief sketch of the history of Quaternions..................Pages 317–329

APPENDIX II.

Hyper-Space 330

GREEK ALPHABET.

Letters.	Names.	Letters.	Names.
$A\ \alpha$	Alpha	$N\ \nu$	Nu
$B\ \beta$	Bĕta	$\Xi\ \xi$	Xi
$\Gamma\ \gamma$	Gamma	$O\ o$	Omicron
$\Delta\ \delta$	Delta	$\Pi\ \pi$	Pi
$E\ \varepsilon$	Epsilon	$P\ \rho$	Rho
$Z\ \zeta$	Zēta	$\Sigma\ \sigma\ \varsigma$	Sigma
$H\ \eta$	Eta	$T\ \tau$	Tau
$\Theta\ \vartheta\ \theta$	Thĕta	$\Upsilon\ \upsilon$	Upsilon
$I\ \iota$	Iōta	$\Phi\ \varphi$	Phi
$K\ \varkappa$	Kappa	$X\ \chi$	Chi
$\Lambda\ \lambda$	Lambda	$\Psi\ \psi$	Psi
$M\ \mu$	Mu	$\Omega\ \omega$	Omega

PART I.

CARTESIAN GEOMETRY.

COÖRDINATE GEOMETRY.

CHAPTER I.

FUNDAMENTAL PRINCIPLES.

Definitions.

1. **Lines** are said to be given when their positions in reference to each other are known or assumed.

2. **A System of Coördinates** is a system of known lines, or of lines and angles, for determining the position of points. The lines constituting a system of coördinates are usually straight, but curved lines may be used.

3. **A Point is known** when its position in reference to a system of coördinates is known.

4. **A Right Line** is one which may be generated by a point moving in one direction only.

5. **Illustrations.**—Let the position of the lines OX and OY in reference to each other, be assumed, so that the position of a point may be determined in reference to them; then will they constitute a system of coördinates. The known angle YOX we will constantly represent by ω. To determine the position of any point p, draw a line pa through p parallel to the line OY; and pb parallel to OX. Then if Oa and Ob are given, the point p will be determined. For by reversing the above process, we lay off Oa a known distance on OX, and draw ap parallel to Ob; then lay off the

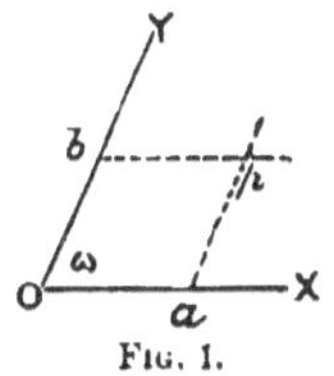

Fig. 1.

1

known distance Ob on OY and draw bp parallel to OX. Since the point will be in both the lines ap and bp, it must be at their intersection p.

The method just described is the one commonly used, but the point p may be determined in other ways in reference to the same lines. Thus, if we know the lengths of the perpendiculars pc and pd, the point becomes known. For, the point will be in the line ep drawn parallel to OY and at a distance from it equal to the perpendicular $Oe = pd$.

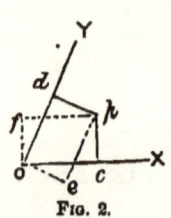
Fig. 2.

Similarly, it will be in the line pf drawn parallel to OX and at a distance from it equal to the perpendicular pc; hence at its intersection with ep at p.

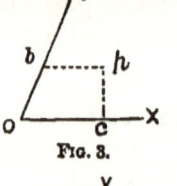
Fig. 3.

Or the perpendicular pc to OX and the parallel pb to the same line may be given.

Or if the perpendicular dp be prolonged to OX, and pc to OY, then will the lines pg and ph determine the point.

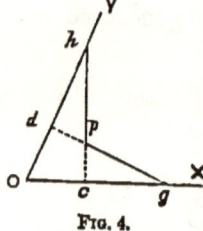
Fig. 4.

Or the lines pc and pd may make any known angles with the lines OX and OY.

The system which is adopted at the beginning of an investigation, must be retained throughout the same.

6. The Coördinate Axes.—When the coördinates consist of right lines intersecting at a point, the right lines thus used are called *coördinate axes*. The line OX is called the *axis of x*, and OY, the *axis of y*. These lines may be prolonged indefinitely in opposite directions from O, dividing the plane which contains them into four parts or angles.

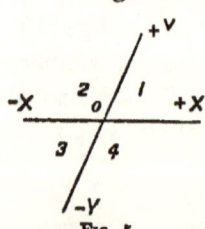
Fig. 5.

The intersection of the axes at O, is called the **origin of coördinates**.

Distances to the right of the axis of y and parallel to the axis of x are called *positive;* those to the left, *negative;* similarly, those parallel to YY and above XX are *positive*, and those below, *negative*.

The angle YOX, marked 1, is called the *first* angle; that marked 2, the *second*; 3, the *third*; and 4, the *fourth*.

The rules in regard to the signs and angles are arbitrary, and may be changed when desired.

7. Kinds of Coördinates.—If the system consists of two or three lines meeting at a point, it is called *rectilinear*.* If the axes form right angles with each other, the system is called *rectangular*. If two axes only are used, it is called *bilinear*. If the axes make oblique angles with each other, the system is called *oblique*.

If the system consists of a fixed line OX, a variable angle pOX, and a variable distance Op, it is called a *polar system*.

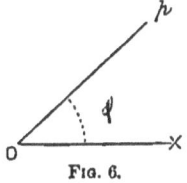
Fig. 6.

The rectilinear and polar systems are the ones most commonly used, though there are many other systems, some of which are used occasionally.

When the system consists of three lines perpendicular respectively to the three sides of a known triangle, it is called *trilinear*. This system forms an important part of Modern Geometry.

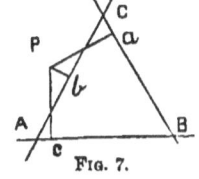
Fig. 7.

The tangential system is in a certain sense the reciprocal of the rectilinear. It begins with a line, and, by the intersection of an infinite number of right lines, determines a point. (See Salmon's *Conic Sections*.) It is also used in Modern Geometry.

The system may consist of a tangent to a given curve, and a radius from the center to the point of tangency. This system is useful for investigating certain properties of the involute.

The system may also consist of the arc of the curve and the angle which the variable tangent makes with a fixed line. This is called the *intrinsic equation* of the curve.

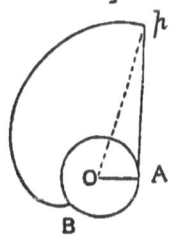
Fig. 8.

Rectilinear System of Coördinates.

8. The Ordinate to a point is the distance from the point to the axis of x, measured on a line parallel to the axis of y. Thus, the ordinate of p_1 is the length of the line p_1a_1, drawn

---

* Sometimes it is called the Cartesian system, from the latinized form of Descartes' name, who first devised the system.

from p_1 parallel to OY and terminated by the axis of x.

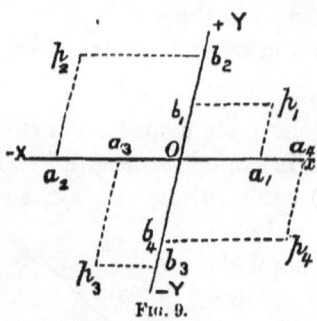
Fig. 9.

Ordinates are generally represented by the letter y, that letter representing *any* ordinate, and when a particular ordinate is known, y may be placed equal to that value. Thus, the ordinate of p_1 is $y = + a_1 p_1 = Ob_1$; of p_2, $y = + a_2 p_2 = Ob_2$; of p_3, $y = -a_3 p_3 = -Ob_3$; of p_4, $y = -a_4 p_4 = -Ob_4$. Assumed ordinates are also designated by accents, or subscripts, thus, y', y'', etc.; y_1, y_2, etc.

9. The Abscissa to a point is the distance from the point to the axis of y measured on a line parallel to the axis of x. The general value of the abscissa is represented by the letter x. Hence, for the abscissa of p_1, we have, $x = + b_1 p_1 = + Oa_1$; of p_2, $x = -b_2 p_2 = -Oa_2$; of p_3, $x = -b_3 p_3 = -Oa_3$; of p_4, $x = +b_4 p_4 = +Oa_4$. We also use x', x'', etc., for known abscissas.

10. The Coördinates to a Point include both the abscissa and ordinate of the point. A point is indicated thus: the point xy, or the point ab, or the point (x, y), or the point $(3, -4)$; in which the value of the abscissa is placed first in order. If the angle between the axes is not given it is supposed to be *right*.

11. The Equations to a Point are equations which express the values of x and y in terms of the known abscissa and ordinate. Thus, if a be a known abscissa, and b a known ordinate, we have for the equations of the point in the

1st *angle*.	2d *angle*.	3d *angle*.	4th *angle*.
$\left.\begin{array}{l}x = + a \\ y = + b\end{array}\right\}$;	$\left.\begin{array}{l}x = - a \\ y = + b\end{array}\right\}$;	$\left.\begin{array}{l}x = - a \\ y = - b\end{array}\right\}$;	$\left.\begin{array}{l}x = + a \\ y = - b\end{array}\right\}$.

EXAMPLES.

1. If the axes are rectangular, locate the points $(5, 7)$; $(-3, -4)$; $(-4, 3)$.

2. If $\omega = 60°$, locate the points $(4, 4)$; $(0, 3)$; $(-4, -3)$.

3. If the axes are rectangular, locate the points (0, − 3); (0, 0); (− 2, 0); (− 0, 2).

4. If ω = 135°, locate the points (3, 2); (3, − 2); (− 3, − 2).

Polar System of Coördinates.

12. The Initial Line is the line from which the variable angle is reckoned. Thus if pOX constitutes a system of polar coördinates in which the variable angle is measured from the fixed line OX, then will OX be *the initial line.*

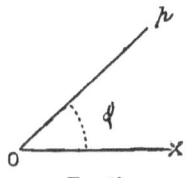

Fig. 10.

13. The Pole is the point about which the line Op is conceived to revolve. Thus O, in the figure, is the pole.

14. The Radius Vector is the line of variable length extending from the pole to the required point. Thus, if p is the point which is to be located, the line Op will be the radius vector. The length of the radius vector we shall generally represent by ρ.

15. The angle between the radius vector and the initial line is called the **variable or vectorial angle**. We will here represent it by φ.

16. Signs of the Coördinates.—*The angle φ is considered positive* when the radius vector revolves *left-handed;* that is, in a direction opposite to that of the hands of a watch; and *negative*, in the opposite direction. Thus, the angle (or arc) ab is positive, and cd, negative.

The radius vector is considered positive when the point and the extremity of the measuring arc are in the same direction *from* the pole, and negative when they are in opposite directions from the pole. Thus, in determining the point p_1, if ab be the measuring arc, Op_1 will be positive, for b and p_1 are in the same direction from O; but if cdi be the measuring arc, the radius vector Op_1 will be negative, for i and p_1 are in opposite directions from the pole.

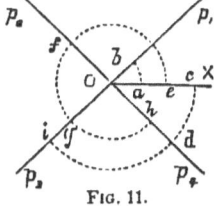

Fig. 11.

17. The Polar Coördinates of a Point involve the position of the fixed line, the vectorial angle, and the radius vector. A point is indicated thus, (φ, ρ).

18. The Polar Equations of a point consist in placing φ and ρ equal to determinate values. Thus the equations of the point p_1 are

$$\left\{\begin{array}{l}\varphi=+ab\\ \rho=+Op_1\end{array}\right\}; \text{ or } \left\{\begin{array}{l}\varphi=+efg\\ \rho=-Op_1\end{array}\right\}; \text{ or } \left\{\begin{array}{l}\varphi=-cdi\\ \rho=-Op_1\end{array}\right\};$$

and of the point p_3,

$$\left\{\begin{array}{l}\varphi=+efg\\ \rho=+Op_3\end{array}\right\}; \text{ or } \left\{\begin{array}{l}\varphi=+ab\\ \rho=-Op_3\end{array}\right\}; \text{ or } \left\{\begin{array}{l}\varphi=-cdi\\ \rho=+Op_3\end{array}\right\}.$$

The points p_2 and p_4 are indicated in a similar manner.

EXAMPLES.

1. Construct the points (60°, 5); (− 45°, − 3); (π, 4).
2. Construct the points (0°, 0); (5π, − 4); (0°, − 3).
3. Construct the points ($\frac{1}{2}\pi$, 1); ($\frac{1}{2}\pi$, − 1); ($\frac{3}{2}\pi$, 1).
4. Construct the points (2π, 2); (2π, − 2); (0, 2).
5. Construct the points (0°, cos 30°); (π, sin 45°); (− $\frac{1}{2}\pi$, tan 45°).
6. Construct the points ($\sin^{-1}\frac{1}{2}$, sec 45°); (π, π); $\left(\cos^{-1}(-\sqrt{\frac{1}{4}}), -\cos\sin^{-1}\frac{1}{4}\right)$.

Distance Between Two Points.

19. Let the Points be given by their Rectilinear Coördinates, those of P being $x'y'$, and of P', $x''y''$.

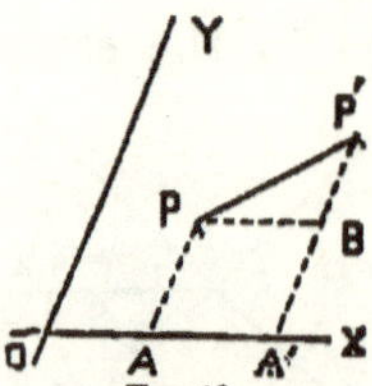

Let $l = PP' = $ the distance between the points, and we have

$$l^2 = (PB)^2 + (BP')^2 - 2PB \cdot BP' \cos PBP'$$
$$= (PB)^2 + (BP')^2 + 2PB \cdot BP' \cos YOX$$
$$= (OA' - OA)^2 + (A'P' - A'B)^2 + 2(OA' - OA)(A'P' - A'B)\cos\omega$$
$$= (x''-x')^2 + (y''-y')^2 + 2(x''-x')(y''-y')\cos\omega,$$

which is the formula required.

Fig. 12.

20. Let the System be Rectangular.

—Then $\omega = 90°$, and the preceding formula becomes

$$l^2 = (x'' - x')^2 + (y'' - y')^2. \qquad (1)$$

If one of the points be at the origin, make $x' = 0$ and $y' = 0$, and we have after dropping the accents

$$l^2 = x^2 + y^2, \qquad (2)$$

which is the square of the distance of *any* point from the origin.

21. The Points being given by Polar Coördinates, to find the distance between them.

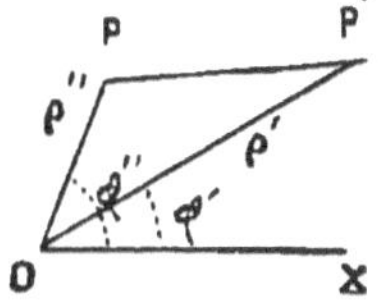

Let

$$OP' = \rho', \quad P'OX = \varphi';$$
$$OP'' = \rho'', \quad P''OX = \varphi'';$$

then

$$P''OP' = \varphi'' - \varphi'.$$

Fig. 13.

From Trigonometry we have

$$(P'P'')^2 = (OP')^2 + (OP'')^2 - 2OP''.OP' \cos P''OP';$$

hence,

$$l^2 = \rho'^2 + \rho''^2 - 2\rho'\rho'' \cos(\varphi'' - \varphi').$$

If $\varphi'' - \varphi' = 90°$, the angle between the lines will be right, and the formula becomes

$$l^2 = \rho'^2 + \rho''^2.$$

EXAMPLES.

1. Find the distance between $(-3, 0)$ and $(3, -5)$.
2. Find the lengths of the three sides of a triangle whose vertices are $(2, 4)$, $(0, 0)$, and $(-3, 5)$.
3. Find the coördinates of the middle point of the line whose extremities are $(5, -1)$, and $(-1, 5)$.
4. If $\omega = 60°$, construct the triangle whose vertices are $(3, 2)$, $(-2, 6)$, $(1, -5)$.
5. Find the distance between the points $(0°, 2)$, and $(-\tfrac{1}{2}\pi, 3)$.
6. Find the distance between the points $(\tfrac{1}{2}\pi, 3)$, and $(\pi, 4)$.

CHAPTER II.

THE RIGHT LINE IN A PLANE.

Definitions.

22. A Locus is a series of positions to which a moving point is restricted by some law. If the successive positions are *consecutive*, the locus is a line, either straight or curved. If any position is isolated, that part of the locus is a point only. The locus of a line may be a surface. Geometrical loci are *real* points, lines, and surfaces. Analytical loci are such as are expressed algebraically, and, in an extended sense, not only represent *real* loci, but also *imaginary* points, lines, and surfaces and loci at infinity.

23. Consecutive Points are such that the distance between them is less than any assignable quantity. Hence, *when compared with finite distances*, the distance between consecutive points may be considered as zero, and in such cases two consecutive points may be treated as one point. The ratio between the distances of different consecutive points may be any finite value.

[In more *familiar language*, we may say that the distance between consecutive points is so exceedingly small that when added to finite or measurable distances, the former may be omitted without *appreciable* error. Thus, 20,000 miles + $\frac{1}{1000}$ of an inch and 20,000 miles + $\frac{1}{100000}$ of an inch are *practically* the same. Their sum is *practically* 40,000 miles, their ratio *practically* 1, and their difference *practically* zero. Hence in regard to the larger numbers, the small fractions may be omitted, and the smaller the fractions the more nearly will the larger part be the true 'value of the expression. When the fraction is less than *any assignable quantity*, it cannot be measured or expressed, but the larger part is measurable, and is called finite. These remarks are true whatever be the actual value of the small fractions. Thus, in the preceding example, the ratio of the first fraction to the second is 10. If the fractions had been $\frac{5}{10000}$ and $\frac{2}{100000}$, their ratio would have been 25 ; while the sum of the entire quantities would have been *practically* 40,000, their difference zero, and their ratio 1. These fractions, however small, may have any conceivable ratio. Quantities which are less than any assignable quantity are called *infinitesimals*. One object of the Differential Calculus is to find the ratio of infinitesimals when related to each other by known laws.]

24. The Equation of a Locus is an equation which expresses the relation between the coördinates of every point of the locus. The distance from the origin to the point where the locus cuts an axis is called the **Intercept** on that axis.

25. Analytical Geometry consists in expressing geometrical quantities by algebraic symbols; establishing an equation of the locus by means of these symbols, then subjecting the quantities and the equation to algebraic operations so as to deduce the properties of the locus, and finally interpreting the result.

26. Coördinate Geometry is that system of geometry in which the points, lines, or surfaces which are to be considered are referred to a system of coördinates, and their properties determined by means of their relations to those coördinates.

Equation of a Right Line.

27. To find the Rectilinear Equation to a Right Line.—Let BP be the line cutting the axis of x at B and the axis y at F. Take any point P on the line and draw PD parallel to OY; then will

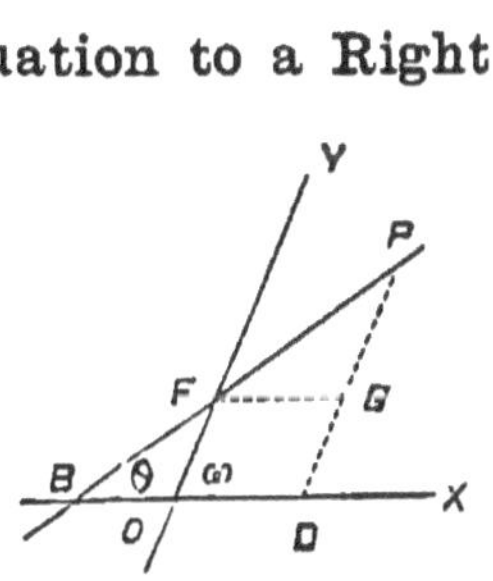

$$x = OD, \quad y = DP.$$
Let $\quad \theta = PBD, \quad b = OF.$

Through F draw FG parallel to OD, then the triangle PFG gives

Fig. 14.

$$FG : PG :: \sin FPG : \sin (PFG = FBX),$$
or $\quad x : y - b :: \sin(\omega - \theta) : \sin \theta$

$$\therefore y = \frac{\sin \theta}{\sin (\omega - \theta)} x + b. \qquad (1)$$

If m' be substituted for the coefficient of x the equation becomes
$$y = m'x + b; \qquad (2)$$
which is the equation sought.

28. Rectangular Equation to a Right Line.—Make $\omega = 90°$ in the equation (1) of the preceding Article, and we have

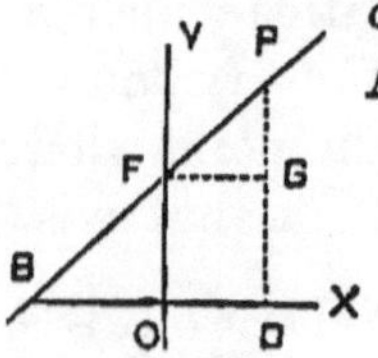

$$y = \frac{\sin \theta}{\cos \theta} x + b;$$

Fig. 15.

or
$$y = \tan \theta . x + b. \qquad (3)$$

Let $m = \tan \theta$, and we have
$$y = mx + b; \qquad (4)$$
which is the required equation.

29. Equation to a Right Line in Terms of its Intercepts.—Equations (2) and (4) are in terms of the intercept (b) on the axis of y and the inclination of the line, which is the more common form. But it may be found in terms of the intercept on both axes.

Let $a = OB$ be the intercept on the axis of x, $OF = b$, the intercept on the axis of y.

In Figs. 14 and 15 we have
$$\frac{OF}{OB} = \frac{\sin OBF}{\sin OFB} = m', \text{ or } m;$$

or
$$\frac{b}{-a} = m' \text{ or } m,$$

where a is negative because B is at the left of the origin. Substituting the value of m' in equation (2), and m in (4) gives

$$\frac{x}{a} + \frac{y}{b} = 1, \qquad (5)$$

which is the required equation.

30. Equation to a Right Line in terms of the Perpendicular upon it from the Origin and the Angle made by the Perpendicular with the Axis of x.

Let $p = OR$ be the perpendicular from the origin upon the line CD; $a = OD$, the intercept on the axis of x; $b = OC$, the intercept on the axis of y. Then

$$a = \frac{p}{\cos \alpha}, \qquad b = \frac{p}{\sin \alpha};$$

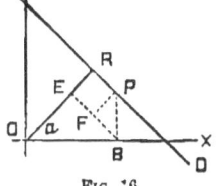

Fig. 16.

and these values substituted in equation (5) give

$$x \cos \alpha + y \sin \alpha = p, \qquad (6)$$

which is the required equation.

If the axes are oblique, the equation becomes

$$x \cos \alpha + y \cos (\omega - \alpha) = p. \qquad (7)$$

31. Of Constants and Variables.

—*A constant* is a quantity which retains the same value during a particular discussion. Constants are of two kinds—*fixed* and *arbitrary*. *Fixed* constants are numerals, as 3, 7, 12, etc. *Arbitrary* constants are those which admit of different values. Thus in equation (4) m may have *any* value *assigned* to it, but when once assigned it becomes *fixed* for that particular discussion; and the same is true of b. Arbitrary constants are usually represented by the first letters of the alphabet.*

Variables are quantities which may have an unlimited number of values in a particular discussion. Thus, in equa-

---

* In higher analysis arbitrary constants are made to vary; or, in other words, certain quantities which are usually considered constant are made to vary. Thus, in the equation of a right line, $y = mx + b$, if b varies while m remains fixed, the line will generate a plane.

Similarly, the parameter of a parabola may increase at a uniform rate while points which generate parabolas shoot out constantly from the vertex with an infinite velocity. Curves thus generated are called "A family of curves."

An arbitrary constant may be defined as a quantity which varies infinitely slow compared with the generator of the curve.

tion (4), (5), or (6), x may have all conceivable values from $+\infty$ to $-\infty$, and the corresponding value of y may be determined from the equation; hence, y may also have an unlimited number of values. The *real* values of the variables in an equation are often confined within *finite limits*, as we shall see in the case of the circle, ellipse, etc.; still, the number of real values within those limits is unlimited.

32. The Absolute Term is that term of an equation which contains no variables. The quantity b in the equation of the right line is the absolute term. *If an equation has no absolute term, the locus passes through the origin;* for the intercepts for that point are both zero.

33. The Discussion of an Equation to a Locus consists in interpreting the results deduced from the equation. The general process consists in solving the equation in reference to one of the variables and assuming fixed values for the other, and interpreting the results.

34. Discussion of the Rectangular Equation to the Right Line.—The equation is (Art. 28),

$$y = mx + b. \qquad (1)$$

1°. Let $x = 0$, then $y = b$.

These are the coördinates of a point on the axis of y at a distance b from the origin. It is the point where the line intersects the axis of y, and the ordinate to this point is the intercept on that axis, (Art. 24).

2°. Let $y = 0$, then $x = -\dfrac{b}{m} = -\dfrac{b}{\tan \theta} = -b \cot \theta$.

These values of x and y are the coördinates of the point where the line intersects the axis of x; and the value of x is the intercept on that axis.

3°. Let $b = 0$, then we have

$$y = mx, \qquad (2)$$

which is the equation to a right line passing through the origin, (Art. 32).

4°. Let $m = 0$, then $\tan \theta = 0$ and $\theta = 0$, and we have

$$y = 0.x + b. \qquad (3)$$

But when $\theta = 0$ the line is parallel to the axis of x; hence this is the equation to a right line parallel to the axis of x and at a distance from it equal to b. In this equation the values of x and y are independent of each other.

5°. Let $m = 0$, and $b = 0$, then will the line coincide with the axis of x, and the equation becomes

$$y = 0.x, \qquad (4)$$

which is the equation of the axis of x. In this equation y is zero for all values of x.

6°. Let $m = \infty$; then $\tan \theta = \infty$, and $\theta = 90°$, and the equation becomes

$$y = \infty.x + b;$$

or
$$\frac{1}{\infty} y = x + \frac{b}{\infty};$$

or
$$0.y = x + 0. \qquad (5)$$

In this equation x is zero for all values of y; hence it is the equation to the axis of y.

The equation to a line parallel to the axis of y is

$$0.y = x + a. \qquad (6)$$

7°. When m and b are both positive, the line lies across the second angle. For, b being positive, the line cuts the axis of y above the origin; and m being positive makes $\tan \theta$ positive, hence the angle θ, above the axis of x and at the right of the line, will be acute.

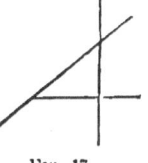

Fig. 17.

8°. When b is positive and m negative, it lies across the first angle.

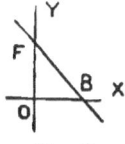

Fig. 18.

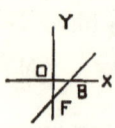

Fig. 19.

9°. When b is negative and m positive, it lies across the fourth angle.

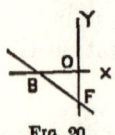

Fig. 20.

10°. When b is negative and m negative, the line lies across the third angle.

[OBS.—It is not necessary for the student to remember all these results, but it is very important that he should be able to make the discussion, and interpret the results. Much practice in interpreting equations is advisable.]

35. Every Equation of the First Degree between Two Variables is the equation to a right line.

The most general form of the equation of the first degree between two variables is

$$Ax + By + C = 0;$$

in which A, B, C, are arbitrary constants. Solving in reference to y gives

$$y = -\frac{A}{B}x - \frac{C}{B}.$$

If $-\dfrac{A}{B}$ be represented by m, and $-\dfrac{C}{B}$ by b, the equation becomes

$$y = mx + b;$$

and therefore represents a right line, (Arts. 27 and 28).

In an algebraic sense, the values of x and y in the preceding equation are indeterminate, and since this is generally true for the equations to loci, this branch of mathematics is sometimes called *Indeterminate Geometry*.

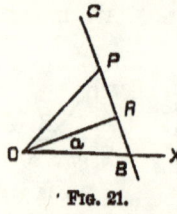

Fig. 21.

36. Polar Equation to a Right Line.—Let BC be the line, OX the initial line, $OR = p$ the perpendicular from the pole O upon the line, the angle $ROX = \alpha$, P any point of the line, $OP = \rho$ the radius vector, and $\varphi = POX$ the variable angle.

The right-angled triangle ORP gives

$$OP \cos POR = OR,$$

or
$$\rho \cos (\varphi - \alpha) = p;$$

which is the required equation.

The angle α is positive when it is generated by a left-handed rotation, or, more generally, from $+x$ towards $+y$.

Discussion.—1°. Let $\varphi = 0$, then $\rho = OB = \dfrac{p}{\cos(-\alpha)} = p \sec \alpha$; which is the intercept on the initial line.

2°. Let $\varphi = \alpha$, then $\rho = OR = \dfrac{p}{\cos 0} = p$.

3°. Let $\varphi = 90° + \alpha$, then $\rho = \dfrac{p}{\cos 90} = \infty$ as it should, since the radius vector becomes parallel to the line, and hence cannot meet it at a finite distance.

4°. Let $\varphi > 90° + \alpha$ and $< 180° + \alpha$, then will $\cos (\varphi - \alpha)$ be negative and the radius vector ρ will also be negative, and the point thus determined will be below B.

5°. Let $\varphi = 180°$, then $\rho = \dfrac{p}{\cos(180° - \alpha)} = \dfrac{-p}{\cos \alpha}$, which gives the point B again.

6°. Let $\varphi = 180° + \alpha$, then $\rho = \dfrac{p}{\cos 180°} = -p$, which gives the point R.

7°. Let $\alpha = 0$, then the equation becomes $\rho \cos \varphi = p$, which is the polar equation of a line perpendicular to the initial line.

8°. Let $\alpha = 90°$, then we have $\rho \sin \varphi = p$ for the equation to a line parallel to the initial line.

9°. If $\varphi = \pi, 2\pi, 3\pi$, etc., then $\rho = \pm p \sec \alpha = OB$. Generally, all the points of the line will be determined once by making φ pass from $0°$ to $180°$; and again by passing from $180°$ to $360°$, and still again, from 2π to 3π and so on indefinitely.

37. To Trace a Right Line.—Two points determine a right line; therefore, assume two different values for one of the variables and find the corresponding values for the other by means of the equation of the line. Construct the points thus found and draw a right line through them.

If the axes are rectilinear, the easiest way is to find the intercepts and lay them off on the respective axes. Therefore, make $y = 0$ in the equation and find x, and lay its value off on the axis of x. Similarly, make $x = 0$, find y and lay it off on the axis of y; the line drawn through the points thus found will be the line required.

If the line is given by its polar coördinates, find the intercept on the axis of x by making $\varphi = 0$ in the equation, and draw the line through it and the extremity of the perpendicular from the origin.

EXAMPLES.

1. Determine three points in the right line $4y = -6x + 3$, whose abscissas are respectively -2, 1, and 2, and draw the right line through the points thus found.

Making $x = -2$ in the equation of the line, we find $y = 3\frac{3}{4}$. With any scale of equal parts lay off on $-x$ two divisions of the scale $= OB$, and at B draw BP parallel to the axis of y and lay off on it $+3\frac{3}{4}$ of the same equal parts.

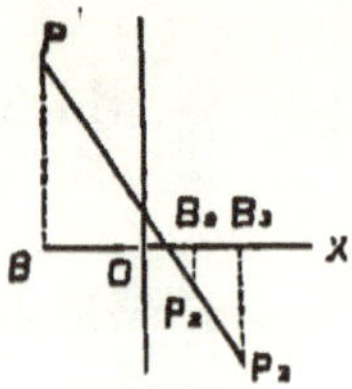

Fig. 22.

Similarly, for $x = 1$ we find $y = -\frac{3}{4}$. For this point, lay off one space $= OB_2$ to the right of the origin, and $B_2 P_2 = \frac{3}{4}$ of a space below the axis of x, which gives P_2 for the required point.

For $x = 2$, we find $y = -2\frac{1}{4}$, by means of which the point P_3 is found.

If the construction is correct the three points will be in a right line.

If $x = 0$, $y = \frac{3}{4}$ which is the *y-intercept*. If $y = 0$, $x = \frac{1}{2}$ which is the *x-intercept*.

2. Construct the line $\frac{1}{2}y = \frac{1}{3}x - \frac{1}{4}$.

[Make $x = 0$, and find $y = -\frac{1}{2}$ which is the *y-intercept*. Similarly the *x-intercept* is $\frac{3}{4}$.]

3. What angle does the line $y = x$ make with the axes? What are its intercepts?

4. What angle does the line $y = -\frac{3}{4}x - 4$ make with the axis of x? What are its intercepts?

5. Trace the line $\frac{1}{2}y\sqrt{3} = -2x\sqrt{\frac{1}{4}} - 1$.

6. Trace the line $3x - 4y + 2 = 0$.

7. What two lines are represented by the equation $y^2 = 4x^2$? What angle do they make with each other ?

8. What two lines are represented by the equation $x^2 = 4 - 4y + y^2$? At what point do they intersect each other ?

9. What are the intercepts for the two lines represented by the equation $(x - 2)^2 = y^2 + 6y + 9$? Construct the lines and find where they intersect each other.

10. Trace the line $\rho \cos (\varphi - 30°) = 6$. Also $\rho \sin \varphi = -5$.

11. Deduce the equation of the line $\dfrac{x}{a} + \dfrac{y}{b} = 1$ from a figure.

12. Deduce the equation $x \cos \alpha + y \sin \alpha = p$ from a figure.

13. Construct the line whose equation is $y = 3x - 5$, when $\omega = 60°$.

14. Interpret the equation $\rho \cos 3\varphi = 0$.

The Point and Right Line.

38. Equation to a Right Line which shall pass through a given point.—Let the given point be (x', y'), and since it is on the right line

$$y = mx + b \qquad (1)$$

it must satisfy the equation of that line, hence we will have

$$y' = mx' + b. \qquad (2)$$

Eliminating b between these equations gives

$$y - y' = m(x - x'); \qquad (3)$$

which is the required equation. The value of m in this equation is indeterminate, and an indefinite number of values may be assigned to it; hence the equation is an expression for the well-known fact that an infinite number of right lines may be passed through a given point.

From (2) we have

$$m = \dfrac{y' - b}{x'},$$

which substituted in (1) gives

$$y = \dfrac{y' - b}{x'} x + b; \qquad (4)$$

which is another form of the equation, but not being so convenient as (3) it is rarely used in practice.

39. Equation of Condition.—Observe that equation (2) of the preceding article is not the equation of a line, for it contains no variables; neither is it the equation of a point. It is *a true equation between constants*, by means of which the values of the arbitrary constants may be determined. The equation is not only true, but is established upon certain *conditions;* the conditions being determined by the geometric forms, or by the equations which represent those forms. Such equations are called *Equations of Condition*.

40. The Equation to a Right Line passing through Two given points.—Let (x', y'), (x'', y'') be the two points through which the right line

$$y = mx + b \qquad (1)$$

must be made to pass. The coördinates of the points must satisfy the equation to the line; hence we have the two *equations of condition*

$$y' = mx' + b, \qquad (2)$$

$$y'' = mx'' + b, \qquad (3)$$

by means of which the arbitrary constants m and b may be completely determined in terms of $x'y'$ and $x''y''$. Subtracting (2) from (3), we have

$$y'' - y' = m(x'' - x');$$

$$\therefore m = \frac{y'' - y'}{x'' - x'}. \qquad (4)$$

Subtracting (2) from (1), and substituting the value of m in the result, gives

$$y - y' = \frac{y'' - y'}{x'' - x'}(x - x'); \qquad (5)$$

which is the required equation. It may be put under the form

$$\frac{y-y'}{x-x'} = \frac{y''-y'}{x''-x'}. \qquad (6)$$

If one of the points, as (x'', y'') be at the origin, then $x'' = 0$, $y'' = 0$, and the equation becomes

$$y = \frac{y'}{x'} x \qquad (7)$$

which is the equation of a right line passing through the origin, and making an angle with the axis of x whose tangent is $\dfrac{y'}{x'}$.

If one of the points is $(0, b)$ we have

$$y = \frac{y'-b}{x'} x + b = mx + b;$$

which is the ordinary equation to the right line.

EXAMPLES.

1. Find the equation to a right line which makes an angle of 45° with the axis of x, and passes through the point (2, 3).

Here $m = \tang 45° = 1$, and equation (3) Art. 38 becomes

$$y = x + 1,$$

which is the required equation.

2. Find the equation to a right line which passes through the point $(-2, 3)$, and whose intercept on y is 1; and find the angle which the line makes with the axis of x.

(Use Eq. (4) Art. 38, or Eq. (5) Art. 40.)

3. Find the equation to a right line which makes an angle of 150° with the axis of x and passes through the point $(-4, -5)$.

4. Find the equation of the right line which passes through the points $(3, -4)$ and $(-2, 6)$, and find tan θ. (Eq. (5), Art. 40.)

5. Form the equations of the sides of the triangle the coördinates of whose vertices are $(-3, 0)$, $(0, -4)$, $(1, 0)$.

6. Find the equation of a right line passing through the points $(0, 0)$, $(-3, 2)$.

Relations between Two Lines.

41. To find the Point of Intersection of Two Lines.—Let the equation of the line BC be

$$y = mx + b,$$

and of EF,

$$y = m'x + b'.$$

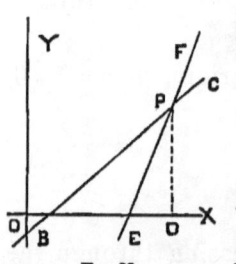

Fig. 23.

If the lines intersect there will be one point in common, and hence the coördinates for that point must satisfy the equations to both lines. To find the values of x and y which will satisfy this condition, we have only to consider the preceding equations as simultaneous, and eliminate x and y successively by the well known rules of algebra. Letting the resulting values of x and y be denoted by x_1 and y_1, we have

$$x_1 = \frac{b' - b}{m - m'} = OD ; \; y_1 = \frac{mb' - m'b}{m - m'} = DP.$$

If $m = m'$ we have

$$x_1 = \infty, \text{ and } y_1 = \infty;$$

hence the lines will not intersect at a finite distance, or, in other words, they will be parallel.
The equation

$$m = m'$$

is *the equation of condition of parallelism of two right lines.*
If $m = m'$ and $b = b'$, we have

$$x_1 = \frac{0}{0}, \; y_1 = \frac{0}{0};$$

both of which are *indeterminate*, hence the lines will coincide throughout.

42. Equation to a Line which is Parallel to a given Line.—Let the *given* line be

$$y = m'x + b'. \qquad (1)$$

The form of the equation of the required line will be

$$y = mx + b; \qquad (2)$$

but, according to the *equation of condition*, $m = m'$, which value in equation (2) gives

$$y = m'x + b, \qquad (3)$$

for the required equation. In this equation the arbitrary constant b is undetermined; hence the equation expresses the well-known fact that *an infinite number of lines may be drawn parallel to a given line*.

43. Equation to a Right Line passing through a given Point and Parallel to a given Line.—Let the given point be (x', y'). Let equation (1) of the preceding article be the given line, then will (3) be parallel to it, and it remains to make (3) pass through the given point. Hence the coördinates of the point must satisfy equation (3), and we have *the equation of condition*

$$y' = m'x' + b\,;$$

which, being subtracted from equation (3), gives

$$y - y' = m'(x - x'),$$

which is the required equation.

44. Angle between Two Lines.—Let the equations be

$$y = mx + b, \text{ (for } BF\text{),}$$
$$y = m'x + b', \text{ (for } EF\text{).}$$

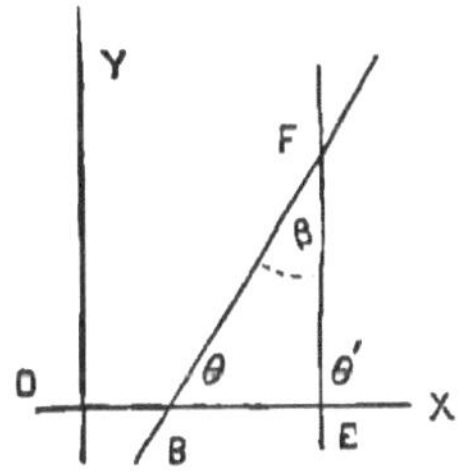

From the figure we have θ' equal to the sum of θ and β, or

$$\beta = \theta' - \theta$$

$$\therefore \tan \beta = \frac{\tan \theta' - \tan \theta}{1 + \tan \theta \tan \theta'}$$

$$= \frac{m' - m}{1 + mm'},$$

Fig. 24.

from which β, the required angle, may be found.

If $m = m'$, then $\beta = 0$, or the lines are parallel, as previously shown.

If the lines are perpendicular to each other, $\beta = 90°$, and $\tan \beta = \infty$, hence we must have

$$1 + mm' = 0,$$

which is *the equation of condition that two right lines shall be mutually perpendicular.*

45. Equation to a Right Line perpendicular to a given one.—Let the given line be

$$y = m'x + b',$$

and let it be required to make the line

$$y = mx + b$$

perpendicular to it. The equation of condition gives

$$m = -\frac{1}{m'},$$

in which m' is known from the given equation, and hence m becomes known; and by substituting its value in the preceding equation, we have

$$y = -\frac{1}{m'}x + b,$$

which is the required equation.

Since the arbitrary constant b remains undetermined, the equation expresses the well-known fact, that *an infinite number of lines may be drawn perpendicular to a given line.*

46. Equation to a Right Line perpendicular to a given line and passing through a given point.—Let the point be (x', y'); and the given line be

$$y = m'x + b;$$

then will the required line be of the form, (Art. 45),

$$y = -\frac{1}{m'}x + b.$$

Since this line is to pass through the point we have *the equation of condition*

$$y' = -\frac{1}{m'}x' + b,$$

which subtracted from the preceding equation, gives

$$y - y' = -\frac{1}{m'}(x - x'),$$

which is the required equation.

47. Equation to a Right Line passing through a given point and cutting a given line at a given angle.—Let the given line be

$$y = m'x + b;$$

and the required line be

$$y - y' = m(x - x');$$

and β the given angle. Then, according to Article 44, we find

$$m = \frac{m' - \tan \beta}{1 + m' \tan \beta};$$

which substituted in the preceding equation gives

$$y - y' = \frac{m' - \tan \beta}{1 + m' \tan \beta}(x - x');$$

for the required equation. If $\beta = 0$ the equation becomes

$$y - y' = m'(x - x').$$

If $\beta = 90°$, the equation becomes, by dividing the numerator and denominator by $\tan \beta$,

$$y - y' = -\frac{1}{m'}(x - x').$$

EXAMPLES.

1. Find the coördinates of the point of intersection of the lines

$$3x - 4y = 1; \; y - 2x = 3.$$

2. Find the coördinates of the point of intersection of the lines

$$\frac{x}{2} - \frac{y}{3} = 1; \; \frac{x}{5} - \frac{y}{7} = 1.$$

3. Find the angle between the lines
$$y = 4x + 5;\ 2y = 6x - 7.$$

4. Find the angle between the lines
$$3y - 4x = 0;\ y = 7x - \tfrac{1}{2}.$$

5. Find the equation of a right line perpendicular to the line $y = 7x - 4$.

6. Find the equation of a right line perpendicular to the line $y = x$, and passing through the point (1, 1) in the line.

7. Find the equation of a right line perpendicular to the line $2y - 4x = 7$, and passing through the point $(-3, -4)$.

8. Find the equation of a right line parallel to $y = -7x + 4$, and passing through the point $(-2, 3)$.

9. Find the perpendicular distance between the parallel lines
$$y = 4x - 3;\ y = 4x + 4.$$

10. Find the perpendicular distance of the line $y = -3x + 5$ from the origin.

11. Find the length of the perpendicular from any point to a right line.

Solution.—Let the point be (x', y') and the line be
$$y = mx + b;$$
then will the equation of the perpendicular be
$$y - y' = -\frac{1}{m}(x - x').$$

Eliminating y and x successively between these equations gives the coördinates of the point of intersection of the line and perpendicular. Letting this point be $x''y''$, we have
$$x'' = \frac{my' - mb + x'}{1 + m^2},$$
$$y'' = \frac{m^2 y' + mx' + b}{1 + m^2}.$$

From these we find
$$x'' - x' = m\frac{y' - mx' - b}{1 + m^2},$$
$$y'' - y' = \frac{-y' + mx' + b}{1 + m^2};$$

and the length of the perpendicular will be
$$\sqrt{(x''-x')^2 + (y''-y')^2} = \frac{y' - mx' - b}{\sqrt{1 + m^2}} = P\,(\text{say}).$$

If the equation of the line be of the form $Ax + By + C = 0$, then

$$P = \frac{Ax' + By' + C}{\sqrt{A^2 + B^2}}.$$

12. Find the distance of the point $(-2, 3)$ from the line $y = 2x - 4$.
13. Find the distance of the point $x'y'$ from the line

$$x \cos \alpha + y \cos (\omega - \alpha) = p.$$

Solution.—The equation of a parallel line passing through the given point will be

$$x' \cos \alpha + y' \cos (\omega - \alpha) = p',$$

and the required distance will be $p' - p$. Subtracting p from the left member, we have for *the length of the required perpendicular*

$$\pm (x' \cos \alpha + y' \cos (\omega - \alpha) - p),$$

in which the plus sign is used when the origin and the point are on *opposite* sides of the line, and *minus* when they are on the *same* side.

If the coördinates are rectangular, the length will be, making $\omega = 90°$,

$$\pm (x' \cos \alpha - y' \sin \alpha - p).$$

14. Given the area (a) and base (b) of a triangle to find the locus of the vertex.

Let the equation of the base be

$$x \cos \alpha + y \sin \alpha - p = 0;$$

then

$$x' \cos \alpha + y' \sin \alpha - p$$

will be the length of the perpendicular from the vertex (x', y') upon the base; which is also $2a \div b$; hence if x' and y' be general variables, we have

$$x \cos \alpha + y \sin \alpha - p = \frac{2a}{b},$$

which is the equation of a right line.

15. To find the equation to a line which shall pass through the intersection of two given lines.

Solution.—The point of intersection may be found as by Article 41, and the equation determined by Article 38. An equation may, however, be written without finding the point of intersection. Let the equations be

$$Ax + By + C = 0, \text{ and } A'x + B'y + C' = 0.$$

Multiplying the former by some constant as m and the latter by another constant as n, and adding the results, we have

$$m(Ax + By + C) + n(A'x + B'y + C') = 0$$

for the required equation. For it is the equation to some line, being of the first degree, and for the point of intersection of the lines both terms will be zero; hence the new line will pass through that point. Since m and n are arbitrary there may be an infinite number of lines through the point. If n be fractional, all possible lines may be represented when $m = 1$.

16. Find the equation to the line which passes through the point of intersection of the lines $3x - 4y - 2 = 0$, and $2x + 7y - 5 = 0$.

17. Find the equation to a right line which bisects the angle between the lines

$$3y + 4x = 12, \text{ and } 5y - 2x = 0.$$

Solution.—Let x', y' be any point on the bisecting line, then will the perpendiculars from this point on the lines be equal to each other. Using the value of P in the answer to the 11th example above, we shall have for the perpendicular upon the first line

$$\frac{4x' + 3y' - 12}{\sqrt{4^2 + 3^2}},$$

and on the second

$$\frac{-2x' + 5y'}{\sqrt{4 + 25}}.$$

Placing these equal to each other and reducing gives

$$31 \cdot 52x' - 8 \cdot 86y' = 64 \cdot 56.$$

But since x' and y' are the coördinates of any point on the line they may be considered as running variables, and the accents may be omitted. Hence the equation of the bisecting line is

$$31 \cdot 52x - 8 \cdot 86y = 64 \cdot 56.$$

The line perpendicular to this line bisects the supplementary angle.

18. Find the equations of the lines bisecting the angles between the lines whose equations are

$$12x + 5y = 8, \text{ and } 3x - 4y = 3.$$

Ans. $21x + 77y - 1 = 0$,
$99x - 27y - 79 = 0$.

CHAPTER III.

TRANSFORMATION OF COÖRDINATES.

48. Certain investigations are more easily made when the locus is referred to a particular system of coördinates than to any other; and it is often convenient to change from one system to another by means of general formulas. This may be done by finding the relations between the coördinates of any point in the two systems. The first system is called the *primitive*, the second, the *new* system; and the required formulas are called the *Equations for the Transformation of Coördinates*.

49. Formulas for passing from one Rectilinear System to another, *the axes being parallel but the origin different.*

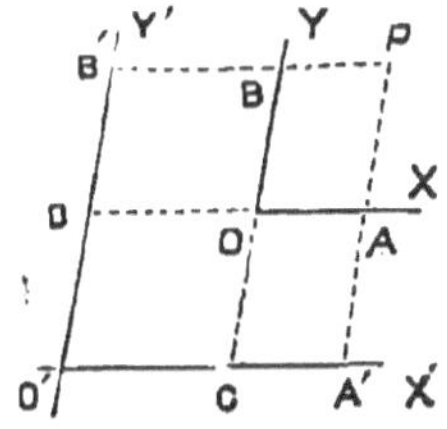

Let P be the point whose coördinates in reference to the system $Y'O'X'$ are

$$O'A' = x_1, \ O'B' = y_1,$$

and in reference to the system YOX, are

$$OA = x, \ AP = y;$$

Fig. 25.

it is required to find the relation between these coördinates.

We have $\quad x_1 = O'A' = O'C + CA' = O'C + OA,$

$\quad y_1 = O'B' = O'D + DB' = O'D + OB.$

Let the coördinates of the origin O in reference to O' be

$$O'C = m, \ CO = n,$$

and the preceding equations become

$$x_1 = m + x,$$
$$y_1 = n + y.$$

50. Formulas for passing from one Rectilinear System to another the origin being the same.

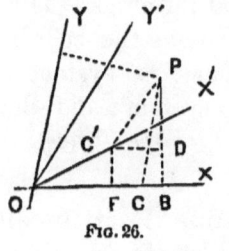
Fig. 26.

Let YOX be the primitive system, and $Y'OX'$ the new system.

Let P be any point. Draw PC' parallel to OY', PC parallel to OY, PB and $C'F$ perpendicular to OX, and $C'D$ parallel to OX. Then

$$OC = x,\ CP = y,\ OC' = x',\ C'P = y'.$$

Also let $\quad \omega = YOX,\ \alpha = X'OX,\ \beta = Y'OX.$

Then $\quad PB = PD + DB = PD + C'F$
$$= y' \sin \beta + x' \sin \alpha.$$

But we also have $\quad PB = CP \sin PCB$
$$= y \sin \omega;$$

$$\therefore\ y \sin \omega = x' \sin \alpha + y' \sin \beta. \tag{a}$$

Dropping perpendiculars from P and C' upon OY, and proceeding as before we find

$$x \sin \omega = x' \sin(\omega - \alpha) + y' \sin(\omega - \beta). \tag{b}$$

51. Formulas for passing from a Rectilinear System YOX to another Rectilinear System $Y'O'X'$ the origin being different.

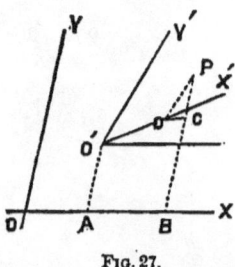
Fig. 27.

The solution consists in transforming the equations from the system YOX to another parallel system whose origin is at O', and then to the system $Y'O'X'$. This may be done by substituting the values of x and y from the equations of Article 49, which are

$$x = x_1 - m,$$
$$y = y_1 - n,$$

in the equations of the preceding article. This being done, and the subscripts dropped, we have

$$\left. \begin{array}{l} (x - m) \sin \omega = x' \sin (\omega - \alpha) + y' \sin (\omega - \beta) \\ (y - n) \sin \omega = x' \sin \alpha + y' \sin \beta \end{array} \right\} ; \quad (c)$$

which are the most *general* equations for the transformation.

52. Formulas for passing from a Rectangular to an Oblique System of Coördinates.

In this case the axes of the primitive system make a right angle with each other; hence, $\omega = 90°$. Substituting this value in the equations of Article 51, we have,

$$\left. \begin{array}{l} x = m + x' \cos \alpha + y' \cos \beta \\ y = n + x' \sin \alpha + y' \sin \beta \end{array} \right\} . \quad (d)$$

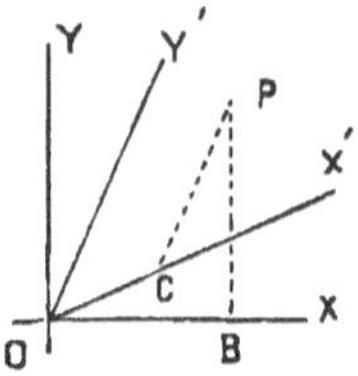

Fig. 28.

If the origin is the same in both systems, we have $m = 0$ and $n = 0$, and the *Equations for passing from Rectangular to Oblique Coördinates the origin being the same*, become

$$\left. \begin{array}{l} x = x' \cos \alpha + y' \cos \beta \\ y = x' \sin \alpha + y' \sin \beta \end{array} \right\} . \quad (e)$$

If the axes OX and OX' coincide, $\alpha = 0$ and the equations become

$$\left. \begin{array}{l} x = x' + y' \cos \beta \\ y = y' \sin \beta \end{array} \right\} . \quad (f)$$

53. Formulas for passing from Oblique to Rectangular Axes, *the origin remaining the same.*—Eliminating y' from equations (e) and finding x'; then eliminating x' and finding y' gives

$$\left. \begin{array}{l} x' = \dfrac{x \sin \beta - y \cos \beta}{\sin (\beta - \alpha)} \\[1em] y' = \dfrac{y \cos \alpha - x \sin \alpha}{\sin (\beta - \alpha)} \end{array} \right\} ; \quad (g)$$

which are the required equations.

54. Formulas for passing from one Rectangular System to another *the origin remaining the same.*—For this case make $m = 0$, $n = 0$, $\omega = 90°$, $\beta = 90° + \alpha$, in the equations of Article 51, and find

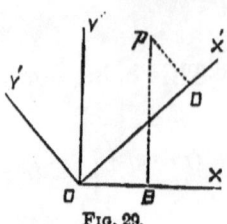
Fig. 29.

$$\left. \begin{array}{l} x = x' \cos \alpha - y' \sin \alpha \\ y = x' \sin \alpha + y' \cos \alpha \end{array} \right\} ; \quad (h)$$

which are the required equations.

If $Y'OX'$ be the primitive system and we pass back to the system YOX, the required formulas will be found by eliminating x' and y' successively from equations (h). We thus find

$$\left. \begin{array}{l} x' = x \cos \alpha + y \sin \alpha \\ y' = - x \sin \alpha + y \cos \alpha \end{array} \right\} . \quad (h')$$

Or these equations may be found by changing α to $-\alpha$, x to x', y to y', x' to x, and y' to y in equations (h). From either sets of equations, we find

$$x^2 + y^2 = x'^2 + y'^2 ;$$

which is the square of the distance of any point from the origin, and is constant in reference to every system of rectangular axes having the same origin.

55. Formulas for passing from Rectangular to Polar Coördinates.

Let OX' be the initial line, making an angle α with the axis of x, φ the vectorial angle, $x = OB$, and $y = PB$. We have from the figure, *the pole being at the origin,*

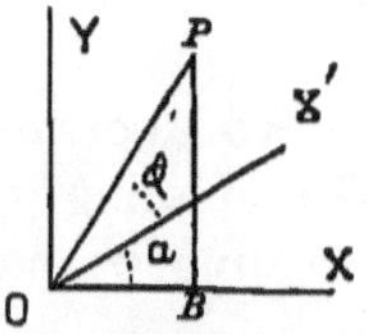

Fig. 30.

$$\left. \begin{array}{l} x = \rho \cos (\varphi + \alpha) \\ y = \rho \sin (\varphi + \alpha) \end{array} \right\} ; \quad (i)$$

which are the required equations.

If the initial line coincides with the axis of x, we have $\alpha = 0$, and

$$\left. \begin{array}{l} x = \rho \cos \varphi \\ y = \rho \sin \varphi \end{array} \right\} . \quad (j)$$

If the pole be not at the origin let its coördinates be x_0, and y_0, and we have

$$x = x_0 + \rho \cos (\varphi + \alpha) \brace y = y_0 + \rho \sin (\varphi + \alpha)}. \quad (k)$$

If the axes are oblique, making an angle ω between them, the formulas become

$$x = x_0 + \frac{\rho \sin [\omega - (\varphi + \alpha)]}{\sin \omega} \brace y = y_0 + \frac{\rho \sin (\varphi + \alpha)}{\sin \omega}}. \quad (l)$$

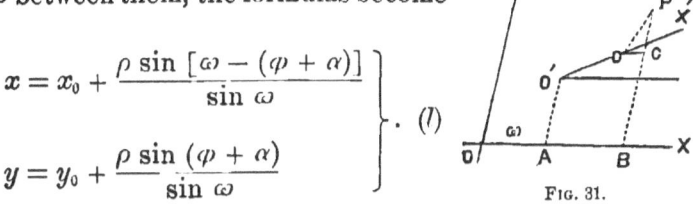

Fig. 31.

To pass from Polar to Rectangular Coördinates, the pole being at the origin, we find from equations (j)

$$x^2 + y^2 = \rho^2 (\sin^2 \varphi + \cos^2 \varphi) = \rho^2 ; \brace \therefore \rho = \sqrt{x^2 + y^2}; \brace \cos \varphi = \frac{x}{\sqrt{x^2 + y^2}}, \sin \varphi = \frac{y}{\sqrt{x^2 + y^2}}}. \quad (m)$$

EXAMPLES.

1. To find the polar equation to a right line.

Solution.—The rectangular equation is

$$y = mx + b.$$

Let the pole be at the origin, then equations (i) give

$$\rho \sin (\varphi + \alpha) = m\rho \cos (\varphi + \alpha) + b;$$

which is the required equation.

Let α be the angle between the axis of x and the perpendicular (p) from the origin, then will

$$m = -\cot \alpha = -\frac{\cos \alpha}{\sin \alpha}, \text{ and } p = b \sin \alpha.$$

Substituting the values of m and b, we find

$$\rho \cos \varphi = p.$$

If the vectorial angle be also measured from the axis of x, and be denoted by φ', then $\varphi = \varphi' - \alpha$, and the preceding equation becomes,—dropping the accent,—

$$\rho \cos (\varphi - \alpha) = p,$$

which is the equation given in Article 36.

2. To find the equation to a right line referred to oblique axes.

Solution.—The equation of the right line referred to rectangular axes is

$$y = mx + b.$$

The equations for transformation the origin remaining the same will be equations (*e*) Art. 52; hence we have

$$x' \sin \alpha + y' \sin \beta = m (x' \cos \alpha + y' \cos \beta) + b \; ;$$

or $\qquad (\sin \beta - m \cos \beta) y' = (m \cos \alpha - \sin \alpha) x' + b,$

which is the required equation.

Let the axis of x' coincide with the axis of x, then $\alpha = 0$, $m = \dfrac{\sin \theta}{\cos \theta}$, and we have

$$(\sin \beta \cos \theta - \cos \beta \sin \theta) y' = \sin \theta . x' + b \cos \theta;$$

or $\qquad \sin (\beta - \theta) \; y' = \sin \theta . x' + b \cos \theta;$

$$\therefore y' = \frac{\sin \theta}{\sin (\beta - \theta)} x' + \frac{\cos \theta}{\sin (\beta - \theta)} b.$$

If $x' = 0$, then $y' = \dfrac{\cos \theta}{\sin (\beta - \theta)} b$, which is the *y-intercept*; call it b', and the equation becomes,—dropping the accents,—

$$y = \frac{\sin \theta}{\sin (\beta - \theta)} x + b,$$

which is the same as the equation in Article 27.

3. Given the point (3, 7), required its coördinates for another parallel system the coördinates of whose origin are (2, — 2). *Ans.* (1, 9).

4. Given the point (— 2, 3) to find its polar coördinates, the pole being at the origin and the initial line making an angle of 45° with the axis of x. *Ans.* $\rho = \sqrt{13}$, $\varphi = \tan^{-1} (-\tfrac{1}{2}) - 45°$.

5. Given the point (3, 4) to find its polar coördinates, the pole being at the point (6, 8) and the initial line making an angle of 30° with the axis of x.

6. Transform to rectangular axes the equation $\rho^2 = c^2 \cos 2\varphi$, the origin being at the pole and the initial line coinciding with the axis of x.

Ans. $(x^2 + y^2)^2 = c^2 (x^2 - y^2)$.

CHAPTER IV.

CONIC SECTIONS,—THEIR EQUATIONS AND PROPERTIES.

56. THE curves known as the Circle, Ellipse, Hyperbola, and Parabola are called Conic Sections, because they may be formed by the intersection of a plane with a cone, as will be shown hereafter. But the investigations in this chapter will have no reference to the cone. The equation of the curve will be deduced from a definition of some property of the curve. These curves are discussed by themselves because of their great importance in the Physical and Astronomical Sciences.

EQUATIONS OF THE CURVES.

The Circle.

57. The Circle is a curve every point of which is at a constant distance from a fixed point called the centre. It may be described mechanically by means of a pencil at some point P of a string OP moving around the fixed point O. The locus of the point P will be a circle.

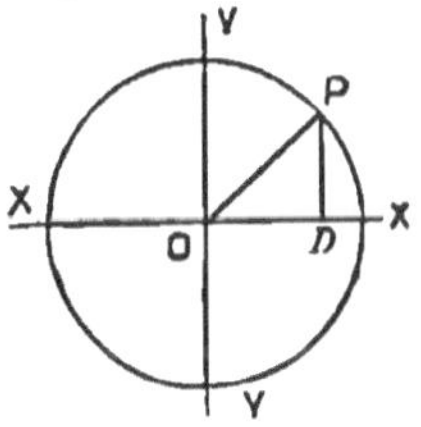

To find the Equation to the Circle the origin being at the centre.

Take any point P in the circumference and drop the perpendicular PD. Let $OP = R$ the radius of the circle, $x = OD$ and $y = PD$, then will the right triangle OPD give

$$x^2 + y^2 = R^2, \qquad (a)$$

Fig. 32.

which is the equation sought. It may be put under the form

$$\frac{x^2}{R^2} + \frac{y^2}{R^2} = 1. \qquad (b)$$

58. General Equation to the Circle *referred to rectangular coördinates.*—Let O be the origin of coördinates, C the centre of the circle, P any point whose coördinates are (x_1, y_1) and (m, n) the coördinates of the centre. According to Article 49, we have, for the transformation of coördinates from C to O,

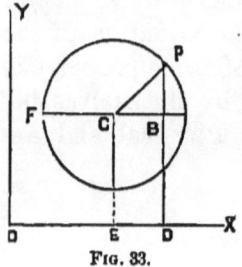

Fig. 33.

$$x = x_1 - m;$$
$$y = y_1 - n;$$

which substituted in equation (a) (Art. 57), and dropping the subscripts, since x_1 and y_1 are now general variables, gives

$$(x - m)^2 + (y - n)^2 = R^2; \qquad (c)$$

which is the equation sought.

59. Equation of the Circle referred to a diameter and tangent at its vertex.—For this case $m = R = OC$, and $n = 0$, and the preceding equation becomes

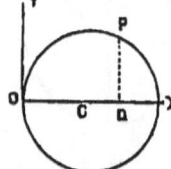

$$y^2 = 2Rx - x^2, \qquad (d)$$

Fig. 34. which is the required equation.

60. Discussion of Equation (a), **Art. 57.**—The equation is $x^2 + y^2 = R^2$. Let $x = 0$, then

$$y = \pm R;$$

that is, the locus cuts the axis of y in two points equidistant from the centre.

Let $y = 0$, then

$$x = \pm R;$$

that is, the locus cuts the axis of x in two points equidistant from the centre.

Solving the equation for y gives

$$y = \pm \sqrt{R^2 - x^2},$$

which shows that for all values of $\pm x$ less than R, y has two equal and opposite values; hence the curve, lies both above and below the axis of x and is symmetrical in respect to that axis.

If x exceeds R the value of y will be imaginary, which shows that no part of the curve is at a distance greater than R to the right or left of the centre. Similar properties may be shown in regard to the axis of y.

61. *To construct the locus whose equation is $x^2 + y^2 = 25$.*
Let $x = 0$, then $y = \pm 5$.

$x = 1, \quad y = \pm \sqrt{24} = \pm 4.9$ nearly.
$x = 2, \quad y = \pm \sqrt{21} = \pm 4.6$ "
$x = 3, \quad y = \pm \sqrt{16} = \pm 4.$
$x = 4, \quad y = \pm \sqrt{9} = \pm 3.$
$x = 5, \quad y = 0.$

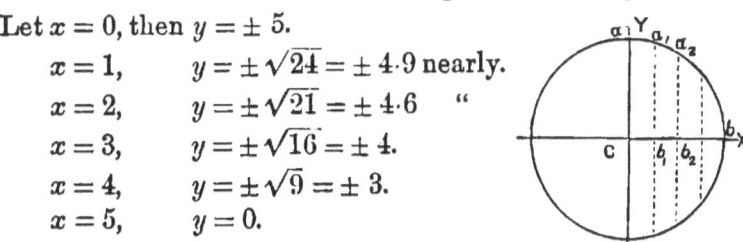

Fig. 35.

Lay off above and below the centre, on the axis of y, a distance Ca equal to 5. Then lay off to the right of the centre a distance Cb_1 equal to 1, and erect an ordinate b_1a_1 above and below the axis of x, equal to 4.9; and so on to $x = 5 = Cb$, which lay off on the axis of x. In this way any number of points may be obtained and the curve may be traced through them.

62. Discussion of Equation (d).—Let $x = 0$, then $y = 0$, hence the curve passes through the origin, (Art. 32).

Let $y = 0$, then $x = 0$ and $2R$, hence the curve cuts the axis of x in two points, one at the origin, and the other at a distance $2R$ from the origin.

Solving for y, gives

$$y = \pm \sqrt{(2R - x)x},$$

hence the curve is symmetrical in reference to the axis of x.

Letting x be negative, we have

$$y = \pm \sqrt{(2R + x)(-x)};$$

which being imaginary shows that no part of the curve lies on the negative side of the origin.

Solving for x gives

$$x = R \pm \sqrt{R^2 - y^2}$$

which shows that the curve is not symmetrical in reference to the axis of y. If y be less than R, there will be two real values for x, but if it exceeds R, x will be imaginary.

63. The Centre of a Curve *is a point which bisects any line drawn through it terminated by the curve.*

64. A Diameter *is a line which bisects a system of parallel chords.* Every diameter passes through the centre, but every line which passes through the centre may not be a diameter, as may be illustrated by higher plane curves. But it will be found that every line which passes through the centre of a conic section, will be a diameter. **An Axis to a Curve** *is a diameter* which is perpendicular to the chords which it bisects. The point where *any* diameter cuts the curve is called **the vertex** of that diameter.

EXAMPLES.

1. What are the coördinates of the centre of a circle whose equation is $(x - 3)^2 + (y + 4)^2 = 16$?

2. Form the equation to the circle whose centre is $(4, -3)$, and radius $= 3$.

3. Construct a circle whose equation is $(y - 3)^2 = 4x - x^2$.

4. In the circle $y^2 + (x - 4)^2 = 16$, what are the ordinates at the point whose abscissa is 7?

5. If the base of a triangle is constant, and the adjacent sides vary in such a way as to preserve a constant angle between them; show that the locus of the apex is a circle.

The Ellipse.

65. An Ellipse is a curve, *the sum of the distances from any point of which to two fixed points is constant.*

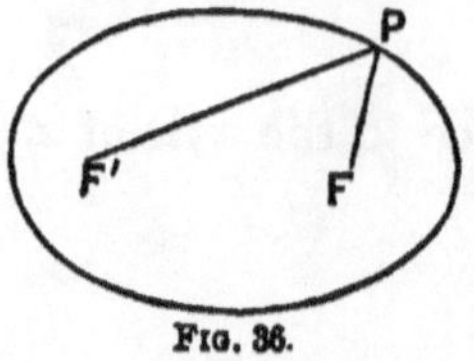

FIG. 36.

66. To trace the curve mechanically, fix two points of a string respectively at the points F and F', and place a pencil point at P. Slide the pencil around, keeping the string constantly stretched; then will the point P describe an ellipse. For, in every position of P, we have

$$F'P + PF = a \text{ constant}.$$

67. Definitions.—The two fixed points F and F' are called the **foci** of the ellipse. The point C midway between the foci is called the **focal centre**, or simply *the centre*. The lines PF' and PF, drawn from any point P of the ellipse to the foci, are called the **focal radii**. The diameter $A'A$, passing through the foci, is the **transverse** or **major axis**, and the diameter BB' perpendicular to AA' is the **conjugate**, or **minor axis**.

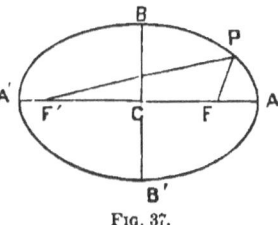

Fig. 37.

68. Construction of the Ellipse by points.—Take any line as $A'A$ for the major axis; bisect it at C and erect the perpendicular BB'. If the length of the minor axis is also assumed, take BC equal to the semi-axis and with B as a centre and AC as a radius describe an arc, cutting $A'A$ in the points F' and F. These points will be the foci, for we have

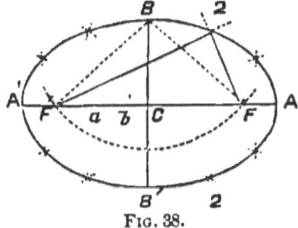

Fig. 38.

$$BF' + BF = A'A = a \text{ constant}.$$

Or assume the foci F' and F, and divide the distance CF' into several parts, Cb, ab, etc., equal or unequal. Take the distance Ab as a radius and F'' as a centre, describe an arc above and below the axis AA'; and with bA' as another radius and F as a centre describe arcs cutting the former ones in 2 and 2. These will be points in the curve. In this manner any number of points may be found through which the curve may be traced. With the radius $F'2$, two other arcs may be described having F as a centre; so that with each radius four arcs may be described, two having F as a centre and two others having F' as another centre. If AC

be one radius, CA' will be the other, and the intersection of the arcs described with them having respectively the centres F' and F'' willgive the extremities B and B' of the minor axis.

An ellipse may also be constructed by points by means of its equation. Thus, if the equation to the ellipse be (see equation (a_1) of the next Article),

$$3x^2 + 4y^2 = 16,$$

points in the locus may be found by assuming values for y and deducing the corresponding values for x. Thus if

then
$$y = \pm 0, \pm 1, \pm 2;$$
$$x = \pm 2\cdot 3, \pm 2, \pm 0;$$

which gives eight points in the curve. Any number of intermediate points may be found in a similar manner and the corresponding points constructed as shown for the circle in Article 61, through which the curve may be traced.

69. Axial Equation to the Ellipse; or in other words, to find the equation to the ellipse when its axes coincide with the coördinate axes.

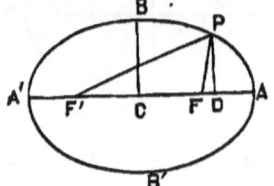

Fig. 39.

Let $\quad 2c = FF'$; $2a = A'A$.

From any point P of the ellipse drop the perpendicular PD; then will

$$x = CD, y = PD,$$
$$FD = x - c, F'D = x + c.$$

From the right-angled triangles FDP and $F''DP$, we have

$$F'P = \sqrt{(x+c)^2 + y^2};$$
$$FP = \sqrt{(x-c)^2 + y^2}.$$

Adding, we have, since $FP + F'P = AA', = 2a$,

$$\sqrt{(x+c)^2 + y^2} + \sqrt{(x-c)^2 + y^2} = 2a.$$

Freeing this equation of radicals, gives

$$(a^2 - c^2) x^2 + a^2 y^2 = a^2 (a^2 - c^2).$$

For the purpose of simplifying this equation, let

$$a^2 - c^2 = b^2,$$

and the equation becomes

$$b^2 x^2 + a^2 y^2 = a^2 b^2, \qquad (a_1)$$

which is the required equation. Dividing through by $a^2 b^2$ gives

$$\frac{x^2}{a^2} + \frac{y^2}{b^2} = 1, \qquad (b_1)$$

or $\qquad a^{-2} x^2 + b^{-2} y^2 = 1,$

which forms are sometimes more convenient than equation (a_1).

70. Discussion of Equation (a_1), or (b_1).—Let $y = 0$, then

$$x = \pm a;$$

hence, the curve cuts the axis of x at two points, A and A', equidistant to the right and left of the centre.

Let $x = 0$, then

$$y = \pm b;$$

hence, the curve cuts the axis of y at two points equidistant from the centre, and the line BB', Fig. 40, is a diameter.

Solving the equation for y gives,

$$y = \pm \frac{b}{a} \sqrt{a^2 - x^2};$$

hence, for all values of x less than a, there are two equal and opposite values for y; therefore the curve is symmetrical in reference to the transverse axis. If $x > a$, y is imaginary. In a similar way we may show that the curve is symmetrical in reference to the minor axis.

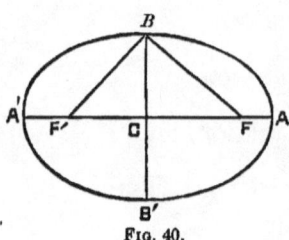

Fig. 40.

The value of b^2 given above is

$$b^2 = a^2 - c^2;$$
$$\therefore \sqrt{b^2 + c^2} = a;$$

that is, the distance from either extremity of the semi-conjugate axis to either focus equals the semi-major axis; a result which was deduced in Article 68 directly from the definition of the curve.

71. A General Equation to the Ellipse, *the coördinate axes being parallel to the axes of the ellipse.*—Let O be the origin of coördinates, C the centre of the ellipse whose coördinates are

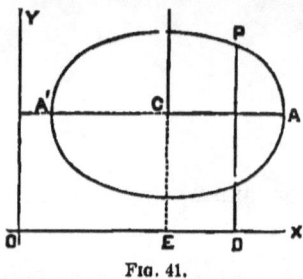

Fig. 41.

$$OE = m,\ EC = n;$$

P any point whose coördinates are $OD = x_1,\ DP = y_1;$

then, according to Article 49, we have for transferring the origin from C to O

$$x = x_1 - m;$$
$$y = y_1 - n;$$

which substituted in equation (a_1) (Art. 69) give, after dropping the subscripts,

$$b^2(x - m)^2 + a^2(y - n)^2 = a^2 b^2; \qquad (c_1)$$

which is the required equation.

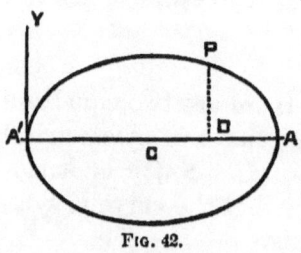

Fig. 42.

72. Rectangular Equation to the Ellipse, the origin being at the vertex of the major axis, *the axis of x coinciding with the major axis.*—In this case we have

$$m = A'C = a,\ n = 0;$$

hence, equation (c_1) becomes,

$$y^2 = \frac{b^2}{a^2}(2ax - x^2), \qquad (d.)$$

which is the required equation.

EXAMPLES.

1. Find the axial equation of the ellipse, the major axis being 7 and the minor axis 3.
2. Construct the equation of the ellipse, the foci being (3, 0), (− 3, 0) and the vertex of the semi-conjugate axis being (0, 4).
3. The vertex of the semi-transverse axis being (5, 0) and one of the foci (3, 0), form the equation to the ellipse.
4. In an ellipse the sum of the focal radii is 18, and the square of the distance between the foci less 8^2 is 36, find the equation to the ellipse.
5. Find the axial equation of an ellipse which shall pass through the points (− 3, 2), (−4, 1).
6. If the origin is at the vertex of the conjugate axis, and the axis of y coincides with that axis, show that the equation to the ellipse becomes $x^2 = \frac{a^2}{b^2}(2by - y^2)$.

7. The lower end of a bar whose length is $2l$ slides upon a horizontal plane while its upper end slides along a vertical plane; what curve will a fly describe that remains upon the bar at a distance d from the centre while the bar slides down?

8. The ordinate of a circle $x^2 + y^2 = r^2$ is increased by a line equal in length to n times the ordinate; show that the locus of the end of the ordinate thus increased is an ellipse.

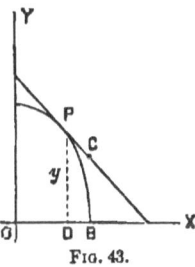
Fig. 43.

The Hyperbola.

73. An Hyperbola *is a curve the difference of the distances of any point of which from two fixed points is constant.*

74. Mechanical Construction.—Let F' and F'' be the fixed points. Let the ruler $F'B$ be pivoted at F'. Take a string whose length BPF is less than $F'B$, and fasten one end at F and the other at some point B on the ruler. With a pencil point at P keep the string pressed against the ruler as the latter

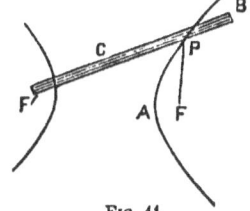
Fig. 44.

is turned about the point F'; then will the point P describe the arc of an hyperbola. For if $PC = PF$ we have

$$BF' - (BP + PC) = CF';$$
or $\quad (BF' - BP) - PC = CF'.$
But $\quad BF' - BP = PF';$
$\quad \therefore PF' - PC = CF',$
or $\quad PF' - PF = CF' = a\ constant;$

hence the locus of P will be an hyperbola. By pivoting the ruler at F, another branch of the curve may be described.

75. Construction by Points.—Join $F'F$ and let $A'A$ be the constant difference between PF' and PF, P being any point on the curve. Make $F'V = A'A$. With F' as a centre and a radius $F'a$ greater than $F'A'$ describe arcs above and below the line AA'. Do the same with F as a centre. Then with a radius equal to Va, and with F as a centre describe arcs intersecting the former ones. Do the same with F' as a centre. The intersection of the corresponding arcs will be points in the curve, for, by the construction, $PF' - PF = F'V = $ a constant. In this way any number of points may be located through which the curve may be traced.

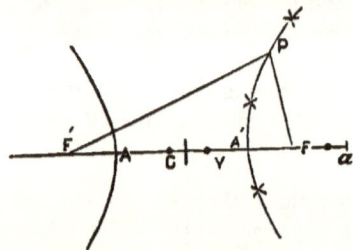
Fig. 45.

76. Definitions.—The fixed points F' and F are called the **foci** of the hyperbola. The line AA' is the **transverse axis.** The lines $F'P$ and FP are the **focal radii.** The points A' and A are the *vertices* of the curve. The point C, midway between the foci, is **the centre,** or *focal centre*, of the curve.

77. A Branch of a Curve is a continuous portion of the curve. The hyperbola has two branches, the ellipse and circle each, one branch. Some curves have many branches.

78. Equation of the Hyperbola referred to its axes.—The origin will be at the centre C, midway between

THE HYPERBOLA.

the foci F and F'. Let the axis of x coincide with the axis CA of the curve, and from any point P of the curve drop the perpendicular PD; then will

$$x = CD\ ;\ y = DP.$$

Let $\quad AA' = 2a\ ;\ FF' = 2c,$

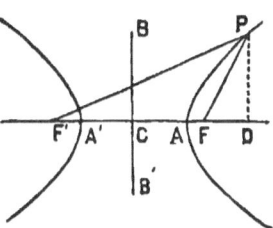

Fig. 46.

then $\quad CA = a\ ;\ CF = c\ ;\ F'D = x + c\ ;\ FD = x - c.$

The right angled triangles $F'DP$ and FDP, give

$$F'P = \sqrt{(x+c)^2 + y^2}\ ;$$
$$FP = \sqrt{(x-c)^2 + y^2}\ ;$$

and from the definition of the curve we have

$$F'P - FP = 2a = \sqrt{(x+c)^2 + y^2} - \sqrt{(x-c)^2 + y^2}.$$

Clearing of radicals and reducing gives

$$(c^2 - a^2) x^2 - a^2 y^2 = a^2 (c^2 - a^2)\ ;$$

and making $\quad c^2 - a^2 = b^2,$
the equation becomes

$$b^2 x^2 - a^2 y^2 = a^2 b^2, \qquad (a_2)$$

which is the required equation. It may be written, by dividing through by $a^2 b^2$,

$$\frac{x^2}{a^2} - \frac{y^2}{b^2} = 1 \qquad (b_2)$$

or, $\quad a^{-2} x^2 - b^{-2} y^2 = 1.$

Equation (b_2) is the equation of the curve in terms of its intercepts.

79. Discussion of Equation (a_2). — Let $y = 0$, then $x = \pm a$; hence the curve cuts the axis of x in two points A and A' equidistant from the origin. AA' is a diameter, and also an axis of the curve, and C is the centre.

Let $x = 0$, then $y = \pm b\sqrt{-1}$, which, being imaginary, shows that the curve does not cut the axis of y. The *position* of this imaginary line corresponds with that of the conjugate axis of the ellipse; hence, by way of analogy, and in an analytical sense, we speak of the *conjugate axis* of the hyperbola and represent its half length by b, the *real* part of the preceding expression.

From the expression $c^2 - a^2 = b^2$, given in the preceding Article, we have

$$c^2 = a^2 + b^2;$$

hence: *The distance from the centre to either focus equals the hypothenuse of the triangle constructed on the semi-axes.*

If $a = b$ in Eq. (a_2) we have

$$x^2 - y^2 = a^2$$

which is called the equation to the *equilateral hyperbola*. It resembles somewhat the equation to the circle.

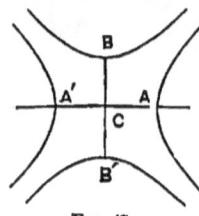

Fig. 47.

If an hyperbola be constructed having its transverse axis coincident and equal to BB', the axis conjugate to AA', it is called a conjugate hyperbola in reference to the former one. Either hyperbola is called a conjugate in reference to the other. The equation to the conjugate hyperbola will be

$$\frac{y^2}{b^2} - \frac{x^2}{a^2} = 1.$$

Solving equation (a_2) in reference to y gives

$$y = \pm \frac{b}{a}\sqrt{x^2 - a^2},$$

which shows that for all values of x less than a, y is imaginary, and for all values greater than a, y has two real values, equal and opposite. Hence the curve lies both above and below the axis of x, and is symmetrical in reference to that axis.

Solving in reference to x, we have

$$x = \pm \frac{a}{b}\sqrt{y^2 + b^2},$$

which is real for all values of y; hence, for every positive or negative value of y, there are two real, equal and opposite values of x, or: *The curve is symmetrical in reference to the axis of y.*

80. A General Equation to the Hyperbola, *the axes of coördinates being parallel to the axes of the curve.*

Let the coördinates of the centre in reference to the new axes be

$$OE = m\;;\; EC = n\;;$$

and of any point P, (x_1, y_1). Then, according to Article 49, we have

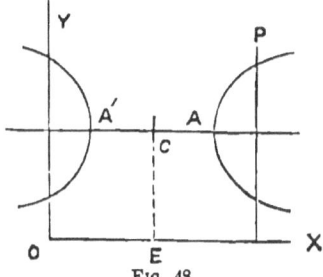

Fig. 48.

$$x = x_1 - m,$$
$$y = y_1 - n\;;$$

which values substituted in equation (a_2) give,—after dropping the subscripts,—

$$b^2(x - m)^2 - a^2(y - n)^2 = a^2 b^2, \qquad (c_2)$$

which is the required equation.

81. Equation to the Hyperbola, the origin being at the left vertex of the Curve, *the axis of x coinciding with the transverse axis.*—We will have $m = a$ and $n = 0$ in equation (c_2), and the equation becomes

$$y^2 = \frac{b^2}{a^2}(x^2 - 2ax). \qquad (d_2)$$

EXAMPLES.

1. Find the axial equation to the hyperbola, the transverse axis being 7 and the distance between the foci being 9.

2. The square of the conjugate axis being -9, and the transverse axis being 4, find the axial equation to the hyperbola.

3. Find the equation to the hyperbola, the origin being at the left vertex, the major axis being 12, and the distance between the foci 16.

4. In a given hyperbola, the difference of the focal radii $=8$, and the difference between the squares of that difference and the distance between the foci $=-9$; find the equation.

5. The ordinate of an hyperbola is prolonged so as to equal the corresponding focal radius; find the locus of the extremity of the prolongation.

6. Show that the equation of the hyperbola whose real axis is the conjugate axis of the hyperbola $b^2x^2 - a^2y^2 = a^2b^2$, is $a^2y^2 - b^2x^2 = a^2b^2$

The Parabola.

82. The Parabola is a curve *every point of which is equally distant from a fixed line and fixed point.* If F is the fixed point and GE the fixed line, then for any point P on the curve we have $PF = PG$.

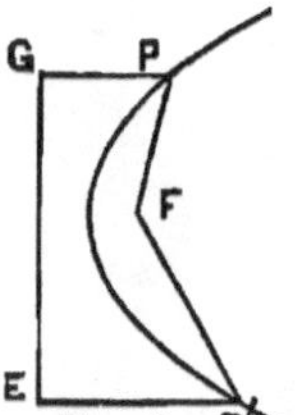

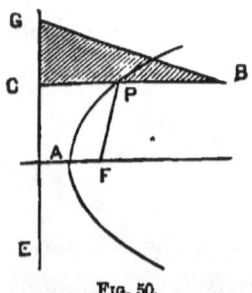

Fig. 49.

Fig. 50.

83. To describe the curve mechanically. Let GE be a fixed line and F a fixed point. Take a board or ruler having a right angle at C, and attach one end of a string at B. Stretch the string along the edge BC, and swing the end C around to the point F and secure it there. With a pencil point P press the string against the edge BC while the ruler is moved along the fixed line, the pencil being permitted to move along the edge BC; the curve traced by the point will be the required curve. For the length of the string will be

$$BP + PC = BP + PF,$$

and, subtracting BP from both sides of the equation, we have

$$PC = PF,$$

which is the condition required by Article 82.

84. Definitions.—The fixed point F is called the **focus**. The fixed line GO is called the **directrix**. The straight line through the focus, perpendicular to the directrix, is called the **axis** *of the parabola;* the point A where the axis cuts the curve is called the **vertex**; and the line FP, drawn from the focus to any point of the curve, is called the **focal radius**.

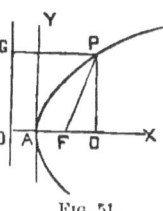

Fig. 51.

85. To Construct a Parabola by Points.—Assume a fixed line GO and a point F. Through F draw a line FO perpendicular to GO. Bisect OF at A, then will A be one point in the curve, for it is equidistant from the fixed line and point. To find another point, assume any radius as FP greater than FA, and with F as a centre describe an arc and intersect it by a line drawn parallel to OG and at a distance from it equal to FP; then will the point of intersection P be a point on the curve. For by the construction we will have $GP = PF$. In a similar manner any number of points may be found.

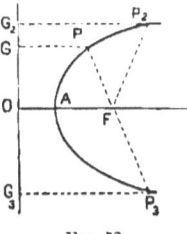

Fig. 52.

The points may also be found by means of the equation of the curve. See equation (d_3) of the following Article, and Article 61.

86. Equation to the Parabola the Origin being at the Vertex of the Curve.—Let the axis of x coincide with the axis of the curve, and the axis of y be tangent to it.

We have, (Fig. 51),

$$x = AD, \ y = PD,$$

and letting p be the constant distance OF, we have

$$OA = AF = \tfrac{1}{2}p;$$
$$GP = OD = x + \tfrac{1}{2}p = PF;$$
$$FD = x - \tfrac{1}{2}p;$$

and the triangle FDP gives

$$FP^2 = FD^2 + DP^2;$$

or
$$(x + \tfrac{1}{2}p)^2 = (x - \tfrac{1}{2}p)^2 + y^2;$$
which reduced gives
$$y^2 = 2px \qquad (d_3)$$
for the required equation.

87. Discussion of Equation (d_3). — If $x = 0$, $y = 0$, hence the curve passes through the origin of coördinates. This also follows from the fact that the equation has no absolute term, (Art. 32).

Solving for y gives
$$y = \pm \sqrt{2px}$$
which is real for every positive value of x, and gives two equal and opposite values for y. Therefore, the curve is symmetrical in reference to the axis of x. If x be negative, y will be imaginary, hence the curve extends only in the positive direction of x. It has only one branch. As x increases indefinitely, y also increases indefinitely, hence the branch is infinite and the curve is not reëntrant. Strictly speaking, therefore, it cannot have a centre, but, for the sake of symmetry in the *form of expression*, we say that *its centre is at an infinite distance from the vertex*. Since all diameters of a curve pass through the centre, it follows that *all diameters of a parabola are parallel to the axis*.

Let $x = \tfrac{1}{2}p$, the abscissa of the focus, then
$$y = p,$$
hence the ordinate at the focus equals twice the distance of the focus from the vertex; and :—*The double ordinate at the focus equals four times the distance of the focus from the vertex.*

88. Equation to the Parabola the Origin being at any Point *and the axis of x parallel to the axis of the curve.*—Let the vertex of the curve in reference to the new origin be (m, n), x and y the coördinates of any point when the origin is at A; x_1, y_1 the corresponding coördinates of the same point when the origin is at O, then according to Article 49, we have
$$x = x_1 - m, \quad y = y_1 - n,$$

which in equation (d_3) give, after dropping the subscripts,

$$(y-n)^2 - 2p(x-m) = 0. \quad (c_3)$$

If the *origin* be at B, the intersection of the axis and directrix, we have $n = 0$, and $m = BA = \tfrac{1}{2}p$, and equation (c_3) becomes

$$y^2 = 2p(x - \tfrac{1}{2}p),$$

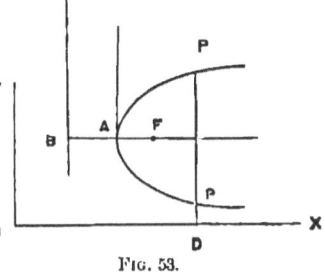

Fig. 53.

which is *the equation to the parabola referred to its axis and directrix*. If the *origin* be removed to the *focus*, the axes remaining parallel, we have

$$y^2 = 2p(x + \tfrac{1}{2}p).$$

EXAMPLES.

1. Find the rectangular equation to a parabola in which the coördinates of the focus are (4, 0), the origin being at the vertex.
2. The origin being at the vertex, find the rectangular equation to a parabola which shall pass through the point $x' = 4$, $y' = 16$.
3. The distance of the focus from the directrix being 8, find the equation of the parabola referred to the axis and directrix.
4. Show that the locus of the centre of the circle which is tangent to the axis of y and passes through a fixed point on the axis of x is a parabola.

89. The equations to the circle, ellipse, and hyperbola have been given when the origin is at the centre of the curve, but no such equation can be given for the parabola, since it has no centre at a finite distance, as shown in Article 87.

The Equations of the Conic Sections Compared.

90. When the Curves are referred to their Axes we have found their equations to be; see equations (b), (b_1), (b_2), of the preceding pages; for

The Circle $\qquad \dfrac{x^2}{R^2} + \dfrac{y^2}{R^2} = 1; \qquad (b)$

The Ellipse $\quad \dfrac{x^2}{a^2}+\dfrac{y^2}{b^2}=1;\quad$ (b_1)

The Hyperbola $\quad \dfrac{x^2}{a^2}-\dfrac{y^2}{b^2}=1;\quad$ (b_2)

The Parabola $\quad$ (*No finite equation.*)

A comparison of these equations shows that, if in the equation of the ellipse $a = b = R$, it becomes the equation of the circle. Also, in the equation of the ellipse, if we make $b^2 = -b^2$ it becomes the equation of the hyperbola.

We shall therefore, when the curves are referred to their axes, determine a required property for the ellipse, and deduce the corresponding property for the circle by making $a = b$ in the result, and for the hyperbola by making $b^2 = -b^2$.

[We may readily conceive, how, for a given major axis these curves pass from one to the other. Thus, in a circle the foci may be considered as consecutive with the centre; and if the foci separate from each other, the major axis remaining constantly equal to the diameter of the circle, we have an ellipse whose minor axis b is constantly diminishing; and when the foci reach the ends of the major axis the ellipse becomes a right line, and when they pass those points, the ellipse changes to an hyperbola in which the conjugate axis will increase as the foci are more and more separated.]

91. When the Curves are referred to their axes and a tangent at the left vertex, we have; see equations (*d*), (d_1), (d_2), (d_3); for

The Circle $\quad y^2 = 2Rx - x^2;\quad$ (*d*)

The Ellipse $\quad y^2 = \dfrac{2b^2}{a}x - \dfrac{b^2}{a^2}x^2;\quad$ (d_1)

The Hyperbola $\quad y^2 = -\dfrac{2b^2}{a}x + \dfrac{b^2}{a^2}x^2;\quad$ (d_2)

The Parabola $\quad y^2 = 2px;\quad$ (d_3)

all of which are included under the general form

$$y^2 = Px + Rx^2. \qquad (D)$$

In this equation R is the ratio of the squares of the semi-axes, and P is called the principal parameter of the curve (Arts. 95, and 145). Hence for the Conics

	the value of the parameter,	and the ratio of the squares of the semi-axes, is
for the circle	$P = 2R,$	$R = -1;$
ellipse	$P = \dfrac{2b^2}{a},$	$R = -\dfrac{b^2}{a^2};$
hyperbola	$P = -\dfrac{2b^2}{a},$	$R = \dfrac{b^2}{a^2};$
parabola	$P = 2p,$	$R = 0.$

A comparison of these equations also shows, that if, in the equation of the ellipse, we make $a = b = R$ it becomes the equation to the circle; and if $b^2 = -b^2$ it becomes the equation to the hyperbola. Hence, when the curves are referred to corresponding vertices and axes, the properties of the circle and hyperbola may be deduced from those of the ellipse, by substituting for the value of b in the results for the ellipse, the values given above for the respective curves.

92. General Equation of the Second Degree.—All the equations of preceding Articles are of the second degree, and it will be shown, (Art. 177), that every equation of the second degree represents a conic.

The general equation of the second degree may be written

$$Ax^2 + 2Hxy + By^2 + 2Gx + 2Fy + C = 0.$$

It is written this way so as to conform to certain modern usage. It is shown in Article 178 that this equation will represent

a circle if $A = B$ and $H = 0$;
an ellipse if $H^2 - AB$ is negative;
an hyperbola if $H^2 - AB$ is positive;
a parabola if $H^2 - AB = 0$.

EXAMPLES.

Determine to what locus the following equations belong:

$$x^2 + y^2 + 3y - 4x + 2 = 0.$$
$$3x^2 + 5xy + 7y^2 - 2x + 4y - 8 = 0.$$
$$2x^2 + 4xy + y^2 + 3x - 2y - 1 = 0.$$
$$5x^2 - 3xy - 2y^2 - 2x - 7 = 0.$$
$$4x^2 + 12xy + 9y^2 - x + 10y - 4 = 0.$$
$$2y^2 - 3x + 4 = 0.$$

Eccentricity.

93. The Eccentricity of a Conic Section is *the ratio of the distance of the focus from the centre to the length of the semi-transverse axis.* In other words, if the semi-transverse

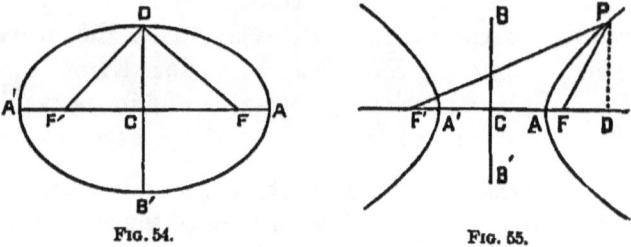

Fig. 54.   Fig. 55.

axis were unity, the eccentricity would be the distance of the focus from the centre.

Let e = the eccentricity;
 c = the distance of either focus from the centre, = $CF = CF'$;

then we have for *the ellipse*,

$$e = \frac{CF}{CA} = \frac{c}{a} = \frac{\sqrt{a^2 - b^2}}{a} = \sqrt{1 - \frac{b^2}{a^2}}; \qquad (1)$$

and changing b^2 to $-b^2$, we have for *the hyperbola*,

$$e = \frac{\sqrt{a^2 + b^2}}{a} = \sqrt{1 + \frac{b^2}{a^2}}; \qquad (2)$$

and making $a = \infty$, (Art. 87), we have for *the parabola*
$$c = 1; \qquad (3)$$
and making $a = b$, we have for *the circle*
$$c = 0.$$

From these we observe that for

The Circle	$e = 0$;
The Ellipse	$e < 1$;
The Hyperbola	$e > 1$;
The Parabola	$e = 1$.

From equation (1) we find
$$b^2 = a^2(1 - e^2), \qquad (4)$$
from which the value of the semi-minor axis may be found in terms of the eccentricity and the semi-major axis.

If $e = 0$, $b = a$.
If $e < 1$, $b < a$.
If $e > 1$, $b = \sqrt{-1}\, a\, (e^2 - 1)^{\frac{1}{2}}$.
If $e = 1$, $b = 0 \times \infty$, which is indeterminate, although in this case, we know from Article 87, that $b = \infty$.

Remark.—Here we have another analogy between the several conic sections. To illustrate, Let F and F' be the foci of an ellipse, C its centre, and AY a tangent at the principal vertex. Suppose that the foci approach uniformly towards C, the curve being of such variable dimensions as to remain constantly tangent to AY; all such curves will be ellipses, but when F and F' become consecutive to C, the curve becomes a circle also tangent to AY, and having zero for its eccentricity, and AC for its radius. Beginning again with the foci, as shown in the figure, let the right focus F' move to the right, the focus F remaining fixed, and, therefore, the distance AF remaining constant; then, for all finite values of FF' the curve will be an ellipse, the eccentricity of which may be determined as shown above. But as the centre C moves to the right, the distances AC and FC approach equality, and for $FC = \infty$, we have $AC = \infty$; hence, ultimately,

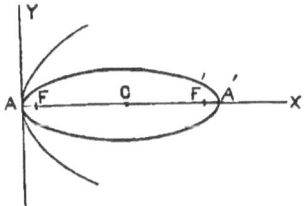

Fig. 56.

$$\frac{FC}{AC} = 1,$$

and the finite portion of the curve which is that in the vicinity of A, becomes a parabola.

When the eccentricity exceeds unity, the vertex of the curve which disappeared at the right of A, will reappear at A', at the left of A, the distance AA' at first being $-\infty$,* but constantly decreasing (numerically) as the eccentricity increases. The finite portion of this curve is an hyperbola. This statement is here made without proof, but it can be shown to be true in a beautiful manner by means of a cone and cutting plane,

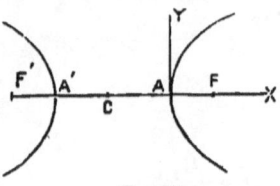

Fig. 57.

(see Art. 185). If A' continues to approach A, and finally coincides with it, the eccentricity becomes infinite, and the hyperbola becomes the straight line AP; for the difference of the distances PF' and PF will be constant, being zero. Hence a straight line is one limit of the hyperbola. If now the foci F' and F approach A and finally coincide with it, the two branches of the hyperbola become any two right lines passing through it; which is another limiting case of the hyperbola.

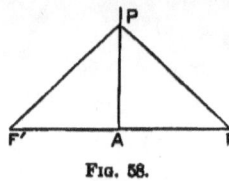

Fig. 58.

93a. **Eccentric Angle.**—If a circle be described on the major axis of an ellipse, and an ordinate BP' be erected, and CP, CP' be drawn, the angle $P'CA$ is called the eccentric angle in reference to the point P.

Let $x' = CB$, $y' = PB$, $\varphi = P'CA$, $CA = a$, $CD = b$,

then $x' = a \cos \varphi$, $\qquad y' = b \sin \varphi$,

the value of x' being deduced directly from the figure, and of y' by substituting the value of x' in the axial equation to the ellipse and solving for y'.

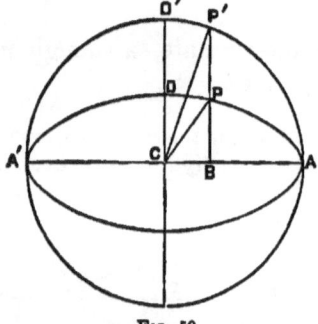

Fig. 59.

The auxiliary angle φ is useful in solving certain problems (see example 64, p. 141).

---

*[There are many cases in which a function passes from $+\infty$ to $-\infty$ for a finite increase of the variable. Thus, in the equation $y = \tan x$, y passes from 0 to $+\infty$, as x increases from 0 to $\frac{1}{2}\pi$; and as x passes $\frac{1}{2}\pi$, y changes from $+\infty$ to $-\infty$. Similar conditions exist for the equations $y = \sec x$; $y = \cot x$; $y = \dfrac{1}{\sin x}$; etc.]

Latus Rectum.

94. The Latus Rectum of a Conic Section *is the double ordinate to the transverse axis through the focus.* This is also called *the principal parameter*, or *parameter of the curve.*

95. Value of the Latus Rectum.—For *the ellipse*, make $x = \pm c$ in equation (a_1), (Art. 69), and solve for y; double the ordinate thus found will be the value sought. We have

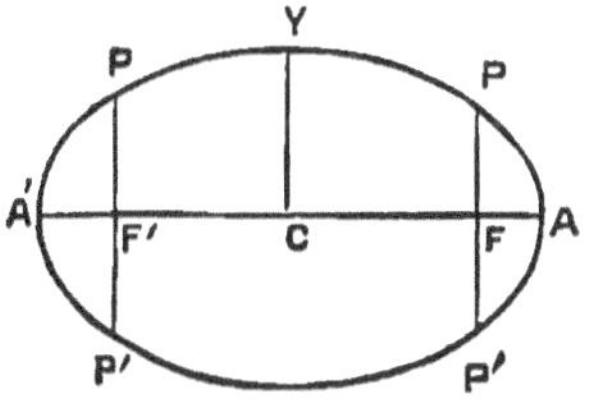

$$y^2 = \frac{b^2}{a^2}(a^2 - c^2)$$

$$= \frac{b^4}{a^2}, \text{ (since } a^2 - c^2 = b^2\text{)};$$

$$\therefore y = \pm \frac{b^2}{a}; \tag{1}$$

FIG. 60.

in which the positive value is the ordinate above the major axis, and the negative value the part below. As these are equal in value, the double ordinate at the focus will be, numerically, twice the value of either;

$$\therefore PP' = \frac{2b^2}{a} = \frac{(2b)^2}{2a} = 2a\,(1 - e^2). \tag{2}$$

Similarly, for *the hyperbola*,

$$y = \mp \frac{b^2}{a};$$

$$\therefore PP' = \frac{(2b)^2}{2a} = \frac{(BB')^2}{A'A} = 2a(e^2 - 1); \tag{3}$$

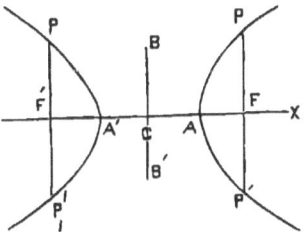

FIG. 61.

that is: *In the ellipse and hyperbola, the latus rectum is a third proportional to the transverse axis and its conjugate.*

These expressions are also true when $a = b = R$, hence, true for the circle; and, by analogy, we may say that the principal parameter of a circle is any diameter.

For the **Parabola**, make $x = \frac{1}{2}p$ in equation (d_3), Article 86, which equation is

$$y^2 = 2px,$$

and we find

$$2y = PP' = 2p = 4OF;$$

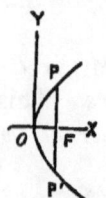
Fig. 62.

that is: *The latus rectum of a parabola equals four times the distance of the focus from the vertex of the curve.*

96. Remark.—We now see that the coefficients of x in the equations of Article 91 are the parameters of the respective curves, and are represented by P in equation (D). We see from inspection that we may write, for the square of the ratio of the semi-axes,

$$R = \frac{P}{2a},$$

in which the parameter P may be constant, while R varies inversely as the semi-major axis a. The equation to the conic may, therefore, be written, the origin being at the principal vertex,

$$y^2 = Px + \frac{P}{2a}x^2 = \frac{P}{2a}(2ax + x^2).$$

If $a = \infty$, $\dfrac{P}{a} = 0$, and we have $y^2 = Px$, which is the equation of the parabola.

EXAMPLES.

1. What is the principal parameter to the curve

$$3x^2 + 4y^2 = 12\ ?$$

2. What is the principal parameter to the curve

$$2x^2 - 7y^2 = 8\ ?$$

3. What is the principal parameter to the curve

$$4y^2 = 12x\ ?$$

Of Ordinates.

97. Let the point P be (x', y'), and P', (x'', y''), then equation (b_1), (Art. 69), gives for the ellipse:

$$y'^2 = \frac{b^2}{a^2}(a^2 - x'^2);$$

$$y''^2 = \frac{b^2}{a^2}(a^2 - x''^2);$$

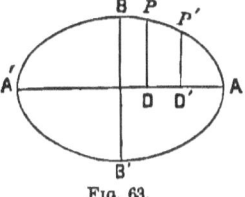

Fig. 63.

hence, $\quad y'^2 : y''^2 :: (a^2 - x'^2) : (a^2 - x''^2)$

$$:: (a + x')(a - x') : (a + x'')(a - x'')$$

$$:: A'D.DA : A'D'.D'A;$$

which is also true for the circle and hyperbola, since it does not contain b. Hence, for the *circle, ellipse,* and *hyperbola: The squares of the ordinates to the major axis are proportional to the products of the corresponding segments into which the axis is divided by the ordinates.*

From the preceding proportion, we find

$$\frac{A'D.DA}{DP^2} = \frac{A'D'.D'A}{D'P'^2},$$

which is a convenient form for memorizing.

If a circle be described on the major axis of an ellipse, and an ordinate $P'D$ be erected, we have, from the equations of the curves,

$$P'D^2 = a^2 - x^2, \ PD^2 = \frac{b^2}{a^2}(a^2 - x^2);$$

$$\therefore P'D : PD :: a : b,$$

or $2a : 2b;$

that is: *If a circle be described on the major axis of an ellipse, and*

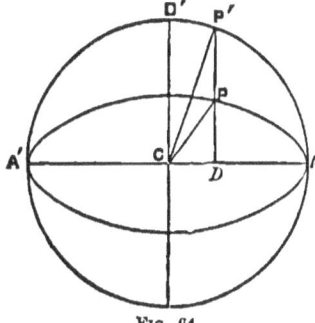

Fig. 64.

a common ordinate be erected to that axis, the ordinate of the circle will be to the ordinate of the ellipse, as the major axis of the ellipse is to its minor axis.

A similar proportion may be found if a circle be described on the minor axis, and an ordinate be erected to that axis.

For the Parabola we have

$$y'^2 = 2px'; \qquad y''^2 = 2px'';$$

$$\therefore y'^2 : y''^2 :: x' : x''$$

that is: *The squares of the ordinates to the axis of the parabola, are as the corresponding abscissas.*

Of Intersections and Tangents.

98. To find the points of Intersection of a Right Line with an Ellipse.—The axial equation to the ellipse is

$$b^2x^2 + a^2y^2 = a^2b^2;$$

and the equation of the right line

$$y = mx + d;$$

(d being used instead of the b heretofore given so as not to confound it with the b in the equation of the ellipse). The coördinates of the points of intersection must satisfy both equations; hence, considering the equations as simultaneous, and eliminating y, we find

$$x = \frac{-a}{b^2 + a^2m^2}\left[amd + \sqrt{a^2m^2d^2 - (b^2 + a^2m^2)(d^2 - b^2)}\right]. \quad (1)$$

If the quantity under the radical be positive, there will be two real points of intersection; if it be negative, the points will be imaginary, or, in other words, the line will not intersect the curve; but if the radical part is zero, there will be only one point, and the line will be tangent to the curve. The last condition requires that we have

$$a^2m^2d^2 = (b^2 + a^2m^2)(d^2 - b^2);$$

$$\therefore d = \sqrt{a^2m^2 + b^2}, \quad (2)$$

which substituted in the preceding equation of the right line, gives

$$y = mx + \sqrt{a^2m^2 + b^2} \ ; \tag{3}$$

and every equation of this form is the equation of a tangent to an ellipse.

For the Hyperbola, this becomes by changing b^2 to $-b^2$,

$$y = mx + \sqrt{a^2m^2 - b^2} \ ; \tag{4}$$

and for **the Circle**, making $a^2 = b^2 = R^2$ in equation (3),

$$y = mx + R\sqrt{m^2 + 1}. \tag{5}$$

For the Parabola we may find

$$y = mx + \frac{p}{2m}. \tag{6}$$

These are called *the Magical Equations* to the tangent.

The magical equation of the tangent to the circle is easily found geometrically.

Let BT be the tangent at S, CS the radius, and

$$m = \tan BTC = \tan BCS = BS \div CS;$$

then

$$\sec SCB = \sqrt{\tang^2 + 1} = \sqrt{m^2 + 1} \ ;$$

$$\therefore BC = R\sqrt{m^2 + 1} \ ;$$

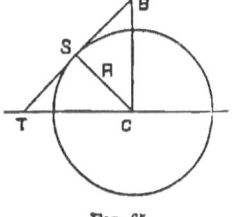

Fig. 65.

which is the intercept on the axis of y; hence the equation of TB, which is of the form

$$y = mx + b,$$

becomes $\qquad y = mx + R\sqrt{m^2 + 1}$

as given above.

Similarly, the points of intersection of any two curves may be found by considering their equations as simultaneous and eliminating first one variable and then the other. Curves, or lines, represented by equations of the second degree are called curves of *the second order*. Two curves of the second order will, in general, intersect each other in four points; for the elimination between two *general* equations of the second degree gives rise to an equation of the fourth degree of which there will be four roots.

EXAMPLES.

1. Find the points of intersection of the lines
$$x^2 + y^2 = R^2;\ y = 3x - 4.$$

2. Find the points of intersection of the lines
$$y^2 = 4x;\ y = -2x + 5.$$

3. Find the points of intersection of the curves
$$y^2 = 9x;\ x^2 + y^2 + 3x - y = 16.$$

4. Find the points of intersection of the parabola $y^2 = 2x + 10$, and the hyperbola $16x^2 - 4y^2 = 8$.

5. Find the equation of a tangent to the ellipse $3x^2 + 7y^2 = 8$, the tangent being inclined 45° to the axis of x.

Of Tangents and Subtangents.

99. A Tangent to a curve is a right line which passes through two consecutive points of the curve, or which touches the curve in one point. Let P and P' be any two points of a curve through which a secant is passed. Let the secant turn about the point P, the point P' remaining in the secant and moving towards P; when P' becomes consecutive to P, or when it falls upon and coincides with P, the line $T'PT$ will be tangent to the curve.*

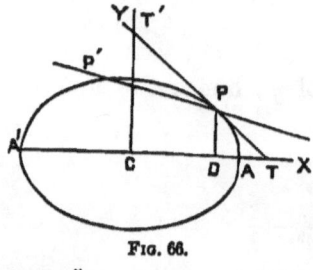

Fig. 66.

---

* In elementary geometry a tangent is defined to be a line which touches a curve in one point only. While this definition answers for many curves,

100. Equation of a Tangent to the Ellipse.—Take the axial equation

$$b^2x^2 + a^2y^2 = a^2b^2;$$

and let a secant be passed through the points (x', y'), (x'', y''). The coördinates of these points must satisfy the equation of the ellipse; hence we have the *equations of condition*

$$b^2x'^2 + a^2y'^2 = a^2b^2;$$

$$b^2x''^2 + a^2y''^2 = a^2b^2.$$

Subtracting one from the other and factoring, we have

$$\frac{y'' - y'}{x'' - x'} = -\frac{b^2}{a^2} \cdot \frac{x' + x''}{y' + y''}. \qquad (1)$$

The equation to a right line passing through these points is, (Art. 40),

$$y - y' = \frac{y'' - y'}{x'' - x'}(x - x');$$

in which substitute the value of the left member of equation (1), and we have

$$y - y' = -\frac{b^2}{a^2} \cdot \frac{x' + x''}{y' + y''}(x - x'). \qquad (2)$$

When the points through which the secant passes become consecutive, we have

$$x' = x'', \text{ and } y' = y'';$$

---

including the conic sections, it is not general; for there are many curves in which a line tangent at one point of a curve may cut it in several other points. The definition in the text involving two consecutive points, is not only the more general, but it is the most useful in making investigations. Students at first, are generally slow to admit that a line passing through two points, though they be consecutive, is a tangent, for they affirm that such a line is a secant. But a tangent may be considered as a special case of a secant, as a circle is a special case of an ellipse; and when the secant passes through *two consecutive points* its position will differ from that tangent which passes through only one of them by less than any assignable quantity. The expressions therefore for the slope of such a secant will be the same as for the tangent, (see Art. 23). We may say then that *a tangent is a secant which passes through two consecutive points of a curve.*

and these values substituted in the preceding equation give

$$b^2 x'x + a^2 y'y = b^2 x'^2 + a^2 y'^2 = a^2 b^2; \qquad (3)$$

for the required equation. Dividing through by $a^2 b^2$, it becomes

$$\frac{x'x}{a^2} + \frac{y'y}{b^2} = 1; \qquad (e)$$

the *form* of which is similar to that of the ellipse.

101. Equation of the Tangent to the Circle, *the origin being at the centre.*

Make $a = b = R$ in equation (e), and we have

$$\frac{x'x}{R^2} + \frac{y'y}{R^2} = 1, \qquad (e_1)$$

or
$$x'x + y'y = R^2;$$

for the required equation.

[This result may easily be deduced directly from the figure. For we have

$$m = \tan PTX = -CT' \div CT = -CD \div PD = -\frac{x'}{y'},$$

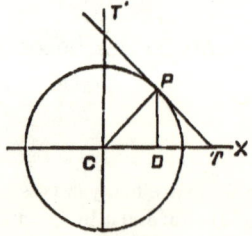

Fig. 67.

and

$$CT' : CP :: CP : PD; \therefore CT' = \frac{R^2}{y'}.$$

The equation of the line TT' will be of the form

$$y = mx + b,$$

$$= -\frac{x'}{y'} x + \frac{R^2}{y'},$$

which becomes
$$xx' + yy' = R^2,$$

as given above.]

102. Equation of the Tangent to the Hyperbola, *the curve being referred to its axes.* — Writing $-b^2$ for b^2, in equation (e) in Article 100, gives

$$\frac{xx'}{a^2} - \frac{yy'}{b^2} = 1, \quad (e_2)$$

which is the required equation.

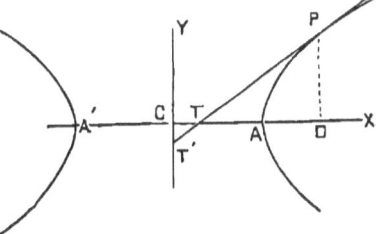

Fig. 68.

103. Equation of the Tangent to the Parabola, *the curve being referred to its axis, and the tangent at its vertex.* — The equation to the curve will be

$$y^2 = 2px;$$

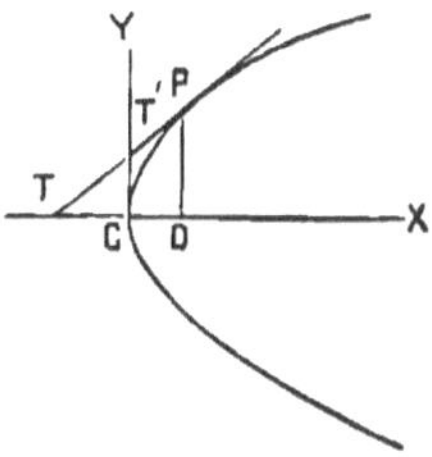

and *the equations of condition* for a secant will be

$$y'^2 = 2px'; \quad y''^2 = 2px'';$$

Fig. 69.

from which we find

$$\frac{y'' - y'}{x'' - x'} = \frac{2p}{y'' + y'}.$$

But the equation of the right line passing through two points is, (Art. 40),

$$\frac{y - y'}{x - x'} = \frac{y'' - y'}{x'' - x'} = \frac{2p}{y'' + y'}, \text{ (from the preceding equation). Making } y'' = y', \text{ we find}$$

$$y'y = p(x + x'), \quad (e_3)$$

which is the required equation.

These equations may be discussed in the same manner as the equation to any other right line.

104. Intercepts of the Tangent to a Conic Section.— *For the intercept on the axis of x,* make $y = 0$ in equations (e),

(e_1), (e_2), (e_3) and we have for

$$\text{the ellipse} \quad CT = \frac{a^2}{x'};$$

$$\text{the circle} \quad CT = \frac{R^2}{x'};$$

$$\text{the hyperbola} \quad CT = \frac{a^2}{x'};$$

$$\text{the parabola} \quad CT = -x'.$$

To find the intercept on the axis of y make $x = 0$ in the same equations, and we find for

$$\text{the ellipse} \quad CT' = \frac{b^2}{y'};$$

$$\text{the circle} \quad CT' = \frac{R^2}{y'};$$

$$\text{the hyperbola} \quad CT' = -\frac{b^2}{y'};$$

$$\text{the parabola} \quad CT' = p\frac{x'}{y'}.$$

EXAMPLES.

1. Find the equation of the tangent to the ellipse $3x^2 + 5y^2 = 10$ at a point whose abscissa is 1; also find its intercepts and construct the line.

2. Find the equation of the tangent to the circle $x^2 + y^2 = 13$ at the point (3, −2), and construct the line.

3. Find the equation of a tangent to the hyperbola $3x^2 - 4y^2 = 12$ at the point where the latus rectum cuts the curve.

4. Find the equation of a tangent to the parabola $y^2 = 4x$, and determine its equation when it is inclined 30° to the axis of x.

(Observe that, in equation (e_3), $\frac{p}{y'} = $ *tan. of the inclination*, or *the slope* as it is sometimes called.)

104a. *To find the equation of the tangent to an ellipse in terms of the eccentric angle.*

The coördinates of the point will be (Art. 93a),

$$x' = a \cos \varphi, \; y' = b \sin \varphi;$$

and these values in the equation of the tangent give

$$a \sin \varphi \cdot y + b \cos \varphi \cdot x = ab,$$

for the required equation.

The intercepts will be

$$x = \frac{a}{\cos \varphi},\ y = \frac{b}{\sin \varphi}.$$

105. The Subtangent *is the distance between the foot of the ordinate of contact and the foot of the tangent. The foot of the tangent* is the point where it intersects either axis of coördinates, but unless otherwise mentioned, the foot will be considered as on the axis of x. Let PT be a tangent at the point P, then will PD be the ordinate of contact, D the foot of the ordinate, T the foot of the tangent, and DT the subtangent. The subtangent is the projection of the tangent on the axis of x. Let $x_2 = CT$, the intercept of the tangent on the axis of x, and $x' = CD$, the abscissa of contact; then will

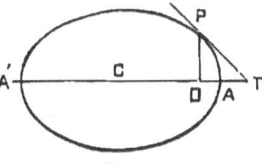

Fig. 70.

$$DT = x_2 - x'.$$

Substituting the value of $x_2 = CT$, from the preceding Article, we have for the *ellipse, hyperbola*, and *circle*,

$$DT = \frac{a^2}{x'} - x' = \frac{a^2 - x'^2}{x'} = \frac{(a+x')(a-x')}{x'}$$

$$= \frac{A'D \cdot DA}{CD}; \quad (1)$$

hence: *The subtangent of the* ELLIPSE, HYPERBOLA *and* CIRCLE *is a fourth proportional to the segments of the major axis formed by the ordinate of contact, and the abscissa to the point of contact.*

In the hyperbola the axis is not divided by the ordinate, but in order to generalize the principle we consider that the axis is prolonged, and define a segment as the distance from the foot of the ordinate to either extremity of the axis.

For *the parabola* we have $CD = x'$, (Fig. 69), and $CT = -x'$, (Art. 104); hence disregarding the sign of CT, we have

$$TD = x' + x' = 2x';$$

hence: *The subtangent of the parabola is bisected at the vertex of the curve.*

[OBS.—Considering CD as positive, CT will be negative, and we have.

$$DT = + CD - CT$$
$$= x' - (-x') = 2x',$$

as before.]

Length of the Tangent.

106. In all the Conics we have, (Figs. 68, 69, 70),

$$TP = \sqrt{TD^2 + DP^2} = \sqrt{TD^2 + y^2}.$$

Also, $$TP = \frac{y'}{\sin T}.$$

EXAMPLES.

1. What is the length of the subtangent to the ellipse $5x^2 + 7y^2 = 35$ at a point whose abscissa is 2? and what is the length of the tangent for the same point?
2. What is the length of the subtangent to the circle $x^2 + y^2 = 25$ at the point $(-3, -4)$? and what is the length of the tangent?
3. What is the length of the tangent, and of the subtangent to the parabola $y^2 = 9x$ at the point $(4, 6)$?

107. In a Conic Section the Acute Angles between the tangent and focal radii, at any point, are equal to each other.—Let P be any point on *the ellipse*, PT the tangent, PF' and PF the focal radii; then will

$$TPF = T'PF'.$$

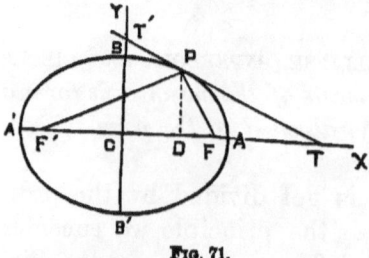

Fig. 71.

The equation to the right line passing through the

points F and P will be, (Art 40),

$$y - y' = \frac{y'' - y'}{x'' - x'}(x - x');$$

in which the point F is ($y'' = 0$, $x'' = c$), and P, (x', y'); and hence the equation becomes

$$y - y' = \frac{y'}{-c + x'}(x - x').$$

Similarly, the equation of the line $F'P$ will be

$$y - y' = \frac{y'}{c + x'}(x - x').$$

The equation of the tangent line PT is, (Eq. (e) Art. 100),

$$y = -\frac{b^2 x'}{a^2 y'}x + \frac{b^2}{y'};$$

hence, according to Article 44, we have

$$\tan F'PT = \frac{-\dfrac{b^2 x'}{a^2 y'} - \dfrac{y'}{c + x'}}{1 + \dfrac{y'}{c + x'}\left(-\dfrac{b^2 x'}{a^2 y'}\right)} = -\frac{b^2(a^2 + cx')}{cy'(a^2 + cx')} = -\frac{b^2}{cy'}; \quad (1)$$

and

$$\tan FPT = \frac{-\dfrac{b^2 x'}{a^2 y'} - \dfrac{y'}{-c + x'}}{1 + \dfrac{y'}{-c + x'}\left(-\dfrac{b^2 x'}{a^2 y'}\right)} = \frac{b^2(a^2 - cx')}{cy'(a^2 - cx')} = \frac{b^2}{cy'}. \quad (2)$$

But from the figure we have

$$\tan T'PF' = \tan(180° - F'PT) = -\tan F'PT;$$

which compared with equations (1) and (2) gives

$$T'PF' = FPT. \quad (3)$$

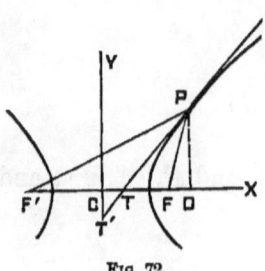

Fig. 72.

For the Hyperbola we also have

$$T'PF' = FPT.$$

In this curve the focal radii lie on opposite sides of the tangent, while for the ellipse they are on the same side.

In the Parabola, the equation of the tangent TP is, (Art. 103, Eq. (e_3)),

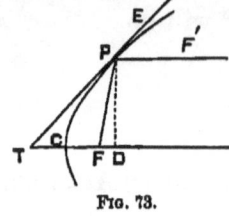

Fig. 73.

$$y = \frac{p}{y'}(x + x'),$$

in which

$$\frac{p}{y'} = \tan PTF.$$

The equation of the line PF is, (Eq. (5), Art. 40),

$$y - y' = \frac{y'' - y'}{x'' - x'}(x - x');$$

in which $y'' = 0$, $x'' = \tfrac{1}{2}p = CF$, $x' = CD$, $y' = DP$; and the equation becomes

$$y - y' = \frac{-y'}{\tfrac{1}{2}p - x'}(x - x');$$

hence, according to Article 44, we have

$$\tan TPF = \frac{\dfrac{-y'}{\tfrac{1}{2}p - x'} - \dfrac{p}{y'}}{1 + \dfrac{p}{y'}\left(\dfrac{-y'}{\tfrac{1}{2}p - x'}\right)} = \frac{p}{y'}; \qquad (4)$$

that is

$$TPF = PTF = EPF'; \qquad (5)$$

PF' being a diameter passing through P. In order that the *wording* of the proposition at the beginning of this Article shall apply strictly to the parabola, it is necessary to consider the diameter PF' as a focal radius, as we have previously done, (Art. 87, and *Remark* in Art. 93).

Since the angle T equals TPF, we have

$$TF = FP, \qquad (6)$$

and the triangle TFP is isosceles.

For the Circle, the focal radii coincide and form the radius, hence the tangent will be perpendicular to the radius, as is well known.

In the ellipse and parabola the tangent bisects the *external* angle formed by the focal radii, but in the hyperbola the *internal* angle is bisected.

108. Construction of the Tangent to a Conic Section.—*A.* LET THE TANGENT BE DRAWN THROUGH A POINT ON THE CURVE. 1°. *By means of focal radii.* On the focal radii, or on one of them and on the other prolonged if necessary, take equal distances

$$PH = PF,$$

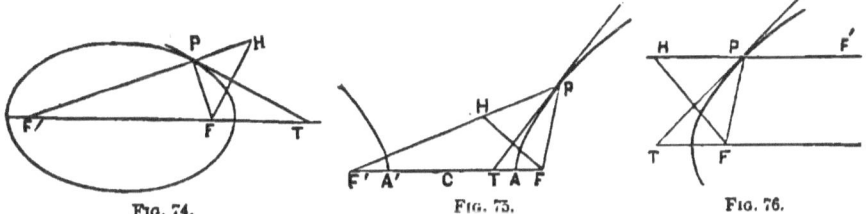

Fig. 74.   Fig. 75.   Fig. 76.

and join F and H. Through P draw a line PT perpendicular to FH, and it will be the tangent required; for it bisects the angle FPH formed by the focal radii.

[The mode of constructing the curves by means of a string leads to the same construction. For the string from one focal radius will be elongated an amount equal to that by which the other is shortened; hence PH will represent the *rate* of shortening of PF', and PF the corresponding *rate* of elongating PF, and the direction of the pencil point will be that of the resultant of these two rates, and will, therefore, bisect the angle between them.]

2°. *By means of the Subtangent.* According to Article 105, the subtangent is independent of the minor axis. To apply the method to the ellipse, draw a circle on $A'A$ as a diameter, and through the given point P erect an ordinate $P'PD$, and at P' in the circle draw a tangent $P'T$.

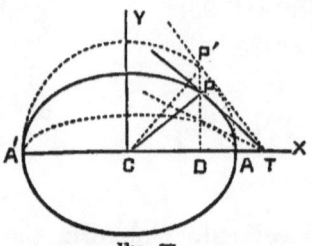

Fig. 77.

The line TP, passing through the points T and P, will be the tangent required.

For **the parabola**, let P be the point; drop the perpendicular PD on the axis CD, and take CT equal to CD on the axis prolonged; then will PT, drawn through P and T, be the tangent required. For the subtangent TD will be bisected at the vertex, (Art. 105).

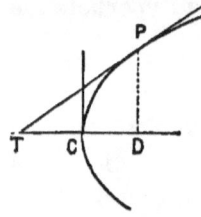

Fig. 78.

3°. *By means of the intercept of the tangent on the axis of x*, especially for **the hyperbola**. This method is equivalent to the former one, for the subtangent is deduced at once by means of the intercept. Let P be the point of which CD is the abscissa $= x'$. With C as a centre, and CD as a radius, describe an arc, and at the vertex A erect an ordinate AM, prolonging it till it intersects the arc DM at M. Join C and M, and with a radius CA, and centre C, describe an arc intersecting CM at N. Drop the perpendicular NT; then will the line PT, drawn through P and T, be the tangent required. For we have

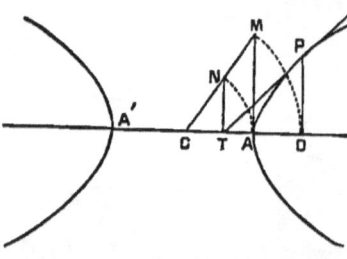

Fig. 79.

$$CM : CA :: (CN = CA) : CT;$$
or
$$x' : a :: a : CT$$
$$\therefore CT = \frac{a^2}{x'};$$

which, according to Article 104, is the intercept of the tangent on the axis of x.

In a manner quite similar, the intercept on the axis of y may be found.

According to the last equation, the value of CT diminishes as x increases, and if $x = \infty$, $CT = 0$; hence, as x increases indefinitely, the intercept approaches the centre C as a limit, and the point T, for the right-hand branch of the curve, can never be at the left of C.

4°. *By means of a Normal.*—See Article 118.

5°. *By means of conjugate diameters.*—See Article 126.

B. To DRAW A TANGENT TO A CONIC SECTION THROUGH A POINT WITHOUT THE CURVE.—For the **ellipse**, with one focus F' as a centre, and a radius equal to the major axis, describe an arc; and with the given point P as a centre, and radius PF equal to the distance of P from the other focus, describe another arc intersecting the former in the points M and N. Join $F'M$ and $F'N$ and the points P' and P'', where these lines intersect the curve, will be the tangent points, and the lines PP' and PP'' will be the tangents required. For, by the construction,

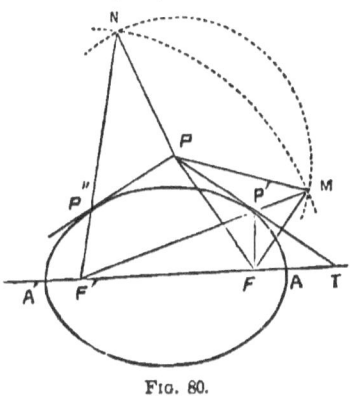

Fig. 80.

$$F'P' + P'M = A'A;$$

and because the point P' is on the curve

$$F'P' + P'F = A'A;$$
$$\therefore P'F = P'M;$$

also $$PF = PM;$$

therefore two points, P and P', are equally distant from M and F, hence the line PP' will be perpendicular to FM, and bisect the angle $FP'M$, and therefore is a tangent required. Similarly, the line PP'' is also a tangent.

For *the hyperbola*, let P be the point. Use the same language and corresponding quantities as just given for the ellipse. The proof is also the same, only observing that $P'M$ becomes negative for the hyperbola.

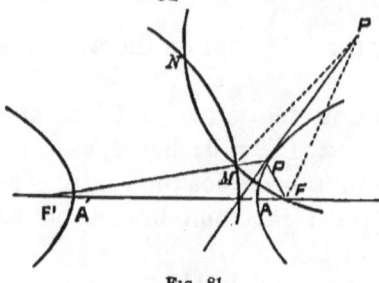

Fig. 81.

For *the parabola* with the given point P as a centre, and radius PF, (F being the focus), describe an arc cutting the directrix MN in the points M and N. Draw MP' and NP'' perpendicular to the directrix, and the points P' and P'' where they intersect the curve will be tangent points, and PP', PP'' will be the required tangents. For by the construction P' is equidistant from M and F, and because P' is on the curve, it is equally distant from the same points, hence PP' will be perpendicular to the line joining M and F, and, therefore, will bisect $MP'F$, which is the required condition. Similarly, PP'' is a tangent.

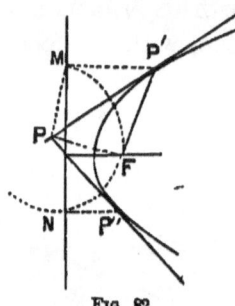

Fig. 82.

(To make this construction appear the same as for the ellipse and hyperbola, it is only necessary to consider one focus as infinitely distant. The arc described from that focus as a centre, will be the straight line MN at the left of vertex, a distance equal to that of the vertex from the focus. And the focal lines drawn through M and N to the remote focus will be parallel to the axis of the curve. In the preceding cases, no tangent will be possible, if the point is within the curve.)

C. To DRAW A TANGENT TO A CONIC SECTION PARALLEL TO A GIVEN LINE.—See Article 126.

Normals and Subnormals.

109. The Normal to a curve *is the perpendicular to the tangent at the point of tangency and limited by one of the axes.*

It will be understood, unless otherwise stated, that the foot of the normal is the point where the normal intersects the axis of x. Thus, PN is the normal at the point P.

The Subnormal is the projection of the normal on the axis of x. Thus, ND is the subnormal.

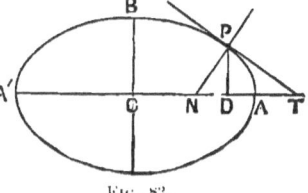

Fig. 82.

110. Equation of the Normal to the Ellipse.—Let $x'y'$ be the point P through which the normal is drawn. The equation to a line passing through this point will be

$$y - y' = m(x - x'). \qquad (1)$$

The equation to the tangent through the same point, Art. 100, Eq. (e), may be written

$$y = -\frac{b^2 x'}{a^2 y'} x + \frac{b^2}{y'}; \qquad (2)$$

in which $-\dfrac{b^2 x'}{a^2 y'}$ is the tangent of the angle which the tangent line makes with the axis of x, (Art. 28), and the condition which will make the former line perpendicular to the latter is (Art. 45),

$$m = -\frac{1}{m'} = \frac{1}{\dfrac{b^2 x'}{a^2 y'}} = \frac{a^2 y'}{b^2 x'}; \qquad (3)$$

and this value in equation (1) gives

$$y - y' = \frac{a^2 y'}{b^2 x'}(x - x'); \qquad (4)$$

which is the required equation. Clearing of fractions and dividing by $x'y'$ gives

$$\frac{a^2 x}{x'} - \frac{b^2 y}{y'} = a^2 - b^2 = c^2; \qquad (5)$$

which is also the equation of the normal, and is similar in form to the equation of the tangent.

111. The equation to the Normal of the Circle becomes, by making $a=b=R$ in equation (5);

$$\frac{x}{x'} - \frac{y}{y'} = 0; \qquad (6)$$

and since it has no absolute term it expresses the well-known fact that *the normal to the circle* (which is the radius) *passes through the centre of the circle.*

112. The equation to the Normal of the Hyperbola becomes, from Eq. (5), (writing $-b^2$ for b^2)

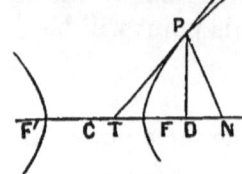

Fig. 84.

$$\frac{a^2 x}{x'} + \frac{b^2 y}{y'} = a^2 + b^2 = c^2. \qquad (7)$$

113. The equation to the Normal of the Parabola is found to be, by a process similar to that in Article 110,

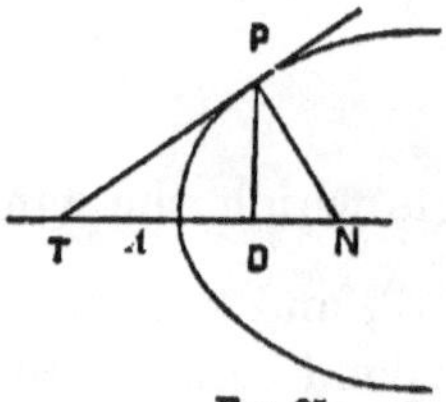

$$y - y' = -\frac{y'}{p}(x - x'). \qquad (8)$$

Fig. 85.

113a.—To find the equation to the normal of an ellipse in terms of the eccentric angle.

The equation to the tangent may be written, (Art. 104 *a*).

$$y = -\frac{b \cos \varphi}{a \sin \varphi} x + \frac{b}{\sin \varphi};$$

hence the equation to the normal passing through the point ($a \cos \varphi$, $b \sin \varphi$) will be

$$y - b \sin \varphi = \frac{a \sin \varphi}{b \cos \varphi}(x - a \cos \varphi);$$

which reduces to

$$a \sec \varphi \cdot x - b \operatorname{cosec} \varphi \cdot y = a^2 - b^2 = c^2.$$

114. Intercepts of the Normal.—To find the intercept on the axis of x, make $y = 0$ in equations (5), (6), (7), (8), and the corresponding value of x will be the intercept required. In this way we find, for the ellipse . . . $CN = \dfrac{c^2}{a^2} x' = e^2 x'$;

the circle. . . . $CN = 0$;

the hyperbola . $CN = e^2 x'$;

the parabola. . . $AN = x' + p$;

in which e is the eccentricity, (Art. 93), and p is one-half the parameter of the axis. The value of CN for the ellipse and hyperbola *appear* to be the same, but they are not really so, since e for the ellipse is less than unity, and for the hyperbola, greater than unity.

The intercept on the axis of y may be found by making $x = 0$ and solving for y in the same equations.

114a. The Length of the Subnormal is the difference between CD and CN. Using only the positive results, since it is the numerical value which is sought, we have

$$DN = x' \sim CN;$$

in which substitute the value of CN from the preceding Article, and we have for

the ellipse . . . $DN = x' - e^2 x' = (1 - e^2) x' = \dfrac{b^2}{a^2} x'$;

the circle. . . . $DN = x'$;

the hyperbola. $DN = \dfrac{b^2}{a^2} x'$;

the parabola. . $DN = p$.

In the ellipse, circle, and hyperbola, the length of the subnormal varies directly as the abscissa, while in the parabola it is constant. In the ellipse if $x' = 0$, $DN = 0$; and if $x' = a$, $DN = \dfrac{b^2}{a}$; hence the subnormal is greatest at the extremity of the major axis, and at that point equals the ordinate through the focus, (Art. 95). For the hyperbola the subnormal is a minimum at the vertex of the curve, at which point $x' = a$, and we have $DN = \dfrac{b^2}{a} =$ half the latus rectum, (Art. 95). From this point it increases indefinitely with x'.

115. Length of the Normal.—In all the curves, we have

$$\text{Normal} = \sqrt{(\text{Subnormal})^2 + (\text{Ordinate})^2}$$
$$= \sqrt{\overline{PD}^2 + \overline{DN}^2}.$$

116. Distance of the foot of the Normal from either Focus.—1°. *The distance from the focus which is on the positive side of the origin.*

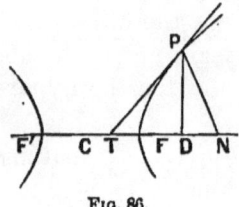

Fig. 86.

The required distance is NF, F being at the focus, and we have

$$FN = CF \sim CN;$$

in which substitute the value of CF (equal ae for all but the parabola, (Art. 93), and equal $\tfrac{1}{2}p$ for the parabola, (Art. 85),) and CN, (Art. 114), and we have the following numerical values for *the distance of the foot of the normal from the focus on the positive side of the origin*, for

the ellipse... $FN = e(a - ex')$, $(e<1)$;
the circle... $FN = 0$, $(e = 0)$;
the hyperbola $FN = e(ex' - a)$, $(e>1)$;
the parabola.. $FN = x' + \tfrac{1}{2}p$, $(e = 1)$.

Similarly, for the distance on the negative side, since x' becomes negative, we have numerically, for

the ellipse... $F'N = e(a + ex')$;
the circle ... $F'N = 0$;
the hyperbola $F'N = e(ex' + a)$;
the parabola.. $F'N = \infty$ on the positive side.

[As there is no *real* second focus to the parabola, this expression is only one of *form*. It is necessarily positive since no part of the curve lies on the negative side of the origin.]

117. In any Conic Section the Normal bisects the

Angle between the focal radii, or between one focal radius and the other prolonged.—The angle between the normal and the tangent is right, and it has been proved in Article 107 that the acute angles between the focal radii and the tangent are equal to each other; hence taking each of the latter from a right angle leaves the remaining angle on one side equal to the remaining angle on the other side of the normal.

It will be observed that in the ellipse and parabola, the normal bisects the *internal* angle formed by the focal radii, and in the hyperbola, the *external* angle.

118. Construction of the Normal.—A. LET THE POINT THROUGH WHICH THE NORMAL IS TO BE DRAWN BE ON THE CURVE.

1°. *Construct a tangent to the curve at that point*, (Art. 108), and at the point erect a perpendicular to the tangent; it will be the required line.

2°. *Bisect the angle between the focal radii*, and it will be either a normal or a tangent, (Arts. 108 and 117). If the latter, erect a perpendicular as stated in the preceding case.

3°. *By means of the subnormal.* For the ellipse and **hyperbola** we proceed as follows:

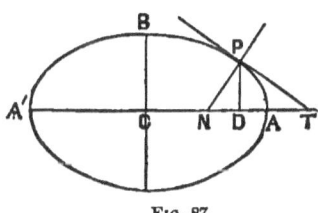

Fig. 87.

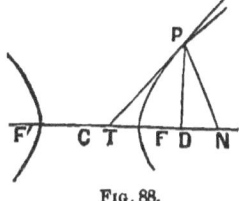

Fig. 88.

We have, (Art. 114a),

$$DN = \frac{b^2}{a^2} x';$$

in which $x' = CD$. To construct this take the lines CA and CB, Fig. 89, making any convenient angle with each other. Make $CA = a$, $CB = b$, and $CD = x'$. Join B and A and draw DJ parallel to AB. Take CG equal to CJ and draw GK parallel to AB; CK will be the required subnormal. For we have

$$\frac{CJ}{CD} = \frac{CB}{CA} = \frac{b}{a}; \therefore CJ = \frac{b}{a}x'.$$

Also $$\frac{CK}{CJ=CG} = \frac{CJ}{CD};$$

$$\therefore CK = \frac{CJ^2}{CD} = \frac{b^2}{a^2}x',$$

which is the value sought. The distance CK being laid off from D, Figs. 87, 88, in the proper direction gives the point N, and PN will be the required normal.

For the parabola let P be the point. Drop the perpendicular PD to the axis of the curve, and lay off DN equal to one-half of the parameter of the curve; then will the line PN be the required normal.

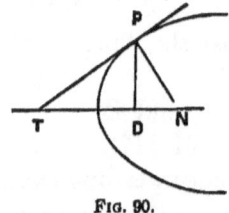

Fig. 90.

[This method also furnishes an easy mode of drawing a tangent to a parabola at a given point. For the tangent will be perpendicular to the normal at the point P.]

4th. *By means of the intercept on the axis of x.*—For the ellipse and hyperbola, the value of the intercept is, (Art. 114),

$$CN = \frac{c^2}{a^2}x';$$

which expression is of the same form as that for DN in the preceding case; hence it may be constructed in the same manner. Making $BC = c$, we have $CK =$ the value of the intercept, which equals CN in Figs. 87, 88, 92.

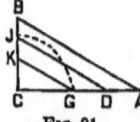

Fig. 91.

For the parabola, we have, Fig. 92,

$$CN = x' + p = CD + DN.$$

5th. *By means of the focal distance of the foot of the normal.* Subtract the value of the intercept as found above, from the distance of the focus from the centre, and the remainder will be the required distance. **The parabola** is the only conic section in which this construction is of special interest. For this curve we have,

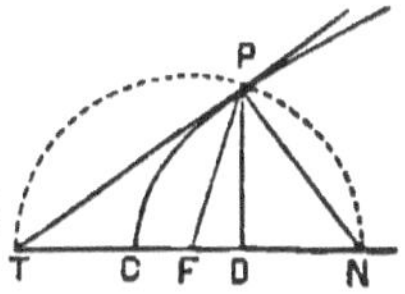

Art. 114a, $DN = p$,

and, Art. 86, $CF = \tfrac{1}{2}p$,

and $CF = DN - CF = \tfrac{1}{2}p$;

also, Art. 105, $TC = CD$;

adding gives $TC + CF = CD + DN - CF$;

or, $TF = FN$;

also, Art. 107, $TF = FP$;

Fig. 92.

hence the points T, P, N, are equi-distant from the focus. Hence with F as a centre, and FP as a radius describe an arc, cutting the axis in the points N and T; the line PN will be the normal, and PT the tangent.

B. LET THE GIVEN POINT THROUGH WHICH THE NORMAL IS TO BE DRAWN, BE WITHIN OR WITHOUT THE CURVE. No general method is known. There are approximate methods, and solutions for special points.

C. NORMAL PARALLEL TO A GIVEN LINE. Construct a tangent perpendicular to the given line, (Art. 126); the point of tangency will be a point of the normal, through which point erect a perpendicular, and it will be the required normal.

EXAMPLES.

1. Find the equation of the normal to the ellipse $2x^2 + 5y^2 = 40$, at a point on the curve whose abscissa is 2.

2. Find the equation of the normal to the parabola $y^2 = 4x$, at a point whose abscissa is 4.

3. Find the intercepts of the normal to the hyperbola $3x^2 - 2y^2 = 16$, at a point whose ordinate is 2.

4. Find the length of the subnormal to the ellipse $\tfrac{1}{2}x^2 + \tfrac{1}{3}y^2 = 5$, at a point whose ordinate is 3.

5. Required the length of the normal to the hyperbola $4x^2 - 7y^2 = 36$, at a point whose ordinate is 4.

6. In the hyperbola $\dfrac{x^2}{9} - \dfrac{y^2}{4} = 1$; required the distance from the centre of the curve to the foot of the normal drawn through the point on the curve whose ordinate is 4.

7. Required the length of the subnormal to the parabola $x = 4y^2$.

8. Required the length of the subnormal in the circle $x^2 + y^2 = 16$, at a point whose ordinate is 2.

119. Linear Equation to the Conic Sections.—Let $F'P = \rho = $ the distance of any point P from the focus F'; $CD = x$: then from the figure we have, as in Article 69,

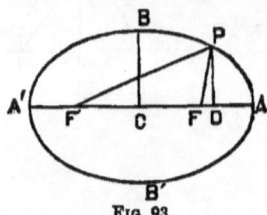

Fig. 93.

$$\rho^2 = (F'C + CD)^2 + PD^2,$$
$$= (ea + x)^2 + y^2.$$

Substituting the value of y^2, Eq. (a_1) Art. 69, and reducing, we have

$$\rho^2 = e^2 a^2 + 2eax + b^2 + \frac{a^2 - b^2}{a^2} x^2$$
$$= a^2 + 2eax + e^2 x^2;$$
$$\therefore \rho = a + ex.$$

Similarly, if the pole be at the focus F on the positive side of the origin, we will find

$$\rho = a - ex;$$

hence, generally, $\qquad \rho = a \pm ex;$

in which ρ will always be positive.

In *the hyperbola* e exceeds unity, and x exceeds a; hence in order that ρ may always appear to be positive in the result, we write for this curve

$$\rho = ex \pm a;$$

the negative sign being used when the pole is at $+c$, and the positive sign, when at $-c$.

For *the parabola* we have, as found in Article 86,

$$FP = \rho = x' + \tfrac{1}{2}p.$$

These equations being of the first degree are called *the linear equations of the conic sections.*

120. Boscovich Definition of a Conic Section.

The linear equation of the ellipse may be written

$$\rho = e\left(\frac{a}{e} + x\right);$$

in which $\frac{a}{e}$ is constant. If a distance $CL = \frac{a}{e}$ be laid off from the centre C, then will

$$MP = LD = LC + CD$$
$$= \frac{a}{e} + x;$$
$$\therefore \frac{\rho}{MP} = \frac{\rho}{\frac{a}{e} + x} = e = a \text{ con-}$$

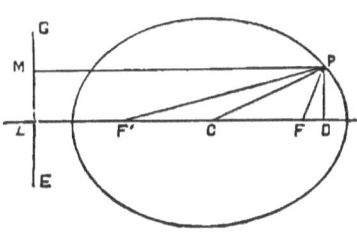

Fig. 94.

stant = the eccentricity of the ellipse.

The abscissa, x, of the point may be negative, in which case the term containing x will be negative.

Similarly, for *the hyperbola*

$$\rho = e\left(x \pm \frac{a}{e}\right).$$

Make $CL = \frac{a}{e}$; then

$$LD = MP = CD - CL$$
$$= x - \frac{a}{e};$$

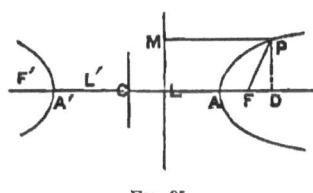

Fig. 95.

and if the directrix be at the left of C, then

$$L'D = x + \frac{a}{e};$$

and in either case we have

$$\frac{FP}{LD} = \frac{\rho}{MP} = e = a \text{ constant} = \text{the eccentricity}$$

of the hyperbola.

For *the parabola*, we have directly from the definition

of the curve, (Art. 82, and Fig. 50),

$$\frac{FP}{CP} = \frac{\rho}{CP} = 1;$$

hence: *A conic section may be defined as a curve such that the ratio of the distances of any point in it from a fixed point and fixed line is constant, and equal to the eccentricity of the curve.*

This is known as Boscovich's definition. The fixed line is the directrix.

121. The principles of the preceding Article furnish a convenient method of describing an hyperbola by a continuous movement. Let GL be the directrix against which moves the triangle GAE. Attach a string at A, stretch it along AE, and swing the point E of the string around to F and fasten it at that point. Place a pencil point at P and keep the string pressed against the edge of the triangle while the triangle moves along the directrix. The curve described will be an hyperbola; for

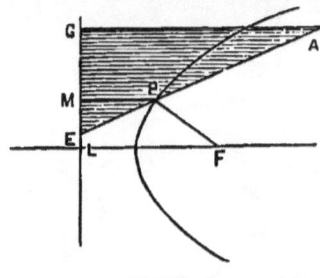

Fig. 96.

$$\frac{PF = PE}{PM} = \frac{AE}{AG} = a \text{ constant}.$$

The ellipse may also be *constructed* from the linear equation, but the method is somewhat complex.

Supplementary Chords and Conjugate Diameters.

122. Supplementary Chords are the chords drawn from any point of a curve to the extremities of any diameter. One of these chords is supplementary in reference to the other. Thus, if $A'A$ is any diameter, then the chords PA and PA' are supplementary in reference to each other. The expression probably came from the

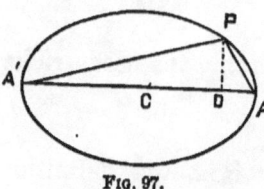

Fig. 97.

relations of these chords to each other in the circle; for in that curve the arc subtended by one chord is the supplement of that subtended by the other.

123. Equation of Condition for Supplementary Chords, in reference to the Transverse Axis.—For the ellipse the equation of the line $A'P$ passing through the point A' will be of the form, (Art. 38),

$$y - y' = m(x - x');$$

in which $y' = 0$, and $x' = -a$; hence the equation becomes

$$y = m(x + a).$$

Similarly, the equation of the line PA will be

$$y = m'(x - a).$$

At the point of intersection of these lines, both equations will be satisfied for the same values of x and y. Multiplying the equations together, we have

$$y^2 = mm'(x^2 - a^2).$$

In order that the point of intersection shall be on the ellipse, the coördinates of that point must satisfy the equation of the ellipse, which is, (Art. 69, Eq. (b_1)),

$$y^2 = -\frac{b^2}{a^2}(x^2 - a^2).$$

Dividing this equation by the preceding one, gives

$$mm' = -\frac{b^2}{a^2};$$

which is the required equation. The sign of the product being negative, it follows that if one angle be acute the other will be obtuse. It will be remembered that the angles are positive for a left-handed rotation measured from $+ x$.

For *the circle,* we have $b = a$, and the equation of condition becomes

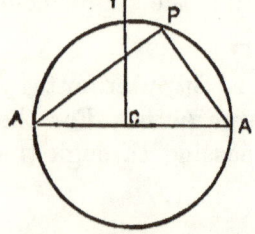

Fig. 98.

$$mm' = -1;$$

which is the condition of perpendicularity; hence: *Supplementary chords in a circle are perpendicular to each other.*

This is another way of stating the principle given in elementary geometry, that *an angle inscribed in a semicircle is right.*

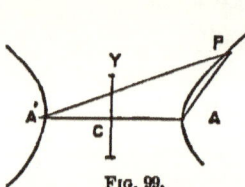

Fig. 99.

For *the hyperbola,* the equation of condition becomes, by changing b^2 to $-b^2$,

$$mm' = \frac{b^2}{a^2};$$

which result being positive shows that both angles are acute or both obtuse.

In *the parabola,* the supplementary chords drawn from P are PA drawn from P to the vertex of the curve, and PA' drawn towards the remote end of the axis. Hence PA' is parallel to the axis and makes with the axis the angle zero; but the angle which PA makes with the axis is $\tan^{-1}\frac{y}{x}$, x and y being the coördinates of the point P in reference to the vertex as an origin.

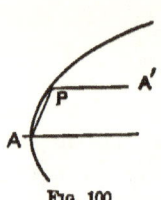

Fig. 100.

124. Two Diameters are Conjugate *when each is parallel to the tangent drawn through the vertex of the other.* — Thus if PP' and QQ' are two diameters such that PP' is parallel to the tangent $T'E'$ passing through Q and QQ' parallel to the tangent at P, then will these diameters be conjugate in reference to each other.

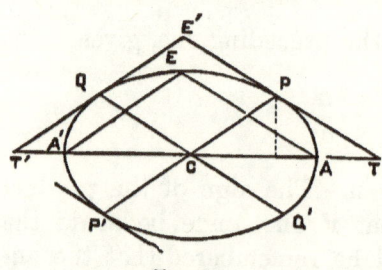

Fig. 101.

125. Equation of Condition for conjugate Diameters.

—For *the ellipse*, the equation of the line CP passing through the origin, will be of the form

$$y = m_2 x,$$

which for the point P, whose coördinates are x' and y', gives the equation of condition

$$y' = m_2 x';$$

$$\therefore m_2 = \frac{y'}{x'}. \tag{1}$$

The equation of the tangent line $E'T$ is, (Art. 100, Eq. (e)),

$$y = -\frac{b^2 x'}{a^2 y'} x + \frac{b^2}{y'},$$

in which $-\dfrac{b^2 x'}{a^2 y'}$ is the tangent of the obtuse angle at T.

Letting this be represented by m_3, we have

$$m_3 = -\frac{b^2 x'}{a^2 y'}; \tag{2}$$

and multiplying (1) and (2) together, we have

$$m_2 m_3 = -\frac{b^2}{a^2}, \tag{3}$$

which is the required equation.

For *the circle*, $a = b = R$, and we have

$$m_2 m_3 = -1; \tag{4}$$

and for *the hyperbola*, $b^2 = -b^2$, and we have

$$m_2 m_3 = \frac{b^2}{a^2}. \tag{5}$$

These values are the same as those found for the conditions of supplementary chords, (Art. 123), therefore we have

$$mm' = m_2 m_3;$$

and if $m = m_2$, then $m' = m_3$; that is:

If one diameter of an ellipse or hyperbola, is parallel to one of two supplementary chords, in reference to the major axis, the conjugate diameter will be parallel to the other chord.

It also follows that:

If a pair of supplementary chords are parallel respectively to a pair of conjugate diameters, they will also be parallel respectively to the tangents passing through the vertices of those diameters.

The parabola, strictly speaking, has no conjugate diameters.

126. To construct a tangent to a conic section by means of conjugate diameters. 1°. *Let the tangent be drawn through a point P on the curve.* Join P with the

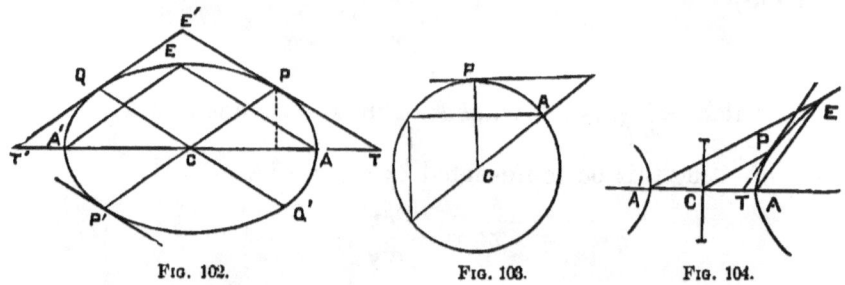

Fig. 102. Fig. 103. Fig. 104.

centre C and through the vertex of the major axis, draw the chord A'E parallel to CP and join E and A; the line PT, parallel to EA will, according to the preceding Article, be the tangent required. The construction is the same for the circle and hyperbola.

2°. *To draw a tangent parallel to a given line.* Let N be

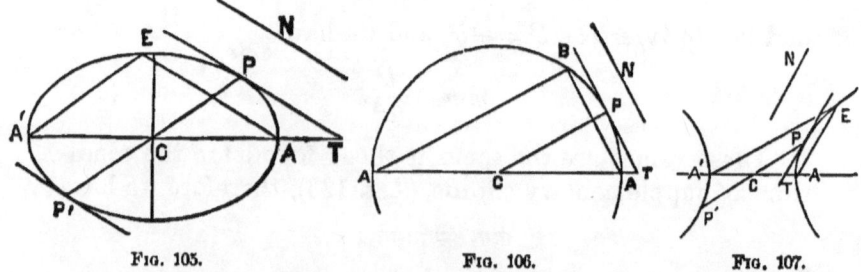

Fig. 105. Fig. 106. Fig. 107.

the line; draw a chord AE through the principal vertex A parallel to the line N, and from E draw the supplementary chord EA'; then from the centre C draw CP parallel to $A'E$, and P, where CP intersects the curve, will be the point of tangency, and PT parallel to N will be the required tangent.

For the parabola, let F be the focus; draw FE parallel to N, and PF making the angle PFE equal to EFX; then will the point P, when PF intersects the curve, be the tangent point, and a line through P, parallel to N, will be the required tangent. For, by the construction, the angle T equals P, and hence $TF = PF$ as it should, (Art. 107).

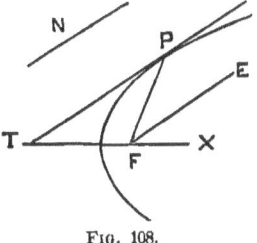

Fig. 108.

The following construction corresponds more nearly in form with that for the other conic sections than does the preceding. Through the vertex of the curve A draw a chord AB parallel to the given line N; from B draw a line BA' towards the vertex infinitely distant, it will be parallel to AX. From the centre draw a line CP parallel to $A'B$, since it will be a diameter it will bisect the chord AB and be parallel to AX, and the point P where it intersects the curve will be the tangent point, and PT, drawn parallel to the line N, will be the required tangent. See also Article 143.

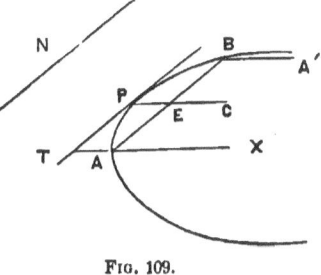

Fig. 109.

127. Equation to the Ellipse referred to Oblique Axes, the origin being at the centre. The equations for changing from rectangular to oblique axes, the origin being the same, are, (Art. 52, Eqs. (e)),

$$x = x' \cos \alpha + y' \cos \beta;$$

$$y = x' \sin \alpha + y' \sin \beta.$$

Substituting these values in the equation of the ellipse

referred to its axes,

$$a^2y^2 + b^2x^2 = a^2b^2,$$

and arranging the result, gives

$$\left\{ \begin{matrix} a^2\sin^2\beta \\ +b^2\cos^2\beta \end{matrix} \right\} y'^2 + \left\{ \begin{matrix} a^2\sin^2\alpha \\ +b^2\cos^2\alpha \end{matrix} \right\} x'^2 + \left\{ \begin{matrix} a^2\sin\alpha\sin\beta \\ +b^2\cos\alpha\cos\beta \end{matrix} \right\} 2x'y' = a^2b^2;$$

which is the required equation, in which α is the angle between the axis x' and the major axis of the ellipse, and β the angle between y' and the same axis. For the hyperbola, change b^2 to $-b^2$.

128. Ellipse referred to Conjugate Diameters.—In order that the new coördinate axes shall coincide with conjugate diameters, they must be subjected to the condition given in Article 125, or

$$m_2 m_3 = -\frac{b^2}{a^2};$$

which becomes

$$\tan\alpha \tan\beta = -\frac{b^2}{a^2};$$

which may be reduced to

$$a^2\sin\alpha\sin\beta + b^2\cos\alpha\cos\beta = 0.$$

This condition causes the coefficient of $x'y'$ in the equation of the preceding Article, to disappear, and that equation becomes

$$(a^2\sin^2\beta + b^2\cos^2\beta)\, y'^2 + (a^2\sin^2\alpha + b^2\cos^2\alpha)\, x'^2 = a^2b^2.$$

In this equation, if $x' = 0$, we have

$$y' = \frac{ab}{\sqrt{a^2\sin^2\beta + b^2\cos^2\beta}};$$

which is the intercept on the axis of y'. Let this be represented by b'. Similarly, the intercept on x' will be

$$x' = \frac{ab}{\sqrt{a^2\sin^2\alpha + b^2\cos^2\alpha}} = a', \text{ (say)}.$$

These values reduce the preceding equation to, after dropping the accents from the variables,

$$\frac{x^2}{a'^2}+\frac{y^2}{b'^2}=1; \text{ or } a'^2 y^2 + b'^2 x^2 = a'^2 b'^2, \qquad (1)$$

which is the equation of the ellipse referred to conjugate diameters.

For *the circle*, $a' = b' = R$, and equation (1) becomes

$$\frac{x^2}{R^2} + \frac{y^2}{R^2} = 1, \qquad (2)$$

which is the same as that previously found for rectangular axes, as it should be, since the conjugate diameters, in this curve, are at right angles with each other.

For *the hyperbola*, $b'^2 = -b'^2$, and equation (1) becomes

$$\frac{x^2}{a'^2} - \frac{y^2}{b'^2} = 1. \qquad (3)$$

These equations are of the same *form* as those deduced for rectangular axes.

129. Discussion of equation (1).—Solving for x gives

$$x = \pm \frac{a'}{b'} \sqrt{b'^2 - y^2};$$

which shows that for every value of y less than b', there are two equal and opposite values for x; hence the curve is *obliquely* symmetrical in reference to the axis of y. In the same manner we find a like symmetry in respect to x. Since the axis of x can be made to coincide with any diameter, it follows that *every diameter bisects a system of chords parallel to its conjugate*.

This equation may be discussed in other respects the same as for rectangular axes. Equation (3) gives corresponding results.

130. Equation of the Tangent referred to conjugate

diameters. The equation to the curve being of the same form as for rectangular axes, the equation to the tangent will be of the same form. We have therefore only to substitute a' for a, and b' for b in the equations previously found. Let P be the point of tangency, $CD = x'$, $PD = y'$; then will the equation of the tangent PT be, (see Art. 100),

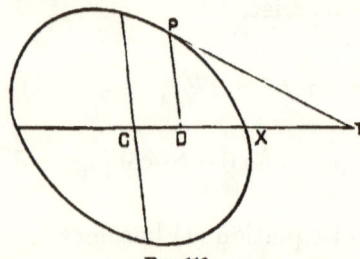

Fig. 110.

for *the ellipse*,

$$\frac{x'x}{a'^2} + \frac{y'y}{b'^2} = 1.$$

For *the hyperbola*, (see Art. 102),

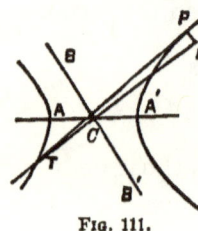

Fig. 111.

$$\frac{x'x}{a'^2} - \frac{y'y}{b'^2} = 1.$$

For the intercept CT make $y = 0$, and we have, for the ellipse, circle, and hyperbola,

$$CT = \frac{a'^2}{x'}.$$

In a similar manner the subtangent DT may be found.

[For a shorter course omit to Article 167.]

131. Transform the equation to the Ellipse from conjugate diameters to its axes.—The formulas for passing from oblique to rectangular axes, the origin remaining the same, are, (Art. 53),

$$x' = \frac{x \sin \beta - y \cos \beta}{\sin (\beta - \alpha)}; \qquad y' = \frac{y \cos \alpha - x \sin \alpha}{\sin (\beta - \alpha)}.$$

These values substituted in the equation to the ellipse referred to conjugate diameters, which is, (Art. 128, Eq. (1)),

$$a'^2 y'^2 + b'^2 x'^2 = a'^2 b'^2,$$

gives

$$\left.\begin{array}{l}(a'^2\cos^2\alpha + b'^2\cos^2\beta)y^2 \\ + (a'^2\sin^2\alpha + b'^2\sin^2\beta)x^2 \\ - 2(a'^2\sin\alpha\cos\alpha + b'^2\sin\beta\cos\beta)xy\end{array}\right\} = a'^2 b'^2 \sin^2(\beta - \alpha).$$

This equation must be reduced to the form

$$a^2 y^2 + b^2 x^2 = a^2 b^2,$$

previously found for the equation to the ellipse referred to its axes, (Art. 69, Eq. (a_1)).

Comparing the two preceding equations, we find:

$$a'^2 \cos^2 \alpha + b'^2 \cos^2 \beta = a^2; \quad (1)$$

$$a'^2 \sin^2 \alpha + b'^2 \sin^2 \beta = b^2; \quad (2)$$

$$a'^2 \sin\alpha\cos\alpha + b'^2 \sin\beta\cos\beta = 0; \quad (3)$$

$$a'^2 b'^2 \sin^2(\beta - \alpha) = a^2 b^2. \quad (4)$$

These are *the equations of condition*, and from them we deduce the following results: Adding the first and second gives

$$a'^2 + b'^2 = a^2 + b^2;$$

or $\qquad 4a'^2 + 4b'^2 = 4a^2 + 4b^2; \qquad (5)$

that is: *The sum of the squares of any pair of conjugate diameters of an ellipse is constant, and equals the sum of the squares of the axes.*

When the axes and one of the conjugate diameters is given, the other conjugate diameter may be found from the preceding equation.

Changing b^2 to $-b^2$, and b'^2 to $-b'^2$ gives

$$4a'^2 - 4b'^2 = 4a^2 - 4b^2; \qquad (6)$$

that is: **For the hyperbola** *the difference of the squares of any two conjugate diameters, equals the difference of the squares of the axes.*

Because the axes are conjugate, we must have, (Art. 128),

$$a^2 \sin\alpha \sin\beta = -b^2 \cos\alpha \cos\beta;$$

$$\therefore \tan\alpha = -\frac{b^2}{a^2}\cot\beta, \qquad (7)$$

from which one of the angles may be found when the other angle and the axes are given.

It appears that if one of these angles is acute the other will be obtuse. From equation (3) we have

$$a'^2 \sin \alpha \cos \alpha = - b'^2 \sin \beta \cos \beta ;$$

which combined with equation (5) by eliminating b', gives

$$a'^2 = \frac{a^2 + b^2}{1 - \dfrac{\sin \alpha \cos \alpha}{\sin \beta \cos \beta}}, \qquad (8)$$

from which one of the conjugate diameters may be found when the axes are given, and the angles α and β have been found. The other axis may be found in a similar manner, or the value of a' may be substituted in equation (5), and b' deduced from the result.

From equation (4) we find

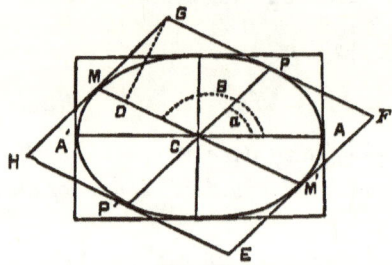

Fig. 112.

$$4a'b' \sin (\beta - \alpha) = 4ab ; \quad (9)$$

the second member of which represents the area of the rectangle constructed on the axes of the ellipse. In the figure

$$PCA = \alpha, \; MCA = \beta ; \; \therefore \; MCP = \beta - \alpha.$$

Construct the parallelogram $EFGH$, having its sides tangent to the ellipse at the extremities of the conjugate diameters PP' and MM'; then will $MCPG$ be one-fourth of the circumscribed parallelogram. Since the sum of the angles of a parallelogram equals four right angles, we have

$$GMC + MCP = 180° ;$$
$$\therefore \sin GMC = \sin (180° - MCP) = \sin (\beta - \alpha),$$
and $(GM = a') \sin (\beta - \alpha) = \text{the perpendicular } GD,$
and $a'b' \sin (\beta - \alpha) = MC . DG = \text{area } MCPG ;$
$$\therefore 4a'b' \sin (\beta - \alpha) = EFGH ;$$

hence: *The area of a parallelogram circumscribing an ellipse in which the sides are tangent to the ellipse at the vertices of a pair of conjugate diameters, is constant and equal to the rectangle constructed on the axes.*

For the hyperbola we substitute $\sqrt{-1}\, b$ for b, and $\sqrt{-1}\, b'$ for b', and obtain precisely the same result, since $\sqrt{-1}$ drops from the equation; hence the preceding conclusion is true for the hyperbola.

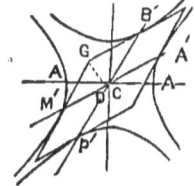

Fig. 113.

EXAMPLE.

If the major axis of an ellipse is 20 and the minor axis is 12, find the conjugate diameters and the angle β, when α is $45°$; deduce the area of the circumscribed parallelogram whose sides are parallel to the conjugate diameters and show that it equals the area of the rectangle on the axes.

132. The Parabola referred to Oblique Axes.—Substituting the values of x and y from equations (d) Article 52, in the rectangular equation of the curve,

$$y^2 = 2px,$$

gives, dropping the accents,

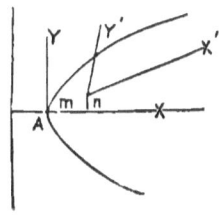

Fig. 114.

$$\left. \begin{array}{l} \sin^2\beta \cdot y^2 + \sin^2\alpha \cdot x^2 + 2\sin\alpha\sin\beta \cdot xy \\ + 2(n\sin\beta - p\cos\beta)y + 2(n\sin\alpha - p\cos\alpha)x \\ + n^2 - 2pm \end{array} \right\} = 0,$$

which is the required equation.

133. Parabola referred to a Diameter and a Tangent at its vertex.—The equation for this case can be deduced directly from the preceding one. Since the new origin will be on the curve, the coördinates of which in reference to the vertex are m and n, we have from the equation of the curve

$$n^2 = 2pm.$$

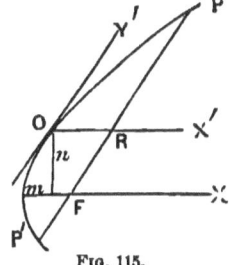

Fig. 115.

Since the new axis of x is to be parallel to the axis of the curve, we have,

$$\alpha = 0 \; ; \; \therefore \; \sin\alpha = 0, \text{ and } \cos\alpha = 1.$$

Because OY' is a tangent, we have, (Art. 103),

$$yn = p\,(x+m),$$

or
$$y = \frac{p}{n}(x+m);$$

$$\therefore \tan\beta = \frac{\sin\beta}{\cos\beta} = \frac{p}{n},$$

or
$$n\sin\beta - p\cos\beta = 0.$$

These several values substituted in the equation of the preceding Article, reduces it to

$$y^2 = \frac{2p}{\sin^2\beta}x;$$

or, if $2p'$ be the coefficient of x, we have

$$y^2 = 2p'x,$$

which is the required equation, and is of the same form as when the curve is referred to its axis and a tangent at its vertex.

Discussion of the equation $y^2 = 2p'x$.

1°. For every positive value of x we have

$$y = \pm\sqrt{2p'x};$$

that is: *The ordinates parallel to the tangent at the vertex of any diameter, are bisected by that diameter;* or the curve is *obliquely* symmetrical in reference to every diameter.

2°. For negative values of x, y is imaginary, hence : *The curve does not extend on the negative side of the tangent.*

3°. For two points on the curve, whose abscissas are respectively x_1 and x_2, we have

$$\frac{y_1^2}{y_2^2} = \frac{x_1}{x_2},$$

or
$$y_1^2 : y_2^2 :: x_1 : x_2;$$

that is: *The squares of the ordinates to any diameter are as their corresponding abscissas.*

4°. The equation referred to a diameter and a tangent at its vertex being of the same form as the rectangular equation previously found, it follows that the equation of the tangent line will be of the same form, hence, if the point of tangency be $x'y'$, we have, for *the equation of the tangent,*

$$yy' = p'(x + x');$$

and if $y = 0$, we have
$$x = -x';$$

that is: *The subtangent to* ANY *diameter is bisected at the vertex of that diameter. Also the two tangents at the extremities of any ordinate intersect on the axis of x.*

134. Value of $\frac{1}{2}\dfrac{p}{\sin^2 \beta}$. Let F be the focus; A' the vertex of the diameter $A'F'$; $x' = AD$, the abscissa of A'; $y' = A'D$, its ordinate; and TA', a tangent. According to Article 103, we find,

$$\tan A'TF = \tan \beta = \frac{p}{y'};$$

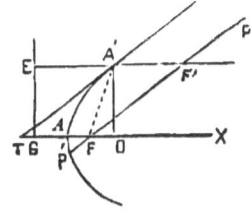

Fig. 116.

$$\therefore \frac{\sin^2 \beta}{1 - \sin^2 \beta} = \frac{p^2}{y'^2} = \frac{p^2}{2px'} = \frac{p}{2x'};$$

hence, solving for $\sin^2 \beta$, gives

$$\sin^2 \beta = \frac{p}{2x' + p};$$

$$\therefore \tfrac{1}{2}\frac{p}{\sin^2 \beta} = x' + \tfrac{1}{2}p.$$

But $TA = AD = x'$, (Art. 104); $AF = \tfrac{1}{2}p$; $TF = FA'$, (Art. 107, Eq. (6));

$$\therefore x' + \tfrac{1}{2}p = TF = FA' = \tfrac{1}{2}\frac{p}{\sin^2 \beta};$$

that is : *The distance of the vertex of any diameter of a parabola from the focus equals the quotient found by dividing the distance of the focus from the principal vertex by the square of the sine of the inclination of the tangent at the vertex of the diameter.*

135. Other relations.—Let EG be the directrix, and PP' a chord through the focus, parallel to the tangent $A'T$. Then, by the construction, $TF = A'F'$, and, according to Article 83, $FA' = A'E$;

$$\therefore TF = FA' = EA' = A'F' = \tfrac{1}{2}\frac{p}{\sin^2\beta}.$$

136. Any double focal ordinate equals $2p \div \sin^2\beta$; in which β is the inclination of the ordinate to the axis of the curve. For, in the equation

$$y^2 = \frac{2p}{\sin^2\beta} x,$$

substitute $x = OR$, (Fig. 115), $= \tfrac{1}{2}\frac{p}{\sin^2\beta}$, and we find

$$y = \pm \frac{p}{\sin^2\beta} = RP', \text{ or } RP.$$

But $PP' = 2y$;

$$\therefore PP' = \frac{2p}{\sin^2\beta};$$

which was to be proved. This value is also called the parameter to any diameter, Article 145.

137. Problem. *Find the point of intersection of a tangent to the parabola with a perpendicular upon the tangent from the focus.*

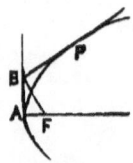

Fig. 117.

Let BP be the tangent, P the point of tangency whose coördinates are x', y', and FB the perpendicular from F upon BP. The equation to the tangent will be, (Art. 103),

$$y = \frac{p}{y'}(x + x');$$

and of the line FB passing through the point $(\tfrac{1}{2}p, 0)$ and perpendicular to BP will be, (Art. 46),

$$y = -\frac{y'}{p}(x - \tfrac{1}{2}p).$$

Combining these equations gives

$$x = 0, \quad y = \tfrac{1}{2}y'.$$

The equation $x = 0$, shows that the intersection will be on the tangent passing through the principal vertex; and conversely:

Perpendiculars drawn to the focal radii at their intersection with the tangent at the principal vertex, will be tangent to the parabola.

The condition $y = \tfrac{1}{2}y'$ furnishes an easy mode of drawing a tangent to the curve; for the distance AB equals one-half the ordinate of the point of tangency. This is substantially the same as in Article 108, 2°.

Geometrical Construction of the Parabola.

138. *Given the focus, and the tangent at the principal vertex to construct the parabola.* Let F be the focus; draw radial lines $F1$, $F2$, etc., and at their intersection with the tangent, 1, 2, etc., draw perpendiculars to the radial lines. The curve drawn tangent to the successive lines will be a parabola. The greater the number of radial lines, the more accurately can the curve be drawn.

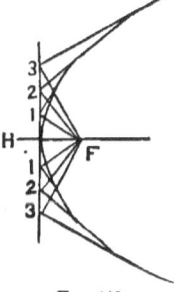

FIG. 118.

139. *To construct a parabola by bisecting subtangents.*—Let TC and TB be the tangents. Join C and B and bisect at A. Draw TA and bisect at E; then will E be a point in the curve. For TA drawn from the intersection of the tangents to the middle of the chord CB, is a diameter, and TA is a subtangent on that diameter, and will be bisected by the curve at E. Join E and C, bisect at G, draw FG and bisect at H, and H will be a point on the

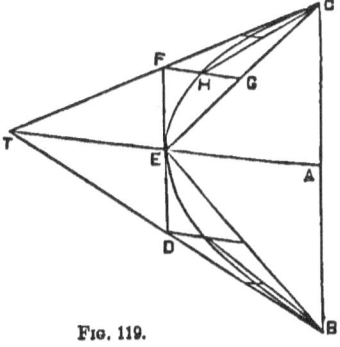

FIG. 119.

curve. In a similar manner any number of points may be found.

140. *By means of enveloping tangents.*—Divide the tangents into the same number of equal parts, and number them in reverse order as shown in the figure, and join the corresponding numbers by straight lines. The curve will be tangent to the *successive* lines. For the intersection of the corresponding tangents will be on a right line and the tangents be equal to each other; hence the subtangents of the corresponding pair of tangents will be equal to each other.

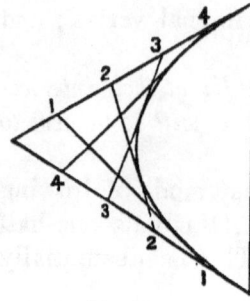
Fig. 120.

141. *By means of bisecting tangents.*—Let TA and TB be two tangents. Bisect them in C and I, and bisect IC in E; then will E be a point in the curve, Art. 139. Bisect CE at F, and CA at H, then will the middle point of FH at J be a point in the curve, and so on.

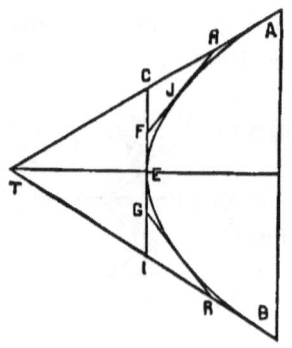
Fig. 121.

142. *By means of intersections.*—Let TC and TD be the tangents. Draw TB to the middle point of CD, and bisect it at 4; this will be one point of the curve. Divide $T4$ and TC into the same number of equal parts. Draw lines from C to the points of division in $T4$, and through the points d, e, f, etc. lines parallel to TB; then will

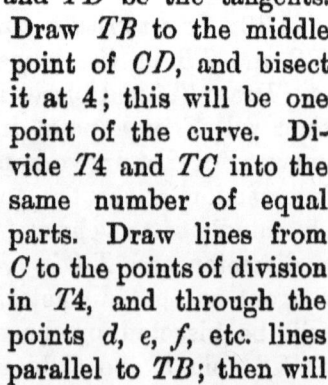
Fig. 122.

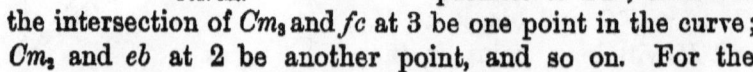
the intersection of Cm_3 and fc at 3 be one point in the curve; Cm_2 and eb at 2 be another point, and so on. For the

distances from d, c, etc. will vary as the squares of Cd, Ce, etc.

Points may also be found by dividing $4B$ into the same number of equal parts as in $T4$ and drawing lines Cn_1 etc. to intersect lines Dm_1 etc.; the points of intersection will be points in the curve.

143. *To draw a tangent parallel to a given line.*—Let N be the given line. Draw two chords AB and DE parallel to N, and bisect them in C and F; then will CFO drawn through F and C be a diameter. Through the vertex O of this diameter draw OT parallel to N, then will OT be the tangent required, (Art. 133, 1°).

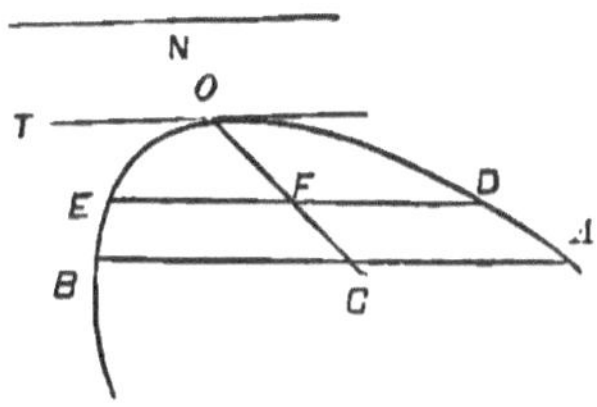

Fig. 123.

144. *To find the axis of a parabola.*—Draw any two parallel chords and bisect them by a line OF'. This line will be a diameter. Draw chords perpendicular to OF', and bisect them by the line AX; then will AX be the axis of the parabola.

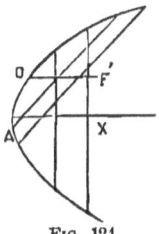

Fig. 124.

Of Parameters.

145. In the Ellipse, (Circle), and Hyperbola, the Parameter in respect to ANY DIAMETER is defined to be *the third proportional of that diameter to its conjugate.* Hence the parameter to the diameter $2a'$ will be

$$2a' : 2b' :: 2b' : parameter;$$

$$\therefore parameter = \frac{(2b')^2}{2a'} = \frac{2b'^2}{a'}.$$

In reference to the major axis, this becomes $\dfrac{2b^2}{a}$, which is the value of *the latus rectum*, (Art. 94).

Similarly, the parameter to the diameter $2b'$ is

$$\frac{(2a')^2}{2b'}.$$

146. In the Parabola the parameter in respect to ANY DIAMETER is defined to be *the third proportional of any abscissa to its corresponding ordinate*, the ordinate being parallel to the tangent at the vertex of that diameter; hence

$$parameter = \frac{y'^2}{x'} = \frac{2p}{\sin^2 \beta}, \text{ (Art. 133)},$$

which equals the double ordinate through the focus, (Art. 136).

If the diameter is the axis of the curve, this value becomes, (since $\beta = 90°$),

$$2p,$$

which is *the latus rectum*, (Art. 95).

Pole and Polar.

147. The Polar of any point in respect to a conic section, is the locus of the intersection of the two tangents drawn at the extremities of any chord passing through the point. It will be shown in Article 152, that this locus is a straight line. Thus, if P be any point in a conic section,

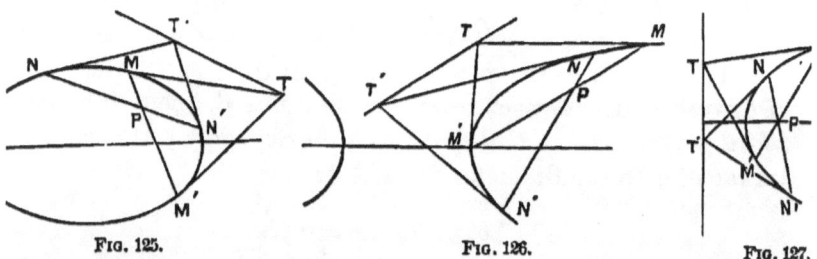

Fig. 125.   Fig. 126.   Fig. 127.

MM' and NN' two chords drawn through this point, then will the point T, the intersection of the tangents TM and TM', be one point in the polar, and T', the intersection of $T'N$ and $T'N'$, another point, and the line TT' passing through these points will be *the polar* required. The chords MM' and NN' are called **chords of contact**, in reference to the tangents drawn from their extremities.

148. The Pole of any right line, with respect to a conic section, is the intersection of the chords of contact in reference to points on the polar. Thus, P is the pole in reference to the polar TT'. A polar point, in reference to the pole, is the point where the diameter through the pole intersects the polar. The *pole* and *polar* are reciprocal terms; neither has any signification without the other.

149. If the pole P is without the curve, the polar will be the chord of contact. For, if any secant PE be drawn, the part EE' will be the chord drawn through the pole. The intersection of the tangents passing through E and E' (but which are not shown in the figure) will be a point on the polar. If now the secant be turned about P, the points E and E' will approach T, and also the intersections of the tangents will continually approach the same point, and when E and E' are consecutive to T, the tangents will intersect at that point. Hence, T is one point. For the same reason, T' is another point. *If the pole is on the* curve, the polar will be tangent to the curve at that point.

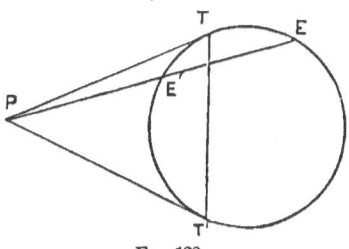

Fig. 128.

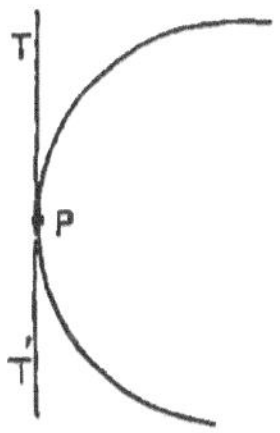

Fig. 129.

150. Equation to the Chord of Contact.—First consider *the ellipse*. The equation of the tangent referred to conjugate diameters, the point of tangency M being $x'y'$, (Fig. 125), will be, (Art. 130),

$$\frac{x'x}{a'^2} + \frac{y'y}{b'^2} = 1;$$

and for the point $x''y''$, or M',

$$\frac{x''x}{a'^2} + \frac{y''y}{b'^2} = 1.$$

But these tangents will intersect in some point as T, which denote by $x_1 y_1$; hence the coördinates of this point must satisfy the equations of both lines, and we have *the equations of condition,*

$$\frac{x'x_1}{a'^2} + \frac{y'y_1}{b'^2} = 1; \qquad \frac{x''x_1}{a'^2} + \frac{y''y_1}{b'^2} = 1.$$

By means of these equations the point of intersection $x_1 y_1$ may be found when the curve and the points of contact are known. Having found this point, let x'' and y'' or x' and y' in the last equations be changed to general variables, and we have

$$\frac{x_1 x}{a'^2} + \frac{y_1 y}{b'^2} = 1, \qquad (1)$$

which, being an equation of the first degree, is the equation of a right line. This line must pass through the point $x'' y''$ because these coördinates satisfy the equation of the line, as shown above; and, for the same reason, it will also pass through $x' y'$; hence it is called *the equation to the chord of contact.*

For *the circle*, make $a' = b' = R$, and we have,

$$\frac{x_1 x}{R^2} + \frac{y_1 y}{R^2} = 1, \qquad (2)$$

for the equation of the chord of contact for the circle.

For *the hyperbola*, we have, for the equation of the chord of contact,

$$\frac{x_1 x}{a'^2} - \frac{y_1 y}{b'^2} = 1. \qquad (3)$$

For *the parabola*, we would find

$$y_1 y = p'(x + x_1). \qquad (4)$$

If all these curves, except the parabola, are referred to their axes, we have $a' = a$, $b' = b$, and in the parabola, if referred to its axis and a tangent at its vertex, $p' = p$. The preceding equations, therefore, remain of the same form whether the curves are referred to their axes or conjugate diameters.

151. Remark.—The changes from variables to constants, and from constants to variables, may appear to the beginner arbitrary and meaningless. We will, therefore, add a few simple illustrations. Take a true numerical expression, as

$$9 \cdot 2 + 4 \cdot 3 = 30;$$

and changing 2 to x, and 3 to y, we have

$$9x + 4y = 30;$$

which is the equation of a right line. This line will pass through the point (2, 3). The above numerical equation is of the form

$$x'x_1 + y'y_1 = 30.$$

If now, in the last equation, we make

$$x' = x, \text{ and } y' = y,$$

we have $\quad x_1 x + y_1 y = 30;$

and making $\quad x_1 = 2, y_1 = 3;$

it becomes $\quad 2x + 3y = 30;$

which is the equation of a right line passing through the point (9, 4), since these values satisfy the equation.

If we have the two numerical equations

$$9 \cdot 1 - 4 \cdot 2 = 1;$$
$$9 \cdot 2 - 4 \cdot 4\tfrac{1}{4} = 1;$$

they may be represented by

$$x_1 x' - y_1 y' = 1; \quad x_1 x'' - y_1 y'' = 1;$$

and if x' and y', or x'' and y'' be made general variables, we have

$$x_1 x + y_1 y = 1;$$

which is the equation of a right line passing through the points (1, − 2) and (2, − 4¼), for the coördinates of both these points satisfy the equation.

In these equations the values of x_1 and y_1 have been assumed arbitrarily; but in *the equation to the chord of contact*, given in the preceding Article, they are determined by means of a certain condition (being the coördinates of the point of intersection of certain tangents), and this condition must be realized so long as x_1 and y_1 remain in the equation.

152. Equation to the Polar.—In regard to the ellipse, we found, in Article 150, that the equation to the chord of contact is

$$\frac{x_1 x}{a'^2} + \frac{y_1 y}{b'^2} = 1. \tag{1}$$

If this line passes through a fixed point P, whose coördinates are x_2, y_2, we will have *the equation of condition*

$$\frac{x_1 x_2}{a'^2} + \frac{y_1 y_2}{b'^2} = 1; \tag{2}$$

If now x_1 and y_1, the coördinates of the intersection of the tangents, be changed to general variables, we have

$$\frac{x_2 x}{a'^2} + \frac{y_2 y}{b'^2} = 1; \tag{3}$$

which will be the equation to the locus of the intersection of the tangents drawn from the extremities of *any* chord passing through the point P. It is the equation of a right line, and is the equation to the *polar*. It is of the *form* of the equation to the tangent line.

For **the circle**, the equation to the polar becomes

$$\frac{x_2 x}{R^2} + \frac{y_2 y}{R^2} = 1; \tag{4}$$

and for **the hyperbola**,

$$\frac{x_2 x}{a'^2} - \frac{y_2 y}{b'^2} = 1. \tag{5}$$

For **the parabola**, when the curve is referred to a diameter and a tangent at its vertex, we would find

$$y_2 y = p'(x + x_2); \qquad (6)$$

in which $p = p'$, if the diameter is the axis.

If all these curves, except the parabola, are referred to their axes, we have

$$a' = a, \text{ and } b' = b.$$

The preceding equations show that the polar of any point in respect to *any* conic section is a right line.

153. Direction of the Polar.—Comparing the equations of the preceding Article with those of Article 130, shows that:

The polar to any point, in respect to a conic section, is parallel to the system of chords bisected by that diameter which passes through the pole.

From this principle it follows directly:

1°. *That for a pole on the axis, the polar will be perpendicular to that axis.*

2°. *For any point, the polar will be parallel to the tangent at the extremity of the diameter passing through the point.*

154. Polar of Special Points. *If the pole be at the centre*, we have, for all but the parabola,

$$x_2 = 0, \ y_2 = 0;$$

which substituted in equations (3) and (5) of Art. 152, give

$$\frac{x}{a'^2} \pm \frac{0}{0} \frac{y}{b'^2} = \frac{1}{0} = \infty;$$

which is the equation of a right line at infinity, whose direction is indeterminate; hence : *The polar of the centre is any right line at infinity.*

If the pole be at the vertex of the major axis, we have, for all but the parabola,

$$x_2 = \mp a, \ y_2 = 0;$$
$$\therefore \ x = 0 \cdot y \pm a,$$

which is the equation of a line parallel to the axis of y, and at a distance from it equal to a. Hence, as the pole passes from the vertex of the major axis to the centre, the polar passes from a distance a from the centre, to an infinite distance.

155. Polar to the Focus.—For the *ellipse* and *hyperbola*, we have
$$x_2 = \pm ae,\ y_2 = 0,$$
which in equations (3) and (5) of Article 152, give
$$x = 0 \cdot y \pm \frac{a}{e};$$
which is the equation to a line parallel to the axis of y, hence: *The polar to either focus is a perpendicular to the major axis and cuts that axis at a distance from the centre equal to $a \div e$, measured on the same side as the focus.*

The polar to the focus is, therefore, *the directrix*, Art. 120.

For the *parabola*, we have for the focus,
$$x_2 = \tfrac{1}{2}p,\ y_2 = 0;$$
which in equation (6), Article 152, gives by reduction,
$$x = 0 \cdot y - \tfrac{1}{2}p;$$
which is the equation of the directrix. Hence: *The polar to the focus of any conic section is the directrix of the curve.*

Tangents at the extremities of a focal chord are called *focal tangents;* hence the two preceding conclusions may be stated as follows:

The locus of the intersection of the focal tangents is the directrix.

EXAMPLES.

1. Required the equation to the polar of the point $x_2 = 2$, $y_2 = 1$, in respect to the ellipse $2y^2 + 3x^2 = 27$.

2. Find the polar of the point $x_2 = -1$, $y_2 = 0$, in respect to the circle $x^2 + y^2 = 9$.

3. Find the equation to the polar of the point $x_2 = 1$, $y_2 = 0$, in respect to the hyperbola $3x^2 - y^2 = 12$.

4. Required the polar of the point $x_2 = 3$, $y_2 = 0$, in respect to the parabola $y^2 = 6x$.

5. Given the polar $y = 2x + 8$, to find the pole in respect to the circle $x^2 + y^2 = 9$.

6. Given the polar $y = x + 2$, to find the pole in respect to the parabola $y^2 = 4x$.

The Hyperbola and its Asymptotes.

156. Definition. An Asymptote *to a curve* is a line towards which the tangent continually approaches as the tangent point moves away from the origin indefinitely.

According to this definition, the curve must have an infinite branch; hence, neither the circle nor the ellipse can have an asymptote.

The parabola has an infinite branch; but it is shown in Article 104, that the intercepts of the tangent increase indefinitely as x' increases indefinitely; hence the tangent does not approach a definite position as a limit.

But in the hyperbola, intercepts on both axes are finite when the abscissa of the point of tangency, x', is infinite. If the origin of coördinates be at the centre of the hyperbola, the intercepts will be zero, when $x = \infty$, (Art. 104), hence the hyperbola has an asymptote passing through the centre. Since the curve is symmetrical in reference to the axis of x, it will have *two asymptotes* passing through the centre.

Many other curves have asymptotes, but their properties are more conveniently investigated by higher analysis.

157. Equations to the Asymptotes of the Hyperbola.—The equation to the tangent line of an hyperbola is, (Art. 102),

$$y = \frac{b^2 x'}{a^2 y'} x - \frac{b^2}{y'}.$$

But from the equation of the curve we have, (Art. 78, Eq. (a_2)),

$$y' = \pm \frac{b}{a}\left[x'^2 - a^2 \right]^{\frac{1}{2}} = \pm \frac{b}{a} x'\left[1 - \frac{a^2}{x'^2} \right]^{\frac{1}{2}};$$

and this value substituted in the preceding equation gives,

$$y = \pm \frac{b}{a} x \frac{1}{\left[1 - \frac{a^2}{x'^2} \right]^{\frac{1}{2}}} \mp \frac{ab}{x'\left[1 - \frac{a^2}{x'^2} \right]^{\frac{1}{2}}}$$

If $x' = \infty$, this becomes

$$y = \pm \frac{b}{a} x, \qquad (1)$$

which is the equation of the asymptotes. The *plus* sign belongs to the asymptote above, and the *minus* sign to that below the axis of x. The asymptotes are equally inclined to the axes of the curve.

158. Equation of Condition for Asymptotes.—Construct a rectangle on the axes, DD' and EG being its diagonals. Let $DCA = \alpha$; then

$$\tan \alpha = \frac{DA}{CA} = \frac{b}{a};$$

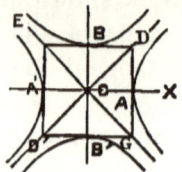

Fig. 129.

and $\tan ACG = -\dfrac{AG}{CA} = -\dfrac{b}{a} = \tan ECA;$

which, compared with the coefficient of x in equation (1), shows that:

The asymptotes of the hyperbola coincide with the diagonals of the rectangle constructed on the axes.

Letting these diagonals be the asymptotes, and $ACG = \beta$, and $c = \sqrt{a^2 + b^2} = CD$; we have

$$\sin \alpha = \frac{b}{c}, \qquad \sin \beta = -\frac{b}{c},$$

$$\cos \alpha = \frac{a}{c}, \qquad \cos \beta = \frac{a}{c};$$

from which we find

$$a^2 \sin^2 \alpha - b^2 \cos^2 \alpha = 0; \qquad (2)$$

$$a^2 \sin^2 \beta - b^2 \cos^2 \beta = 0; \qquad (3)$$

$$a^2 \sin \alpha \sin \beta - b^2 \cos \alpha \cos \beta = -\frac{2a^2 b^2}{a^2 + b^2}, \qquad (4)$$

which are the required equations.

159. Equation to the Hyperbola referred to its Asymptotes.—The equation to the hyperbola referred to oblique axes, the origin being at the centre, is found by changing b^2 into $-b^2$ in the equation of Article 127, and hence, after dropping the accents, is

$$\left.\begin{array}{l}(a^2\sin^2\beta - b^2\cos^2\beta)\,y^2 + (a^2\sin^2\alpha - b^2\cos^2\alpha)\,x^2 \\ + (a^2\sin\alpha\sin\beta - b^2\cos\alpha\cos\beta)\,2xy\end{array}\right\} = -a^2 b^2. \quad (5)$$

Combining this equation with the equations of condition given in the preceding article, we have

$$xy = \tfrac{1}{4}(a^2 + b^2), \quad (6)$$

which is the required equation. It appears that the rectangle under the coördinates is constant. Let the area of this rectangle be represented by k^2, then we have

$$xy = k^2. \quad (7)$$

in which x and y may be both positive or both negative.

For the *conjugate hyperbola*, we have, (since x or y becomes negative),

$$xy = -k^2. \quad (8)$$

For the *equilateral hyperbola*, $a = b$, and we have

$$xy = \pm \tfrac{1}{2}a^2. \quad (9)$$

In these equations $x = 0$ for $y = \infty$, and $y = 0$ for $x = \infty$.

The asymptotes are sometimes called *self-conjugates;* and equation (7) the equation of the hyperbola referred to its self-conjugates.

160. Angle between the Asymptotes in terms of the eccentricity.—We have found for the eccentricity, (Art. 93),

$$e = \frac{c}{a},$$

and, from Article 158,

$$\cos\alpha = \frac{a}{c};$$

$$\therefore \sec\alpha = e;$$

in which $\alpha = DCA$. But $DCG = 2DCA$; and if $DCG = \varphi$, we have
$$\varphi = 2 \sec^{-1} e. \tag{10}$$

For the equilateral hyperbola, this becomes
$$\varphi = 2 \sec^{-1} \sqrt{2} = 90°,$$

that is: *The asymptotes of an equilateral hyperbola are mutually perpendicular.*

161. Problem.—*To find the area of the parallelogram constructed on the coördinates of any point of an hyperbola referred to its asymptotes.*

Let P be the point whose coördinates are $x = CG$, $y = GP$, and let $ECG = \varphi$; then will the area of the required parallelogram $ECGP$ be

$$CG \sin \varphi \cdot GP = xy \sin \varphi$$
$$= \tfrac{1}{4}(a^2 + b^2) \sin \varphi.$$

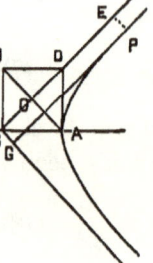

Fig. 130.

Construct the rectangle $ADBC$ on the semi-axes; its area will be ab. Draw the diagonal AB; it will be parallel to CG and EP. The area of the triangle ABC will be

$$\tfrac{1}{2} AB \cdot CO \sin COA,$$

or $\qquad \tfrac{1}{2} \cdot \sqrt{a^2 + b^2} \cdot \tfrac{1}{2} \sqrt{a^2 + b^2} \sin \varphi,$

or $\qquad \tfrac{1}{4}(a^2 + b^2) \sin \varphi;$

which, compared with the expression above, shows that it is equal to the area of the parallelogram $CGPE$, but it is also equal to one-half the area of the rectangle $ADBC$; therefore

$$xy \sin \varphi = \tfrac{1}{2} ab;$$

that is: *The area is constant and equal to one-eighth the rectangle on the axes.*

162. Equation to any Chord referred to the Asymptotes.—Let $x'y'$ and $x''y''$ be the extremities of the chord PP', then will the equation of the chord be, (Art. 40),

$$\frac{y-y'}{x-x'} = \frac{y''-y'}{x''-x'}.$$

But the equation of the curve gives

$$x'y' = x''y'' \therefore y'' = \frac{x'y'}{x''};$$

which, substituted in the preceding equation, gives

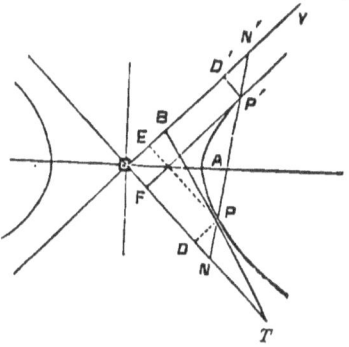

Fig. 131.

$$y - y' = \frac{\dfrac{x'y'}{x''} - y'}{x'' - x'}(x - x');$$

or $$y - y' = -\frac{y'}{x''}(x - x'); \qquad (1)$$

which is the required equation. It may also be written

$$\frac{x-x'}{x''} + \frac{y}{y'} = 1. \qquad (2)$$

163. Equation to the Tangent of the Hyperbola referred to its Asymptotes.—Let the chord pass through P' continually as that point moves along the curve and finally becomes consecutive to P; then, ultimately, will $x' = x''$, and $y' = y''$; and equations (1) and (2) of the preceding Article become

$$y - y' = -\frac{y'}{x'}(x - x'), \qquad (3)$$

or $$\frac{x}{x'} + \frac{y}{y'} = 2; \qquad (4)$$

either of which is the required equation. It is also the equation of the tangent to the conjugate hyperbola.

164. Intercepts of the tangent referred to Asymptotes.—In equation (4) make $y = 0$, and we have

$$CT = 2x' \quad \therefore \quad CD = DT;$$

or: *The subtangent is bisected by the ordinate of contact.*

If $x = 0$ in equation (4), we have

$$CB = 2y' \quad \therefore \quad CE = EB,$$

or: *The intercept on the axis of ordinates is bisected by the ordinate of contact.*

Also: *The portion of the tangent included between the asymptotes is bisected at the point of contact.*

By comparing these results with that found in Article 161, we find that: *Twice the area of the parallelogram constructed on the coördinates of any point equals the area of the triangle of which the tangent to the point is one side, and the intercepts on the asymptotes the other two sides of the triangle.*

165. Problem.—*To find the relation of the segments NP and $N'P'$ of the secant NN', (Fig. 131), contained between the asymptotes and the curve.*

In equation (2) of Article 162, if $x = 0$, we have, (observing that $x'y' = x''y''$),

$$CN' = y = y'' + y';$$

but
$$D'N' = CN' - CD';$$

$$\therefore D'N' = y'' + y' - y'' = y' = DP.$$

In a similar manner we may prove that

$$DN = D'P'.$$

But the triangles DPN and $D'P'N'$ are mutually equiangular; hence

$$PN = P'N';$$

or: *The segments of any chord contained between the curve and its asymptotes are equal.*

This principle furnishes an easy mode of constructing an hyperbola when the asymptotes and one point of the curve are given. Let Tv_3 and T-1 be the asymptotes, and a, any point on the curve. Draw radial lines $a1$, $a2$, $a3$, etc., and prolong them to Tv_3. Lay off $v_3 a_3$ equal to $a3$, $v_2 a_2 = a2$, etc., then will a_3, a_2, etc., be points on the curve.

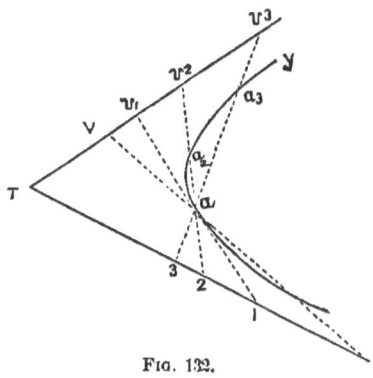

Fig. 132.

166. Tangents at the extremities of Conjugate diameters meet on the asymptotes.—Let CA be any diameter, its equation, referred to the axes x and y, will be, (Art. 40, Eq. (4)),

$$y'x - x'y = 0, \qquad (CA)$$

in which x' and y' are the coördinates of A. The equation of the tangent NA passing through the same point is, (Art. 163, Eq. (4)),

$$\frac{x}{x'} + \frac{y}{y'} = 2. \qquad (AN)$$

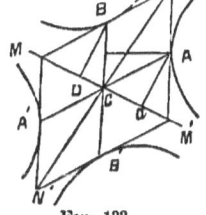

Fig. 133.

To find the equation of the tangent BN, it is necessary to find the coördinates of B. The diameter CB being conjugate to CA, will be parallel to AN, (Art. 124), and hence its equation will be of the same form as that of (AN), but since it passes through the origin, its absolute term will be zero, hence the equation of CB will be

$$\frac{x}{x'} + \frac{y}{y'} = 0; \qquad (CB)$$

which combined with the equation of the conjugate hyperbola

$$xy = -k^2 = -x'y',$$

gives, by elimination,

$$x = CD = \mp x', \qquad y = DB = \pm y'; \qquad (B)$$

which shows that $DC = -Ca$, and $Aa = BD$.

The equation of the tangent BN will be of the same form as that of NA; hence substituting in equation (AN) the values of the coördinates of the point B, as found above, that is, $y' = y'$ and $x' = -x'$, we have

$$-\frac{x}{x'} + \frac{y}{y'} = 2, \qquad (BN)$$

which is the required equation.

To find the intersection of the tangents AN and BN, combine equations (AN) and (BN), and eliminate x and y successively, and we find

$$y = 2y' = CN, \text{ and } x = 0.$$

This value of x shows that the intersection is on the axis of y, that is, it is on the asymptote.

This analysis also shows that:

The diagonals of all parallelograms constructed on a pair of conjugate diameters coincide with the asymptotes of the hyperbola.

The value of y shows that the diagonal CN of the parallelogram constructed on the semi-diameters, is double the ordinate aA of the extremity of the conjugate diameter $A'A$.

Polar Equations of the Conic Sections.

167. General Equations.—Let P be any point on the curve, F the focus, which is also taken as the pole, $2p$ the *latus rectum*, BE the directrix, e the eccentricity, $\rho = FP =$ the radius vector, and $\varphi = PFA$, the variable angle measured from the nearest vertex. Then according to Article 120,

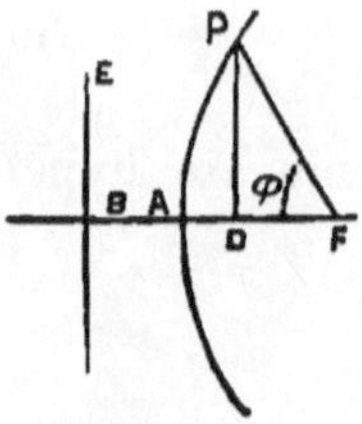

$$FP = e \cdot BD = e(BF - FD) = e\left(\frac{p}{e} - FD\right).$$

But $\quad FD = PF \cos PFD = \rho \cos \varphi,$

therefore, $\quad FP = \rho = e\left(\frac{p}{e} - \rho \cos \varphi\right);$

Fig. 134.

$$\therefore \rho = \frac{p}{1 + e \cos \varphi}, \qquad (1)$$

which is *the polar equation of any conic section, the pole being at the focus and the variable angle measured from the nearest vertex.*

If φ be measured from the remote vertex, $\cos \varphi$ in the preceding equation becomes negative, and we have, (designating the angle by φ'),

$$\rho = \frac{p}{1 - e \cos \varphi'}. \qquad (2)$$

If the angle φ be measured from a line which passes through the focus F and makes an angle β with the axis AF of the curve, (β being positive in reference to the line of reference), we have

$$\rho = \frac{p}{1 + e \cos(\varphi - \beta)}, \qquad (3)$$

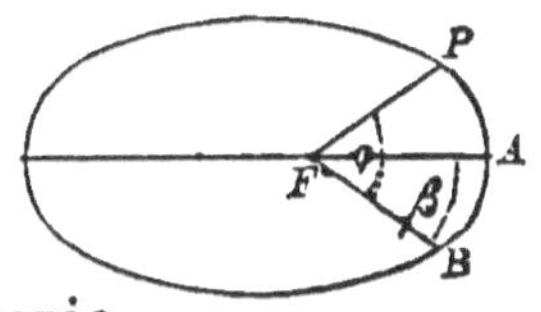

which is a more general equation of a conic.

Fig. 135.

168. Polar Equation to the Parabola.—For this curve, $e = 1$, and if the pole be at the focus, the initial line coinciding with the axes, equation (1) becomes

$$\rho = \frac{p}{1 + \cos \varphi}, \qquad (4)$$

which is the required equation.

169. Polar Equation to the Ellipse.—Let the pole be at one of the foci, and φ measured from the nearest vertex. We have $e < 1$, and $2p = 2a(1 - e^2)$, (Art. 95, Eq. (2)), and equation (1) becomes

$$\rho = \frac{a(1 - e^2)}{1 + e \cos \varphi}, \qquad (5)$$

which is the required equation.

170. Polar Equation to the Hyperbola.—The conditions being the same as for the ellipse, except that $e > 1$, we have

$$\rho = \frac{a(e^2 - 1)}{1 + e \cos \varphi}, \qquad (6)$$

for the required equation.

171. Discussion of Equation (4).—If $\varphi = 0$, $\rho = \tfrac{1}{2}p$, which is the distance from the focus to the vertex of the parabola. .... If $\varphi = 90°$, $\rho = p$, which is one-half the ordinate through the focus and is one-half the *latus rectum*, as it should be. .... If $\psi = 180°$, $\rho = \infty$, hence the curve does not cross the axis in that direction. .... If $\varphi = 180° - i$, i being any value however small, ρ will have a definite value; hence if a line be drawn from the focus in the opposite direction from the vertex, making any angle however small with the axis, it will meet the curve at some point. .... If $\varphi = 270°$, $\rho = p$, as it should; and if $\varphi = 360°$, $\rho = \tfrac{1}{2}p$, which is the same as for $\varphi = 0°$, as it should be.

172. Discussion of Equation (5), or $\rho = \dfrac{a(1-e^2)}{1+e\cos\varphi}$.

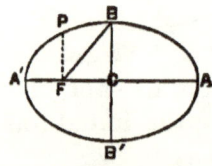

Fig. 136.

If $\varphi = 0°$, $\rho = a(1-e) = a - ea = A'C - FC = A'F$. .... If $\varphi = 90°$, $\rho = a(1-e^2) = FP$. .... For $\varphi = \cos^{-1}\left(-\dfrac{FC}{FB}\right) = \cos^{-1}-\dfrac{ea}{\rho}$, we find $\rho = a$, as it should. ....
If $\varphi = 180°$, $\rho = a(1+e) = FA$. .... If $\varphi = 360°$, $\rho = a(1-e)$ which is the same as for $\varphi = 0°$.

173. Discussion of Equation (6), or $\rho = \dfrac{a(e^2-1)}{1+e\cos\varphi}$, for the hyperbola. Since $e > 1$, the numerator is essentially positive, and the denominator will be negative when $\cos\varphi$ is negative and > 1. For $\varphi = 0°$, $\rho = a(e-1) = F''A'$. .... For $\varphi = 90°$, $\rho = a(e^2-1) = F'P$.

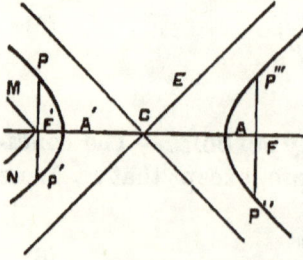

Fig. 137.

.... When $\varphi = \cos^{-1}\left(-\dfrac{1}{e}\right)$ the radius vector becomes parallel to the asymptote, and the equation gives $\rho = \infty$; hence the positive radius vector will not cut the curve. As $\varphi > \cos^{-1}\left(-\dfrac{1}{e}\right)$ the radius vector becomes negative, and the branch $P''A$

will be described. . . . . . If $\varphi = 180°$, $\rho = -a(e+1)$ $= F'A$, and from this point the branch AP''' will be described until $\varphi = 180° + \cos^{-1}\dfrac{1}{e}$ when the radius vector will be parallel to the asymptote CE, and will be infinite. When it passes this value it again becomes positive and will trace the branch $P'A'$. . . . . . For $\varphi = 270°$, $\rho = a(e^2-1)$ $= F'P'$. . . . . For $\varphi = 360°$, $\rho = a(e-1) = F'A'$.

174. Polar equation to the Ellipse, the Pole being at the centre.—The equations for transformation from rectangular to polar coördinates, the pole being at the origin, are, (Art. 55),

$$x = \rho \cos(\varphi + \alpha), \qquad y = \rho \sin(\varphi + \alpha);$$

and if the initial line coincides with the axis of x, these become

$$x = \rho \cos \varphi, \qquad y = \rho \sin \varphi.$$

These values substituted in the axial equation to the ellipse, (Eq. (a_1), Art. 69), give

$$a^2 \rho^2 \sin^2 \varphi + b^2 \rho^2 \cos^2 \varphi = a^2 b^2;$$

$$\therefore \rho = \dfrac{ab}{\sqrt{a^2 \sin^2 \varphi + b^2 \cos^2 \varphi}},$$

which is the required equation.

175. Polar Equation to the Hyperbola, the Pole being at the centre.—Changing b to $b\sqrt{-1}$ in the preceding equation, gives

$$\rho = \pm \dfrac{ab}{\sqrt{-a^2 \sin^2 \varphi + b^2 \cos^2 \varphi}};$$

which is the required equation.

If $\varphi = 0°$, then $\rho = \pm a$;

$\varphi = 90°$, $\qquad \rho = \pm b \sqrt{-1}$;

$\varphi = \tan^{-1}\dfrac{b}{a}$; $\rho = \infty$;

the last value of φ being the inclination of the asymptote.

EXAMPLES.

1. What is the polar equation of the parabola whose rectangular equation is $y^2 = 8x$? What is the length of the radius vector for $\varphi = 0°$, 45°, 60°, 90°, 120°?

$$\text{Ans. } \rho = \frac{4}{1 + \cos \varphi} \; ; \; 2; \; 8 - 4\sqrt{2}; \; \frac{8}{3}; \; 4; \; 8.$$

2. Required the polar equation of an ellipse whose axes are 8 and 6 respectively, the pole being at one of the foci.

3. What is the polar equation of an hyperbola whose transverse axis is 8, and the distance between the foci is 12? What will be the value of ρ for $\varphi = 0°$, 90°, 120°, 180°?

4. If a comet moves in a parabolic orbit having the sun at the focus, and is 150,000,000 miles from the sun when the radius vector makes an angle of 90° with the axis; how near will it approach the sun?

Ans. 75,000,000 miles.

[For a shorter course omit to Chapter VI.]

CHAPTER V.

GENERAL DISCUSSION OF THE EQUATION OF THE SECOND DEGREE.

176. Rectangular Equation to a Conic Section having any position in a plane.—Let P be any point on the arc of a conic section whose coördinates are $OK = x$, and $KP = y$; F the focus, and CB the directrix. Let fall the perpendicular PE upon the directrix, then, according to Article 120, we have $FP \div PE = e = $ the eccentricity. It is required to find FP and PE in terms of x and y and known quantities. Let the coördinates of the focus be $OH = m$, and $HF = n$, the distance OL of the directrix from the origin be d, and the angle which the axis AF of the curve makes with the axis of x, which equals LOX, be θ; then

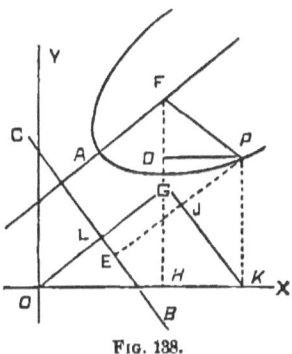

Fig. 188.

$$DP = OK - OH = x - m;$$
$$FD = FH - DH = n - y;$$
$$\therefore FP^2 = (x-m)^2 + (y-n)^2. \quad (1)$$

Draw KG parallel to CB, and note the point J where it cuts PE, then

$$PE = PJ + OG - OL = y \sin JKP + x \cos GOK - d$$
$$= y \sin \theta + x \cos \theta - d.$$

But, according to Boscovich's definition of a Conic, (Art. 120), we have
$$FP^2 = e^2 \cdot PE^2;$$
hence equation (1) becomes
$$(x-m)^2 + (y-n)^2 = e^2(y\sin\theta + x\cos\theta - d)^2;$$
expanding and reducing gives
$$\frac{1-e^2\cos^2\theta}{m^2+n^2-e^2d^2}x^2 - \frac{2e^2\sin\theta\cos\theta}{m^2+n^2-e^2d^2}xy + \frac{1-e^2\sin^2\theta}{m^2+n^2-e^2d^2}y^2$$
$$+ 2\frac{e^2d\cos\theta - m}{m^2+n^2-e^2d^2}x + 2\frac{e^2d\sin\theta - n}{m^2+n^2-e^2d^2}y + 1 = 0; \quad (2)$$
which is the required equation.

177. Every Equation of the Second Degree between two variables may represent a Conic Section.—The general equation of the second degree may be written, (Art. 92),
$$Ax^2 + 2Hxy + By^2 + 2Gx + 2Fy + C = 0;$$
but since this equation may be divided by any one of its coefficients, thereby making the coefficient of that term unity, it will be equally general if we make C unity. Hence we have for the general equation,
$$Ax^2 + 2Hxy + By^2 + 2Gx + 2Fy + 1 = 0;$$
in which there are *five* arbitrary constants, independent of each other. In the equation of the preceding Article there are also *five* arbitrary constants, viz., m, n, d, e, and θ, independent of each other. If now the coefficients of the corresponding terms of the preceding equation and of equation (2) of the preceding Article be placed equal to each other, we have
$$A = \frac{1-e^2\cos^2\theta}{m^2+n^2-e^2d^2}; \qquad B = \frac{1-e^2\sin^2\theta}{m^2+n^2-e^2d^2};$$
$$H = -\frac{e^2\sin\theta\cos\theta}{m^2+n^2-e^2d^2}; \qquad G = \frac{e^2d\cos\theta - m}{m^2+n^2-e^2d^2};$$
$$F = \frac{e^2d\sin\theta - n}{m^2+n^2-e^2d^2}.$$

If any values whatever be assigned to A, B, H, G, F, the values of m, n, d, e, and θ may be found by means of these five equations, and the latter quantities determine the character and position of a conic section; hence, *Every equation of the second degree represents some conic section.*

178. General Test.—Subtracting the product of A times B, of the preceding Article, from the square of H, gives

$$H^2 - AB =$$
$$[e^4 \sin^2\theta \cos^2\theta - (1 - e^2\cos^2\theta)(1 - e^2\sin^2\theta)] \frac{1}{(m^2 + n^2 - e^2 d^2)^2}$$
$$= [e^4 \sin^2\theta \cos^2\theta - 1 + (\sin^2\theta + \cos^2\theta)e^2 - e^4\sin^2\theta\cos^2\theta] \frac{1}{(m^2 + n^2 - e^2 d^2)^2}$$
$$= \frac{e^2 - 1}{(m^2 + n^2 - e^2 d^2)^2};$$

the denominator of which is essentially positive, and therefore the sign of the second member will be the same as that of the numerator of the fraction. When $e > 1$ it will be positive, and negative when $e < 1$. Hence, according to Article 93, we have

for the ellipse .. $e < 1$, and $H^2 - AB < 0$;
for the parabola .. $e = 1$, and $H^2 - AB = 0$;
for the hyperbola $e > 1$, and $H^2 - AB > 0$.

The species of the locus represented by the general equation of the second degree, therefore, depends upon the coefficients A, B, and H, and is an ellipse, parabola, or hyperbola according as $H^2 - AB$ is *negative, zero,* or *positive.*

179. To cause the term containing xy to disappear from the general equation.—The coefficient of xy is, (Art. 177),

$$2H = -\frac{2e^2 \sin\theta \cos\theta}{m^2 + n^2 - e^2 d^2},$$

which will reduce to zero for $\theta = 0°$ or $90°$. The former value makes the axis of x parallel to the axis of the curve, and the latter, perpendicular to it. But changing the coördinate axes cannot change the character of the locus; hence by transforming the equation so that the axis of x will be parallel or perpendicular to the axis of the curve, the general equation becomes

$$Ax^2 + By^2 + 2Gx + 2Fy + 1 = 0,$$

and since this transformation is always possible, this equation includes all varieties and species of conic sections.

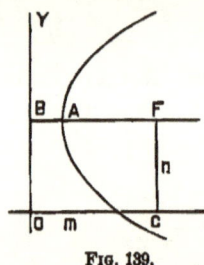

Fig. 139.

Take the axis of x *parallel* to the axis of the curve; then $\theta = 0$, and the values of the coefficients, (Art. 177), become

$$A = \frac{1-e^2}{m^2 + n^2 - e^2 d^2}; \quad B = \frac{1}{m^2 + n^2 - e^2 d^2}.$$

$$H = 0;$$

$$G = \frac{e^2 d - m}{m^2 + n^2 - e^2 d^2}; \quad F = \frac{-n}{m^2 + n^2 - e^2 d^2}$$

180. To cause the coefficient of y to disappear.—This condition requires that F be zero in Article 177,

$$\therefore \theta = 0, \text{ and } n = 0,$$

which makes the axis of x coincide with the axis of the curve. This condition also reduces H to zero, hence the general equation becomes

$$Ax^2 + By^2 + 2Gx + 1 = 0;$$

which involves *all varieties of conic sections*. The values of the coefficients become

$$A = \frac{1-e^2}{m^2 - e^2 d^2}; \quad B = \frac{1}{m^2 - e^2 d^2}; \quad G = \frac{e^2 d - m}{m^2 - e^2 d^2}.$$

181. Remark.—When H and F are zero, *no other coefficients can* GENERALLY *be zero*. The value of A cannot also, generally, be zero, for if it is, e must be unity, which is true only for the parabola; B cannot be zero unless m or d is *infinite*, neither of which would represent a finite curve; and G cannot generally be zero, for if it were we would have $d = m$ in the parabola, which would cause the directrix to pass through the focus, a condition which is not true of the common parabola.

Varieties of Conic Sections.

182. Varieties of the Ellipse.—For this case we have to determine all the *forms of loci* represented by the equation $Ax^2 + By^2 + 2Gx + 1 = 0$, when $H^2 - AB < 0$.

Since H is zero in this equation, we have $-AB < 0$, hence A and B must be finite and have the same signs. The origin may be so taken that $m = e^2 d$, in which case $G = 0$, and the equation becomes

$$Ax^2 + By^2 + 1 = 0, \qquad (1)$$

which involves all the varieties of the ellipse.

If A and B are essentially negative, this equation becomes

$$Ax^2 + By^2 = +1, \qquad (2)$$

which is the equation of the ellipse referred to its axes, (Art. 69). If A and B are essentially positive, we have

$$x = \sqrt{\frac{-1 - By^2}{A}}, \qquad (3)$$

which is *imaginary* and the equation represents no real locus, but we may say that it represents an imaginary locus. If $A = B$ and both are negative, we have $Ax^2 + Ay^2 = 1$, which is the equation of the circle, (Art. 57). Substituting $m = e^2 d$ in the values of A and B of Article 180, we have

$$A = -\frac{1 - e^2}{e^2 d^2 (1 - e^2)} = -\frac{1}{e^2 d^2}; \quad B = -\frac{1}{e^2 d^2 (1 - e^2)}. \qquad (4)$$

Substituting these values in equation (1), multiplying through by $(1 - e^2)$, and representing the coefficients of x^2 and y^2 respectively by A' and B', gives

$$A'x^2 + B'y^2 + (1 - e^2) = 0.$$

If now the absolute term be zero, or $e^2 = 1$, the equation reduces to the form
$$A'x^2 + B'y^2 = 0.$$
But an examination of equation (4) shows that
$$A' = -\frac{1-e^2}{e^2d^2}$$
which is zero when $e^2 = 1$, hence the equation becomes
$$0 \cdot x^2 + B'y^2 = 0,$$
which is satisfied for $\pm x$ *indeterminate* and $y = 0$; and hence is the equation of the axis of x; hence *a straight line* is a particular case of an ellipse. But when $e = +1$, the locus is a parabola, (Art. 93); hence the ellipse and parabola approach the same right line as *one* of the limits of those curves.

There are, therefore, *five* varieties of loci embraced in the general equation of the second degree, which fulfil the condition $H^2 - AB < 0$, and hence are called varieties of the ellipse; viz., the *Ellipse proper*, the *Circle*, the *Point*, the *Right Line*, and an *Imaginary Locus*.

183. Varieties of the Hyperbola. — These are included in the equation $Ax^2 + By^2 + 2Gx + 1 = 0$, (Art. 180), under the condition $H^2 - AB > 0$. Since $H = 0$ in this equation, A and B must have contrary signs so that their product shall always be negative, and hence $-AB$ be always positive. Taking the origin so that $m = e^2 d$; in which case $G = 0$, the equation becomes
$$Ax^2 - By^2 - 1 = 0;$$
in which A and B must have the same sign, since the sign of one term is changed. If A and B are essentially negative, this equation becomes
$$Ax^2 - By^2 = +1, \qquad (1)$$
which is the equation to the *ordinary hyperbola* (or *x-hyper-*

bola), (Art. 78). If A and B are essentially positive, this equation becomes

$$Ax^2 - By^2 = -1,$$

which is the equation to the *conjugate hyperbola* (or *y-hyperbola*). If $A = B$, we have

$$Ax^2 - Ay^2 = \pm 1;$$

which is the equation to the *equilateral hyperbola*. Substituting the value of $m = e^2 d$ in the values of A and B, (Art. 180), they become the same as for the ellipse. Substituting their values in equation (1), multiplying through by $e^2 d^2 (1 - e^2)$, and representing the coefficients by A' and B' (observing that they have contrary signs), and making $d = 0$, e and A' will be indeterminate, and we have

$$A'x^2 - B'y^2 = 0; \quad \therefore x = \pm \sqrt{\frac{B'}{A'}}\, y,$$

which is the equation of *two right lines* passing through the origin. If d be zero or finite, and $e = \pm 1$, in which the upper sign is characteristic of the parabola, the value of A' for both values of e becomes zero, and the equation becomes

$$0 \cdot x^2 - B'y^2 = 0;$$

which is satisfied for $+ x$ *indeterminate* and $y = 0$, and hence is the equation to the axis of x. This same *right line* is, therefore, a limit both of the hyperbola and parabola. Hence, the condition $H^2 - AB > 0$, in the general equation of the second degree, gives *four* varieties of loci, and are called varieties of the hyperbola; viz., the *Common Hyperbola*, the *Equilateral Hyperbola*, *Two Right Lines* intersecting each other, and *One Right Line*.

184. Varieties of the Parabola.—Resuming equation

$$Ax^2 + By^2 + 2Gx + 1 = 0,$$

and the analytical condition for the parabola, which is

$$H^2 - AB = 0;$$

we observe that, since H is zero in the former equation, either A or B, or both A and B, must be zero.

If $A = 0$, we have

$$By^2 + 2Gx + 1 = 0.$$

But, for the parabola, $e = 1$, and the equations in Article 180 become

$$A = 0; \quad B = \frac{1}{m^2 - d^2}; \quad G = \frac{d - m}{m^2 - d^2} = -\frac{1}{m + d}.$$

Substituting these values in the preceding equation and multiplying by $m^2 - d^2$, gives

$$y^2 + 2(d - m)x + m^2 - d^2 = 0.$$

The absolute term is zero when

$$m^2 = d^2; \text{ or } m = \pm d.$$

For $m = + d$, the preceding equation becomes

$$y^2 + 0 \cdot x = 0, .$$

which is the equation of the axis of x. But when $m = + d$ the directrix passes through the focus, (Arts. 176 and 179), and this condition reduces the parabola to *a straight line passing through the focus.*

If $m = - d$, the absolute term vanishes, and the coefficient of x becomes $4d$, and the equation becomes

$$y^2 + 4dx = 0; \text{ or } y^2 = - 4dx.$$

If d be positive, y will be real for negative values of x, and imaginary for positive values.

If d be negative, it may be written

$$y^2 = 4dx,$$

which is *the equation of the common parabola.*

The condition that $m = - d$, places the directrix as far from the origin in one direction as the focus is in the opposite direction, which agrees with Article 85.

If $B = 0$, we have the *form*

$$Ax^2 + 2Gx + 1 = 0;$$

$$\therefore x = -\frac{G}{A} \pm \sqrt{\frac{G^2}{A^2} - \frac{1}{A}},$$

and y *indeterminate;* but to make $B = 0$, m or d must be infinite, either of which reduces A and G to zero, hence the preceding value of x becomes $x = \infty$, hence the line is parallel to y, but infinitely distant.

Substituting $e = 1$, in the equations of Article 180, we find that if $B = \infty$ we have $m^2 = d^2$; $\therefore A = \frac{0}{0}$; and also $G = \frac{0}{0}$; that is, both are indeterminate.

If $A = 0$ and $B = 0$, we have

$$2Gx + 0 \cdot y^2 + 1 = 0,$$

which is the equation of a right line parallel to the axis of y, and at a distance from it equal to $-\frac{1}{2G}$.

In the general equation

$$Ax^2 + 2Hxy + By^2 + 2Gx + 2Fy + 1 = 0,$$

if $A = B = \pm H$, we have $H^2 - AB = 0$ (which characterizes the parabola), and the equation becomes

$$A(x + y)^2 + 2Gx + 2Fy + 1 = 0.$$

Making $A = B = \pm H$ in the equations of Article 177, and at the same time making $e = 1$, we find that $\sin \theta = \cos \theta$; $\therefore \theta = 45°$; hence the axis of the curve cuts both coördinate axes at an angle of $45°$. Transforming the origin of coördinates to a point such that $F = 0$, and $G = 0$, we find from the last equations of Article 177 that

$$m = n, \; d = \sqrt{2}\, n, \text{ and } A = \infty.$$

If $F = G$ we have

$$A(x+y)^2 + 2G(x+y) = -1;$$

$$\therefore y = -x - \frac{G}{A} \pm \sqrt{\frac{G^2}{A^2} - \frac{1}{A}},$$

which is the equation of two real straight lines if $A < G^2$; of two imaginary lines if $A > G^2$, and of one real line if $A = \infty$.

There are, therefore, *five* varieties of the parabola, viz., the *Common Parabola*, the *Imaginary Parabola*, the *Right Line*, two *Parallel Right Lines*, and two *Imaginary Right Lines*.

185. Illustration.—It will be shown hereafter, (Art. 252), that the intersection of a plane with a cone will be a parabola when the plane is parallel to an element of the cone, as LR; an ellipse if the plane cuts all the elements of the cone, as AE; an hyperbola if it cuts both nappes of the cone, as $DA'AH$. Taking these for granted, we may easily illustrate all the varieties of the conic sections.

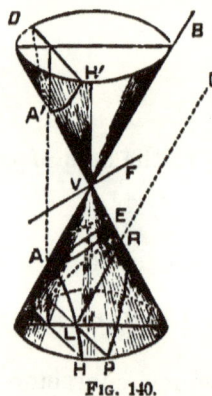

Fig. 140.

If the plane of the ellipse be turned until it is perpendicular to the axis of the cone, the line of intersection will constantly be an ellipse, but when the plane becomes perpendicular to the axis the intersection will be *a circle*. If the plane pass through the apex it will cut all the elements, for they all pass through that point, and the intersection becomes *a point*. If it be turned until it is tangent to the cone along AB, the intersection becomes *a straight line*, which is one of the special cases of an ellipse. If the apex of the cone be infinitely remote, the cone becomes a cylinder, and the intersection AE is still an ellipse.

If the plane $DA'AH$ pass through the vertex, it will intersect both nappes, but the intersection will be *two straight lines* intersecting at the apex. If the plane be turned about an axis through the vertex, the two lines approach the *single line VB* as a limit, and this is a particular case of the hyperbola.

If the plane LR passes through the apex and parallel to an element the intersection will be *a right line, AB*. If the apex of the cone be infinitely distant, the surface becomes a cylinder, and the intersection of a plane parallel to an element will be *two right lines* unless the plane be tangent to the surface, in which case it will be limited to one *right line*. If the plane be parallel to an element, but does not cut the surface of the cylinder, the intersection will be imaginary, and will represent the two imaginary right lines.

186. Problem.—*Pass a conic through any five points in a plane, and show that only one such conic can be passed.*

Let (x_1, y_1), (x_2, y_2), (x_3, y_3), (x_4, y_4), and (x_5, y_5) be the five points. Substituting these values successively in the general equation of the second degree, (Art. 177), we have the five *equations of condition*

$$Ax_1^2 + 2Hx_1y_1 + By_1^2 + 2Gx_1 + 2Fy_1 + 1 = 0;$$

$$Ax_2^2 + 2Hx_2y_2 + By_2^2 + 2Gx_2 + 2Fy_2 + 1 = 0;$$

$$Ax_3^2 + 2Hx_3y_3 + By_3^2 + 2Gx_3 + 2Fy_3 + 1 = 0;$$

$$Ax_4^2 + 2Hx_4y_4 + By_4^2 + 2Gx_4 + 2Fy_4 + 1 = 0;$$

$$Ax_5^2 + 2Hx_5y_5 + By_5^2 + 2Gx_5 + 2Fy_5 + 1 = 0;$$

from which the value of the five arbitrary constants may be found, which values substituted in the general equation $Ax^2 + 2Hxy + By^2 + Gx + Fy + 1 = 0$ will give the required equation. Since the unknown quantities A, H, etc., are of the first degree, only one set of values can be found, and since the five points give five conditions, they can generally be definitely determined.

EXAMPLE.

1. Determine the equation of a conic section which shall pass through the points $(2, 1)$, $\left(\dfrac{5}{3}, 2\right)$, $\left(-\dfrac{2}{3}, -\dfrac{1}{3}\right)$, $\left(-\dfrac{2}{3}, -3\right)$, $\left(\dfrac{5}{3}, -\dfrac{2}{3}\right)$; find its species, its eccentricity, the position of its directrix, and locate the curve in reference to the coördinate axes.

The equations of condition will be

$$4A + 4H + B + 4G + 2F + 1 = 0;$$

$$\frac{25}{9}A + \frac{20}{3}H + 4B + \frac{10}{3}G + 4F + 1 = 0;$$

$$\frac{4}{9}A + \frac{4}{9}H + \frac{1}{9}B - \frac{4}{3}G - \frac{2}{3}F + 1 = 0;$$

$$\frac{4}{9}A + 4H + 9B - \frac{4}{3}G - 6F + 1 = 0;$$

$$\frac{25}{9}A - \frac{20}{9}H + \frac{4}{9}B + \frac{10}{3}G - \frac{4}{9}F + 1 = 0.$$

Solving these equations, give

$$A = -1, \quad B = -\frac{1}{3}, \quad 2H = \frac{2}{3}, \quad 2G = \frac{4}{3}, \quad 2F = -\frac{2}{3};$$

and the equation of the locus will be

$$-x^2 + \frac{2}{3}xy - \frac{1}{3}y^2 + \frac{4}{3}x - \frac{2}{3}y + 1 = 0;$$

in which the absolute term must remain positive since it is positive in Article 177.

To determine the species of this locus, we have

$$H^2 - AB = \frac{1}{9} - \frac{1}{3} = -\frac{2}{9}, \text{ which is } < 0,$$

hence the locus is an ellipse. We now have, according to Article 177,

$$\frac{1 - e^2\cos^2\theta}{m^2 + n^2 - e^2 d^2} = -1, \quad (1) \qquad \frac{1 - e^2 \sin^2\theta}{m^2 + n^2 - e^2 d^2} = -\frac{1}{3}, \quad (2)$$

$$\frac{e^2 \sin\theta \cos\theta}{m^2 + n^2 - e^2 d^2} = -\frac{1}{3}, \quad (3) \qquad \frac{e^2 d \cos\theta - m}{m^2 + n^2 - e^2 d^2} = \frac{2}{3}, \quad (4)$$

$$\frac{e^2 d \sin\theta - n}{m^2 + n^2 - e^2 d^2} = -\frac{1}{3}. \quad (5)$$

From (1) and (2) we have

$$1 - e^2(1 - \sin^2\theta) = 3 - 3e^2 \sin^2\theta;$$

$$\therefore \sin^2\theta = \frac{2 + e^2}{4e^2}. \quad (6)$$

From (2) and (3) we have

$$e^2 \sin \theta \cos \theta = 1 - e^2 \sin^2 \theta.$$

Squaring, substituting $\sin^2 \theta$ from (6), we find

$$e^2 = 2(\sqrt{2} - 1); \qquad (7)$$

$$\therefore e = 0.910 +. \qquad (8)$$

This value of e in (6) gives

$$\sin \theta = \tfrac{1}{2}\sqrt{2 + \sqrt{2}} = 0.9238 + ; \therefore \theta = 67° 30'. \qquad (9)$$

Equation (2) now gives

$$m^2 + n^2 - e^2 d^2 = -3 (1 - e^2 \sin^2 \theta)$$

$$= -\tfrac{3}{2}(2 - \sqrt{2}); \qquad (10)$$

and (4) and (10) give

$$e^2 d \cos \theta - m = - (2 - \sqrt{2}); \qquad (11)$$

and (5) and (10) give

$$e^2 d \sin \theta - n = \tfrac{1}{2} (2 - \sqrt{2}). \qquad (12)$$

Let $A = \tfrac{1}{2}(2 - \sqrt{2}) = \tfrac{1}{2} \sqrt{2} (\sqrt{2} - 1)$;

then, (Eq. (7)), $\qquad e^2 = 2 \sqrt{2} A$;

and from Eq. (9) $\qquad \cos \theta = \tfrac{1}{2} \sqrt{2A}$;

and, therefore, $\qquad \sin \theta = \tfrac{1}{2} \sqrt{2} \sqrt{2 - A}$;

and equations (10), (11), (12), become

$$m^2 + n^2 - 2\sqrt{2} A d^2 = -3A; \qquad (13)$$

$$2A^{\tfrac{3}{2}} d - m = -2A; \qquad (14)$$

$$2Ad \sqrt{2 - A} - n = A. \qquad (15)$$

Eliminating d and n from these equations, we find

$$(A - \tfrac{1}{4}\sqrt{2})m^2$$

$$+ (2A^3 - 4A^2 - A^2 \sqrt{2A - A^2} + \sqrt{2}A)m$$

$$= \frac{3}{2}A^4 - \frac{11}{2} A^3 - 2A^3 \sqrt{2A - A^2} + \sqrt{2}A^2.$$

Introducing the value of A and reducing, we find

$$m^2 - m = 0.68 +, \text{ nearly};$$
$$\therefore m = 1.46 + \text{ or } - 0.46 +.$$

These values in (14) give

$$d = 2.76 \text{ or } - 3.31;$$

and these in (15) give

$$n = 1.81 \text{ or } - 2.82.$$

Making $OH = m = 1.46$, and $HF = n = 1.8 +$, locates the focus F, and $OE' = n = + 2.8 +$, $E'F'' = m = - 0.46$ locates the focus F'; and the line AA' drawn through the foci locates the major axis. It should make an angle $FGX = 67° 30'$.

The distance $OL = d = 2.7 +$, parallel to AA', gives the distance of the directrix LB from the origin, and the line LB, drawn through L, perpendicular to AA' gives one directrix. Similarly, $OL' = d = -3.3 +$, determines the other directrix. The centre C is midway between F and F', and

$$\frac{CF}{AC} = \frac{AF}{AB} = e = 0.91 +.$$

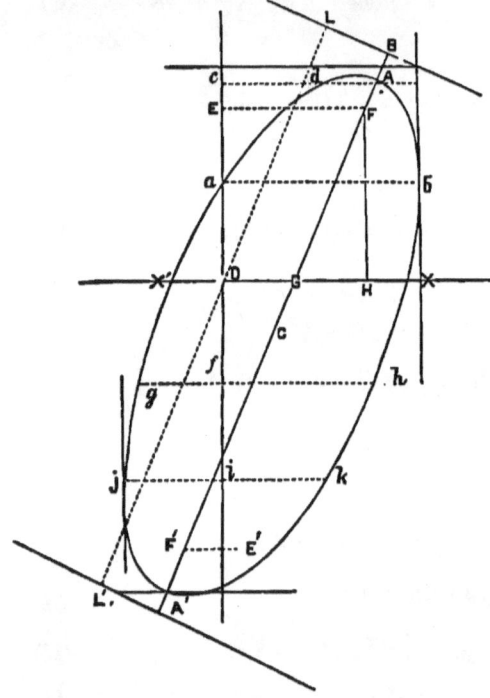

Fig. 141.

187. *Another Mode of Discussion.*—The process of determining the position and dimensions of a curve by the preceding process is often very lengthy, so that instead of resorting to it, the curve, in practice, is simply constructed by points. Thus in the preceding example, in which the equation is

$$x^2 - \frac{2}{3}xy + \frac{1}{3}y^2 - \frac{4}{3}x + \frac{2}{3}y - 1 = 0,$$

we assume values for y and find x. We thus find

if $y = 0$, $x = 1.86 + = OX$, or $-0.53 + = OX'$;
   $y = 1 = Oa$, $x = 2 = ab$, or 0;
   $y = 2 = Oc$, $x = 1.66 + = cA$, or $+1.00 = cd$;
   $y = -1 = Of$, $x = 1.53 + = fh$, or $-0.86 + = fg$;
   etc.;   etc.;

through which points the curve may be traced.

A general discussion may be made without introducing the eccentricity. Resuming the general equation

$$Ax^2 + 2Hxy + By^2 + 2Gx + 2Fy + 1 = 0, \quad (1)$$

and solving for y, we have

$$y = -\frac{Hx+F}{B} \pm \frac{1}{B}\sqrt{(H^2-AB)x^2 + 2(HF-BG)x + F^2 - B},$$

which generally gives two values for y, one of which exceeds the ordinate given by the rational part, $-\frac{Hx+F}{B}$, as much as the other is less than that part. If the rational part be represented by y_1, then

$$y_1 = -\frac{Hx+F}{B} \quad (2)$$

is the equation of a diameter. This diameter will be conjugate to the diameter parallel to the axis of y.

187a. To remove the coefficients F and G from the general equation.—Transform the axes to a new origin, the axes remaining parallel. For this case we have, (Art. 49),

$$x = x_1 + m, \quad y = y_1 + n, \quad (3)$$

(since m and n may be either positive or negative), and substituting these values in equation (1) we find

$$Ax_1^2 + 2Hx_1y_1 + By_1^2 + 2(Am + Hn + G)x_1 + 2(Bn + Hm + F)y_1 + C' = 0, \quad (4)$$

in which C' is the sum of the absolute terms. Making the coefficients of x_1 and y_1 equal to zero, we find

$$m = \frac{BG - HF}{H^2 - AB}, \qquad n = \frac{AF - HG}{H^2 - AB}; \qquad (5)$$

from which the values of m and n may be found, except when $H^2 - AB = 0$, in which case m and n are either infinite, or indeterminate. The finite values of m and n from equations (5), substituted in equation (4), give, after dropping the subscripts,

$$Ax^2 + 2Hxy + By^2 + C' = 0; \qquad (6)$$

hence, if the coefficients of the first powers of both x and y in the general equation can be removed at the same time, it can be done by changing the origin to a point whose coördinates are given by equations (5).

If, in equation (6), we substitute $-x$ for x, and $-y$ for y, it will remain of the same form and value; hence the locus will have *a centre* and the origin of coördinates will be at that centre. Hence, also, for every locus of the second order which has a centre, the origin of coördinates may be taken at such a point that the terms containing the first powers of the variable shall disappear.

Comparing the value of $H^2 - AB$ with Article 178, we see that the parabola has no centre, but that the ellipse and hyperbola have each a centre.

187b. To remove the term containing xy **from the General Equation.**—Turn the new axes through an angle α. For this we have, (Art. 54),

$$x = x' \cos \alpha - y' \sin \alpha,$$

$$y = x' \sin \alpha + y' \cos \alpha;$$

and these values substituted in equation (6) give

$$\left. \begin{array}{l} (A\cos^2\alpha + B\sin^2\alpha + 2H\sin\alpha\cos\alpha)x'^2 \\ + (A\sin^2\alpha + B\cos^2\alpha - 2H\sin\alpha\cos\alpha)y'^2 \\ + [2(B-A)\sin\alpha\cos\alpha + 2H(\cos^2\alpha - \sin^2\alpha)]x'y' + C' = 0 \end{array} \right\}. \quad (7)$$

Making the coefficient of $x'y'$ equal to zero, we find

$$\frac{2\sin\alpha\cos\alpha}{\cos^2\alpha-\sin^2\alpha}=\tan 2\alpha=\frac{2H}{A-B}, \qquad (8)$$

and equation (7) will reduce to the form

$$A'x'^2 + B'y'^2 + C' = 0;$$

in which A' represents the coefficient of x'^2; and B', of y'^2; or, dropping the accents, it becomes

$$Ax^2 + By^2 + C = 0. \qquad (9)$$

Hence the term containing xy in the general equation of the second degree may be made to disappear by turning the rectangular axes through an angle α, the value of which is given by equation (8).

After removing the term containing xy, the general equation becomes

$$Ax^2 + By^2 + 2Gx + 2Fy + C = 0, \qquad (10)$$

and when the locus has not a centre we have $H^2 - AB = 0$; and since H is zero in the preceding equation, we must have either A or B equal to zero, and the equation becomes

or $$\left.\begin{array}{r}Ax^2 + 2Gx + 2Fy + C = 0, \\ By^2 + 2Gx + 2Fy + C = 0.\end{array}\right\} \qquad (11)$$

The origin of coördinates may now be changed to such a point as to reduce these equations to the form

or $$\left.\begin{array}{r}A_1 x^2 + 2F_1 y = 0, \\ B_1 y^2 + 2G_1 x = 0.\end{array}\right\} \qquad (12)$$

which are the well-known equations of the parabola.

EXAMPLES.

1. Required the equation of the conic passing through the five points (0, 0), (2, 3), (8, −6), (18, −9), (32, 12), and find its species.

$$\text{Ans. } y^2 = \frac{9}{2}x.$$

2. Find the equation of a conic passing through $(-2, -4)$, $(1, 2)$, $(0, 0)$, $(3, -6)$, and $(-1, 2)$.

Ans. $y = \pm 2x$.

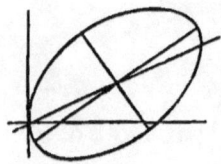

Fig. 142.

3. $y^2 - 2xy + 2x^2 - 2x = 0$, is the locus of what? Does it pass through the origin? Does it cut both coördinate axes?

ADDITIONAL EXAMPLES.

Of Points and Lines.

1. FIND the coördinates of the point which bisects the line joining the points whose coördinates are $(2a, -b)$, $(-a, 2b)$.

2. Show that the area of the triangle inclosed by the lines

$$x + 2y - 5 = 0, \quad 2x + y - 7 = 0, \quad y - x + 1 = 0,$$

is $\frac{3}{2}$.

3. Show that if the whole area included between the lines $x + y = c$, and $bx + ay = ab$ and the coördinate axes, be bisected by the line which joins the origin with their point of intersection, then c is a geometric mean between a and b.

4. Prove that the equations to the lines bisecting the angles between the lines whose equations are

$$12x + 5y = 8, \text{ and } 3x - 4y = 3,$$

will be

$$99x - 27y = 79, \text{ and } 21x + 77y = 1.$$

5. Find the condition of perpendicularity of the straight lines represented by the equations

$$x + (a + b)y + c = 0,$$
$$a(x + ay) + b(x - by) + d = 0.$$

6. Find the equation to the line passing through the origin and perpendicular to the line $x + y = 2$.

7. Find the perpendicular distance of the point $(1, -2)$ from the line $x + y - 3 = 0$.

8. Find the equation to the line which passes through the point (a, b), and through the intersection of the lines

$$\frac{x}{a} + \frac{y}{b} = 1, \quad \frac{x}{b} + \frac{y}{a} = 1.$$

Ans. $\frac{x}{a^2} - \frac{y}{b^2} = \frac{1}{a} - \frac{1}{b}$.

9. Given the coördinates of the vertices of a triangle, to find the equation to the line which joins the middle points of two sides.

10. Find the tangent of the angle between the lines
$$y - mx = 0, \text{ and } my + x = 0,$$
when referred to oblique axes.

Ans. $\dfrac{m^2+1}{m^2-1} \tan \omega$.

11. Show that the tangent of the angle between the two lines contained in the equation $Ay^2 + Bxy + Cx^2 = 0$, is $\dfrac{\sqrt{(B^2-4AC)}}{A+C}$.

12. Show that $4x^2 - 12xy + 9y^2 - 4x + 6y - 12 = 0$ represents two parallel right lines.

13. Show that $9x^2 - 12xy + 4y^2 - 2x + y - 3 = 0$ does not represent right lines, and find what value must be assigned to the coefficient of x in order that it may.

14. Find the equation to the line passing through the intersection of $x \cos \alpha + y \sin \alpha = p$, and $x \cos \beta + y \sin \beta = p'$, and cutting at right angles the line $x \cos \gamma + y \sin \gamma = p''$.

15. The lines drawn from the angles of a triangle to the middle of the opposite sides meet in one point. (For a solution by Quaternions, see Articles 335-7.)

16. Show that the lines which bisect the angles of a triangle intersect in one point. (See Articles 341-3.)

17. Show that the altitudes of a triangle meet in a point.

18. Substitute numerical coefficients in the equation $Ax^2 + 2Hxy + By^2 + 2Gx + 2Fy + C = 0$, so that the result shall represent two right lines.

19. Find the polar equation to the line passing through the points (3, 30°), and (2, 60°).

20. Construct the line $r \cos (\varphi - \pi) + r \sin (\varphi - \tfrac{1}{2}\pi) = 2$.

21. Find the polar equation to a line perpendicular to the line $p = r \cos (\theta - 60°)$.

22. Construct the line $5 = 3r \cos \theta - 6r \sin \theta$.

23. Show that the angle between the lines $a = 4r \cos \theta + 3r \sin \theta$, and $b = 3r \cos \theta - 4r \sin \theta$ is right.

24. Find the polar equation to a line passing through the point (6, 45°) and making an angle of $\tfrac{1}{3}\pi$ with the initial line.

25. Find the point of intersection of the lines $p = r \cos (\theta - \tfrac{3}{4}\pi)$, and $3r \cos \theta + 4r \sin \theta + 5 = 0$.

Of the Circle.

26. Find the equations to the straight lines joining the centre of the circle $x^2 + y^2 = a^2$, with the points in which the line $2(x+y) = a$ meets it.

Ans. $y = \tfrac{1}{3}(\pm \sqrt{7} - 4)x$.

EXAMPLES.

27. Determine the centre of the circle $x^2 + y^2 + 4x + 4y - 1 = 0$.

28. Find the equations of the two circles which touch the straight lines $y \pm x = 0$, and pass through the point (h, k).

Ans. $x^2 + y^2 - 2y (2k \pm \sqrt{2k^2 - 2h^2}) = \frac{1}{2}(2k \pm \sqrt{2k^2 - 2h^2})^2$.

29. Show that the locus of a point whose distance from a fixed line is equal to its distance from a fixed circle, is a parabola.

30. Find the points of intersection of the circle $x^2 + y^2 = 12$, and the line $3x - 2y = 6$.

31. Find the equation to the tangent of the circle $x^2 + y^2 - 2y - 3x = 0$, passing through the origin.

Ans. $2y + 3x = 0$.

32. Find the equation to the circle which passes through the origin and the intercepts $x = 4$ and $y = 5$.

33. Show that the portion of the line $x + y = 3a$, which is between the coördinate axes, is trisected at the points where it cuts the circle $x^2 + y^2 + a^2 = 2a(x + y)$.

34. $ACB = 2R$ is the diameter of a circle, CP, CQ are perpendicular radii; show that the locus of the intersection of AP and BQ is a circle whose centre is in the given circle, and radius is $\sqrt{2}R$.

35. A point moves so that the sum of the squares of its distances from the four sides of a square is constant; show that the locus of the point is a circle.

36. Show that the locus is a circle when the point moves so that the sum of the squares of its distances from the sides of an equilateral triangle is constant.

37. If a point moves so that the sum of the squares of its distances from any number of fixed points is constant, show that the locus will be a circle.

38. Find the equation to the circle when the axes are inclined at an angle $\omega = 120°$, the intercepts being $(0, h)$, and $(0, k)$.

39. Required the radius of the circle $x^2 + y^2 + 2xy \cos \omega - hx - ky = 0$.

Ans. $\dfrac{\sqrt{(h^2 + k^2 - 2hk \cos \omega)}}{2 \sin \omega}$.

40. Find the locus of the middle points of chords drawn from any point of the circle.

41. Given the base of a triangle and the sum of the squares on its sides ; to find the locus of its vertex.

42. ABC is an equilateral triangle, find the locus of a point P, such that $PA = PB + PC$.

43. Find the polar equation to the circle, the initial line being a tangent.

44. Show that the locus

$r = A \cos (\varphi - \alpha) + B \cos (\varphi - \beta) + C \cos (\varphi - \gamma) + $ etc.

is a circle.

45. If the centre be at the pole, show that the polar equation to the chord of a circle which subtends an angle 2β at the centre is $r = c \cos \beta \sec(\varphi - \alpha)$ when α is the angle between the initial line and the line from the centre which bisects the chord.

46. Determine the radius and centre of the circle $r^2 + 2a(\cos\varphi + \sin\varphi)r - c^2 = 0$.

47. A line is drawn from a fixed point O, meeting a fixed circle in P; in OP a point Q is taken so that $OP \cdot OQ = k^2$; find the locus of Q.

48. Find the bilinear equation to a circle, the axes being a pair of tangents and making an angle of 45° between them.

49. Find the equations to the common tangents of the two circles

$$x^2 + y^2 - x - y + 4 = 0, \quad x^2 + y^2 + x + y - 4 = 0.$$

50. If PQ be the diameter of a circle, the polar of P with respect to any circle that cuts the first at right angles, will pass through Q.

Of The Ellipse.

51. What is the eccentricity of the ellipse $3x^2 + 4y^2 = c^2$?

52. If the normal to an ellipse at the extremity of the latus rectum passes through one end of the minor axis, show that the eccentricity will be $\sqrt{\frac{1}{2}(\sqrt{5} - 1)}$.

53. If the tangent to an ellipse at the extremity of the latus rectum meets the minor axis produced in K, and the normal at the same point meets the major axis in G, and if the angle between the normal and the major axis is $\cot^{-1} e^2$, find $CG + CK$, C being at the centre.

54. Find the locus of the middle point of a focal chord.

55. Given the base of a triangle and the product of the tangents of the angles at the base; prove that the locus of the vertex is an ellipse.

56. Show that the equation to the normal of an ellipse expressed in terms of the tangent of the angle which the line makes with the axis of x is $y = mx - m\dfrac{a^2 - b^2}{\sqrt{a^2 + m^2 b^2}}$.

57. In the ellipse $x^2 + 4y^2 = 9$, show that the locus of the middle points of all chords passing through the point $(a, \frac{3}{2}a)$ is $(x - \frac{1}{2}a)^2 + 4(y - \frac{3}{4}a)^2 = \frac{5}{8}a$.

58. If the ordinate of a circle whose equation is $x^2 + y^2 = r^2$ be increased in length by the corresponding abscissa, show that the locus of the extremity of the ordinate thus increased is an ellipse.

59. Find the locus of the intersection of a normal to an ellipse and a perpendicular upon it from the centre.

Ans. $(x^2 + y^2)^2 (a^2 y^2 + b^2 x^2) = c^4 x^2 y^2$.

EXAMPLES. 141

60. Find the locus of the intersection of a normal to an ellipse and a perpendicular upon it from one focus.

Ans. $(a^2 y^2 + b^2 x^2)(y^2 + x^2 - cx)^2 = c^4 x^2 y^2.$

(This curve consists of two loops one inside the other.—*Ed. Times.* Reprint, Vol. XXIV., p. 26.)

61. Find the locus of the intersection of a perpendicular from the centre upon a tangent.

Ans. $\rho^2 = a^2 \cos^2 \varphi + b^2 \cos^2 \varphi.$

62. Show that the lengths of two conjugate semi-diameters, in terms of the eccentric angle, are $a'^2 = a^2 \cos^2 \varphi + b^2 \sin^2 \varphi$, and $b'^2 = a^2 \sin^2 \varphi + b^2 \cos^2 \varphi.$

63. Find the locus of the intersection of tangents drawn through the extremities of a pair of conjugate diameters.

Ans. $a^2 y^2 + b^2 x^2 = 2a^2 b^2$ (*The Wittenberger*, 1878, p. 40).

64. Find the locus of the poles of the normals of an ellipse. (Reprint. *Ed. Times*, Lond., Vol. XXVI., p. 98.)

Solution.—A solution is easily effected by means of the eccentric angle. The equation to the normal may be written, (Art. 113*a*),

$$\frac{a \sec \varphi}{c^2} x - \frac{b \csc \varphi}{c^2} y = 1.$$

Let (x_1, y_1) be the pole of the normal; then, according to Article 150, equation (1), the equation of the normal will be

$$\frac{x_1}{a^2} x + \frac{y_1}{b^2} y = 1,$$

which must be identical with the preceding equation;

$$\therefore \frac{x_1}{a^2} = \frac{a \sec \varphi}{c^2}, \quad \frac{y_1}{b^2} = -\frac{b \csc \varphi}{c^2}.$$

Eliminating φ gives (dropping the subscripts),

$$a^6 y^2 + b^6 x^2 = c^4 x^2 y^2$$

which is the required equation.

(Let the solution also be made by means of the rectangular equation to the normal.)

65. If (x_1, y_1) is the pole of the chord PQ, normal at P to the conic $\frac{x^2}{a^2} + \frac{y^2}{b^2} = 1$, prove that the equation to the normal at Q is $\frac{a^2}{b^2 - y_1^2} \cdot \frac{x}{x_1} + \frac{b^2}{a^2 - x_1^2} \cdot \frac{y}{y_1} = \frac{c^4}{b^2 x_1^2 + a^2 y_1^2}.$ (*Ibid.*, p. 99.)

[Sug.—The values of x_1 and y_1 are given in the preceding example.

The coördinates of the point Q, found by combining the equation of the normal PQ with that of the ellipse, will be

$$\frac{[ac^4 - ab^4 \operatorname{cosec}^2 \varphi] \sec \varphi}{a^4 \sec^2 \varphi + b^4 \operatorname{cosec}^2 \varphi}, \quad \frac{[bc^4 - a^4 b \sec \varphi] \operatorname{cosec} \varphi}{a^4 \sec^2 \varphi + b^4 \operatorname{cosec}^2 \varphi};$$

in which substitute the values of $\sec \varphi$ and $\operatorname{cosec} \varphi$ from the preceding example, and find

$$\frac{b^2 - y_1^2}{b^2 x_1^2 + a^2 y^2} x_1 c^2, \quad -\frac{a^2 - x^2}{b^2 x_1^2 + a^2 y^2} y_1 c^2;$$

then find the equation of the normal through this point.]

66. Show that the focal radii of any point of an ellipse, in terms of the eccentric angle, are $r = a\,(1 - e \cos \varphi)$ and $r' = a\,(1 + e \cos \varphi)$.

67. Two radii vectors at right angles to each other, are drawn from the centre of an ellipse; show that the locus of the intersection of tangents drawn at their extremities is an ellipse whose equation is

$$\frac{x^2}{a^2} + \frac{y^2}{b^2} = \frac{1}{a^2} + \frac{1}{b^2}.$$

68. If P be any point of an ellipse, y its ordinate, A and A' the extremities of the major axis, show that $\tan APA' = -\dfrac{2b^2}{ae^2 y}$.

69. The locus of the middle point of the normal to an ellipse, intercepted between the curve and the major axis, is an ellipse; show that if the eccentricity of the given ellipse be e, that of the locus will be

$$e' = \left[1 - \frac{1 - e^2}{(1 + e^2)^2} \right]^{\frac{1}{2}}.$$

70. Show that the equation to the locus of the middle points of all chords in an ellipse whose length is $2l$, is

$$l^2 \frac{a^2 y^2 + b^2 x^2}{a^4 y^2 + b^4 x^2} + \frac{x^2}{a^2} + \frac{y^2}{b^2} - 1 = 0.$$

71. Find the polar equation to the ellipse, the pole being at the principal vertex, and the major axis being the initial line.

72. Find the polar equation to the ellipse, the pole being at the point (x_0, y_0) in reference to the centre, and the initial line making an angle α with the major axis.

73. The polar equation to the tangent, the pole being at the left focus, the major axis being the initial line, and (ρ', φ') the tangent point, is

$$\rho = \frac{a(1 - e^2)}{\cos (\varphi - \varphi') - e \cos \varphi}.$$

74. Find the polar equation to the curve given in example 60.

75. Given the relative values r, r', r'', of three radii vectors drawn from the focus of an ellipse, and the angles between them; required the relative value of the major axis, the eccentricity of the ellipse, and the position of the major axis.

SUGGESTION.—Let φ be the angle between the major axis and r, measured from the nearest vertex, α the angle between r and r', and β the angle between r and r''. Then equation (5), page 115, gives

$$r = \frac{a(1-e^2)}{1 + e \cos \varphi};$$

$$r' = \frac{a(1-e^2)}{1 + e \cos(\varphi + \alpha)};$$

$$r'' = \frac{a(1-e^2)}{1 + e \cos(\varphi + \beta)};$$

in which r, r', r'', α, and β being known, a, e, and φ may be found by elimination; the last of which gives the position of the major axis in reference to r. We find

$$e = \frac{r - r'}{r \cos \varphi - r' \cos (\varphi + \beta)}$$

$$\tan \varphi = \frac{a(r - r'') - (r - r')(r - r'' \cos \beta)}{r''(r - r') \sin \beta - r'(r - r'') \sin \alpha}.$$

Having found φ from the last equation, its value in the preceding equation gives e, and from these a may be found in terms of r.

[REMARK.—By observing the apparent diameter of the sun on three different dates, and the angles passed over between these dates, the eccentricity of the earth's orbit, the position of its major axis, and the relative value of the major axis may be found. It is difficult to observe the diameter of the sun, so it is better to observe the apparent angular velocity, and use the principle that, The relative distances of the sun are to each other as the *square roots* of the sun's apparent angular velocity.]

The Parabola.

76. A circle radius r touches a parabola whose latus rectum is $4p$ at the vertex, show that if it cuts the parabola, the distance from the vertex to the straight line joining the points of intersection will be $2r - 4p$.

77. Find the equation to the axis of the parabola whose equation is $(x - a)^2 = a(x + y)$.

78. Show that the vertex of the parabola $(3x - 4y)^2 - 50ax + 25a^2 = 0$, is $\left(\dfrac{2}{5}a, -\dfrac{3}{5}a\right)$.

79. QOQ' is a fixed chord in a parabola, $P'OP$ another chord parallel to a given straight line. Op is taken on OP, produced if necessary, such that $OP \cdot Op = OQ \cdot OG'$; show that the locus of p is a parabola.

80. A normal is drawn at the point (x', y'); find the point where it again cuts the parabola.

81. Show that the equation to the parabola referred to a pair of tangents at the extremities of the latus rectum is $\sqrt{x} + \sqrt{y} = \sqrt{p\sqrt{2}}$.

82. From any point there can no more than three normals be drawn.

83. Find the points of intersection of the parabolas $y^2 = 2px$, and $x^2 + 4py = 0$.

84. Through the point (x', y') on a parabola a normal is drawn; find the equation of a tangent parallel to the normal, and the point of contact of the tangent.

85. Find the equation to the normal passing through the point of contact of the tangent in the preceding example.

86. Find the locus of the middle points of focal chords to a parabola.

87. Show that the locus of the centre of a circle which touches a given line and given circle is a parabola.

88. Find the locus of the centre of a circle inscribed in a variable sector of a given circle, one of the bounding radii being fixed.

89. Find the locus of the intersection of mutually perpendicular normals to a parabola.

(SOLUTION.—The equations to the normals will be

$$y - y' = -\frac{y'}{p}(x - x'),$$

$$y - y'' = -\frac{y''}{p}(x - x'').$$

Because they are mutually perpendicular, we have

$$y'' = -\frac{p^2}{y'}, \qquad (1)$$

and the equation to the curve gives

$$x' = \frac{y'^2}{2p}, \qquad x'' = \frac{y''^2}{2p};$$

and the equations become, by substitution, clearing of fractions, and multiplying the first by $2y$,

$$2yy'^3 - 4pyy'x = 4p^2y^2 - 4p^2yy',$$

$$2yy'^3 + 2p^2y'^2 = 2pxy'^2 - p^4.$$

Subtracting and solving for y', we have

$$y' = y \pm \sqrt{y^2 - \frac{p(4y^2 + p^2)}{2(p-x)}}.$$

This gives two points from which normals may be drawn, and these normals must be mutually perpendicular, so that if one be y' the other will be y'' and we have from (1)

$$y + \sqrt{y^2 - \frac{p(4y^2 + p^2)}{2(p-x)}} = \frac{-p^2}{y - \sqrt{y^2 - \frac{p(4y^2 + p^2)}{2(p-x)}}},$$

which reduced gives

$$y^2 = \tfrac{1}{4}p(x - \tfrac{3}{2}p);$$

hence the locus is a parabola having one-fourth the parameter of the given curve, and whose vertex is on the axis of the given curve at a distance from that vertex equal to three times the distance of the given focus from the vertex of the given curve.)

90. Prove that the equation to the normal in terms of the tangent of the angle which it makes with the axis of the curve, is $y = mx - pm - \tfrac{1}{2}pm^3$.

(Obs.—The preceding example may be solved by means of this equation.)

91. If two parabolas whose axes are mutually perpendicular intersect in four points, prove that the four points are in the circumference of a circle.

92. If two parabolas, having a common vertex, and axes at right angles to each other, intersect in the point $x'y'$; then

$$2p : x' :: y' : 2p'.$$

[Obs.—This property enables one to insert graphically two geometric means between two given lines, and for this reason has been used in solving the celebrated problem of the Duplication of the Cube.]

93. Parabolas have their axes parallel and all pass through two given points; prove that their foci lie in a conic section. (Tait & Kelland's *Quaternions.*)

94. Prove that the locus of the poles of normals to a parabola is $(x+p)y^2 + \tfrac{1}{4}p^3 = 0$. (*Educational Times*, Reprint, 1876, Vol. XXVI., p. 99.)

95. An ellipse and a parabola have a common focus, and the other focus of the ellipse moves on the directrix of the parabola; show that the points of contact of a common tangent subtend a right angle at the common focus. (*Math. Visitor*, 1878, p. 45.)

EXAMPLES.

General Equations.

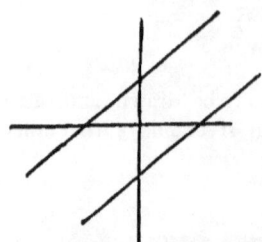

Fig. 143.

96. The equation

$$y^2 - 2xy + x^2 - 1 = 0,$$

is the locus of what curve?

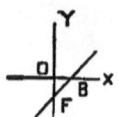

Fig. 144.

97. Show that the equation

$$y^2 - 2xy + x^2 + 2y - 2x + 1 = 0,$$

is the locus of a single right line.

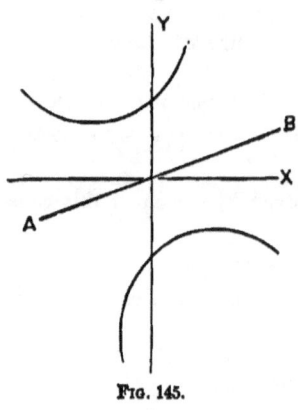

Fig. 145.

98. The equation

$$y^2 - 2xy - x^2 - 2x - 2 = 0,$$

is the locus of what? Does it cut either or both the coördinate axes? What is the inclination of the diameter with the axis of x?

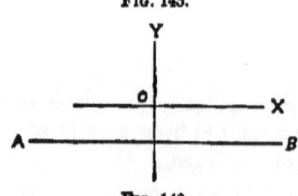

Fig. 146.

99. $y^2 + y + 1 = 0$, is the locus of what?

(AB is the diameter and the lines imaginary.)

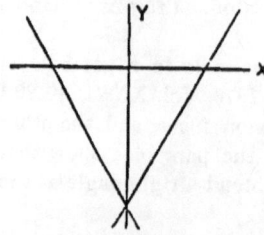

Fig. 147.

100. Determine the character and position of the locus

$$y^2 - 2x^2 + 2y + 1 = 0.$$

Is it symmetrical in reference to the axis of y?

EXAMPLES. 147

101. The equation

$$y^2 - 2xy + x^2 + x = 0,$$

is the locus of what ?
(AB is a diameter whose equation is $y = x$.)

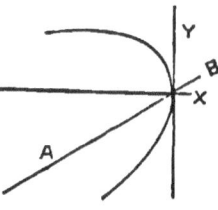

Fig. 148.

102. The equation

$$y^2 - 2xy + x^2 + 2y + 1 = 0,$$

is the locus of what curve? Is the curve tangent to the axis of y? What is the slope of the diameter?

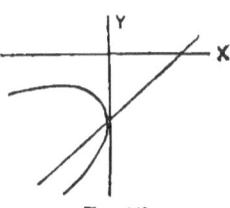

Fig. 149.

103. The equation

$$y^2 - 2xy + x^2 - 2y - 1 = 0,$$

is the locus of what curve?

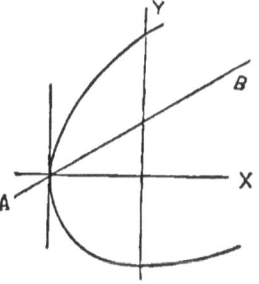

Fig. 150.

104. What locus is represented by the equation

$$y^2 - x^2 + 2x - 2y + 1 = 0 ?$$

Does it cut the axis of y? Does it cut the axis of x? Has it a centre?

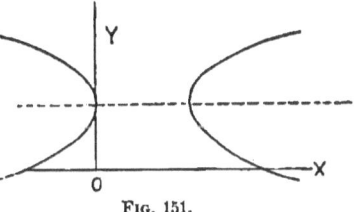

Fig. 151.

105. If A and B are fixed points in the axes Ox, Oy, and a, b, are always taken on Ox, Oy, so that

$$\frac{1}{Oa} - \frac{1}{OA} \text{ varies inversely as } \frac{1}{Ob} - \frac{1}{OB} ;$$

show that the locus of the intersection of Ab and Ba is a conic section touching Ox, Oy, in A and B.

CHAPTER VI.

LOCI IN SPACE.

Of the Point, Right Line and Plane.

188. THE loci previously considered have been confined to one plane, and the coördinate axes have been taken in that plane. When loci are referred to three intersecting planes, they are called *loci in space.*

189. Definitions. — *Coördinate planes* are three planes intersecting each other, to which loci are referred. The three planes will intersect in a common point, and the planes taken two and two will intersect in right lines passing through the common point. *Coördinate axes* are the lines of intersection of the coördinate planes, as OX, OY, and OZ. The *origin* O is the common point of intersection of the coördinate axes. *Rectangular coördinates* are those in which the coördinate axes make right angles with each other. All coördinates not rectangular are *oblique.* The coördinate planes are assumed to be indefinite in extent, and divide all space into eight parts or solid angles, and if the axes are rectangular, the eight parts will be equal to each other. The plane YOX is generally assumed to be horizontal, and the other two, vertical, but as they may have any position in space, this assumption is entirely arbitrary.

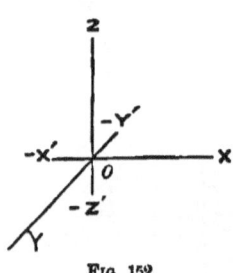

FIG. 152.

190. Axis of a Plane.—Any line perpendicular to a plane may be considered as the axis of that plane. When the axes are rectangular, the coördinate axis OZ is the axis of the plane YOX, since it is perpendicular to that plane, and similarly for the other planes.

191. Order of the Angles.—The angle Z-YOX, which is above the horizontal plane YOX, at the right of the vertical plane ZOY, and in front of the vertical plane ZOX, is called the *first angle*. Similarly,

Z-XOY' is called the 2d angle,
Z-$Y'OX'$ " " " 3d "
Z-$X'OY$ " " " 4th "
$-Z$-YOX " " " 5th " ;

and so on.

In analysis, however, the angles are determined by the *signs* of the coördinates. Thus, OX is plus x, OX', minus x; OZ, plus z, OZ', minus z; OY, plus y, and OY' minus y.

The Point.

192. Coördinates of a Point.—The position of a point is determined by its distance from the coördinate planes measured on lines parallel to the coördinate axes. Let p be the point. Through p draw pG parallel to the axis of x, and let G be the point where it intersects the plane ZOY; then will pG be the x-*ordinate* of the point. Similarly, pE is the y-*ordinate*, and pD the z-*ordinate*. Three coördinates determine a point; for

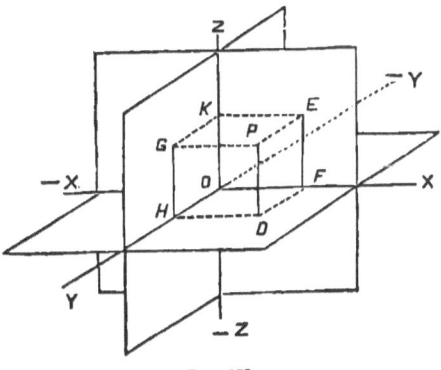

Fig. 153.

three right lines can intersect in one point only. If the coördinates are rectangular, they will be perpendicular to the respective planes, that is, pG will be perpendicular to the plane ZOY, and similarly for the others.

The coördinates of a point may be expressed algebraically by the equations

$$x = a, \quad y = b, \quad z = c,$$

and these are called equations to the point. Three determinate equations of the first degree are therefore necessary and sufficient for determining a point in space.

We have, for a point in the

first angle	$x = +a, \ y = +b, \ z = +c,$
second angle	$x = +a, \ y = -b, \ z = +c,$
third angle	$x = -a, \ y = -b, \ z = +c,$
fourth angle	$x = -a, \ y = +b, \ z = +c,$
fifth angle	$x = +a, \ y = +b, \ z = -c,$
sixth angle	$x = +a, \ y = -b, \ z = -c,$
seventh angle	$x = -a, \ y = -b, \ z = -c,$
eighth angle	$x = -a, \ y = +b, \ z = -c.$

192a. *Coördinates of particular Points.*— For a point at the origin, we have

$$x = 0, \quad y = 0, \quad z = 0.$$

For a point on the axis of x, we have

$$x = \pm a, \quad y = 0, \quad z = 0,$$

and similarly for a point on any other axis.

For a point in the plane xy, we have

$$x = \pm a, \quad y = \pm b, \quad z = 0,$$

and similarly for a point in any other coördinate plane.

192b. *Equations of the Axes.*— For the axis of x, we have

$$z = 0, \quad y = 0, \quad \text{and } x \text{ *indeterminate*.}$$

The value of x being indeterminate, will represent every point of the axis in succession.

For the axis of y, we have

$$z = 0, \quad x = 0, \quad \text{and } y \text{ *indeterminate;*}$$

and for the axis of z,

$$x = 0, \quad y = 0, \quad \text{and } z \text{ indeterminate.}$$

193. Distance between two Points.—Let the coördinates of one point p, be x_1, y_1, z_1, and of the other point O, be x_2, y_2, z_2; then will the distance Op between the points be the diagonal of a parallelopiped, hence

$$Op = l = \sqrt{(x_1 - x_2)^2 + (y_1 - y_2)^2 + (z_1 - z_2)^2}.$$

If the point O be at the origin, then $x_2 = 0$, $y_2 = 0$, $z_2 = 0$, and

$$l = \sqrt{x_1^2 + y_1^2 + z_1^2},$$

which is the distance of any point from the origin.

194. Projections.—The foot of a perpendicular from a point upon a plane is the projection of the point on that plane. Thus D is the projection of the point P upon the plane xy; E, upon xz; G, upon zy. The lines PD, PE, PG, passing through the point perpendicular to the respective planes are called *projecting lines*, or *lines of projection*.

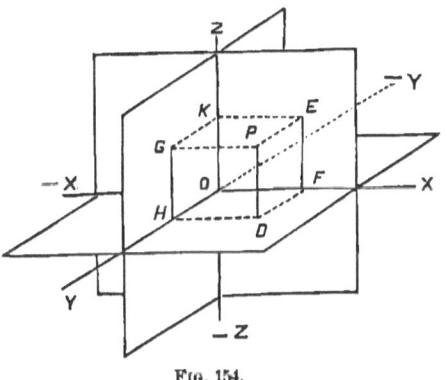

Fig. 154.

The *projection of a line* is the projection of every point of the line. This is true of curved as well as of right lines. If a curved line be in the plane of two projecting lines, its projection will be a right line, but in other cases its projection will be curved.

The Right Line.

195. Equations to a Right Line.—A line is given by its projections. For, if perpendiculars be erected from the

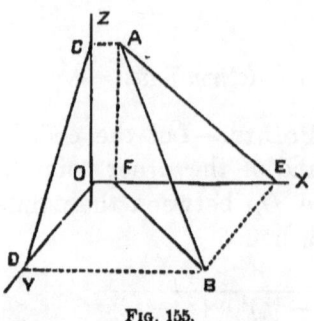

Fig. 155.

corresponding points of the projections, they will be the projecting lines. But the projecting lines intersect in space on the given line, and hence determine its position.

Let AB be the line in space. The projection of the point B on the plane zy is D; similarly, C is the projection of A; and CD will be the projection of the given line on the plane zy. Also, AE will be its projection on zx; and FB, its projection on xy.

Let m be the tangent of the angle which the line AE makes with the axis of z, and n that of CD with the same axis. Also let $a = OE$ the intercept of AE on the axis of x, and $b = OD$ the intercept of CD on the axis of y. Then, according to Article 28, we have for the equation of AE,

$$x = mz + a, \qquad (1)$$

and for CD, $$y = nz + b; \qquad (2)$$

which are the required equations. Two equations of the first degree, each having two of the required variables, are necessary and sufficient for determining a right line in space.

Eliminating z from these equations gives

$$\frac{x-a}{m} = \frac{y-b}{n}; \qquad (3)$$

which is the equation of FB, the projection of the given line on the plane xy.

196. Intersection of two right lines in space.—Let the lines be

$$x = mz + a, \text{ and } x = m'z + a',$$
$$y = nz + b, \qquad y = n'z + b'. \qquad (1)$$

Considering the equations as simultaneous, and eliminating x and y, we have

THE RIGHT LINE.

$$(m - m')z = a' - a\ ;\ (n - n')z = b' - b$$

$$\therefore z = \frac{a' - a}{m - m'}\ \text{and}\ z = \frac{b' - b}{n - n'}. \quad (2)$$

Eliminating z between equations (2), gives

$$\frac{a' - a}{m - m'} = \frac{b' - b}{n - n'}\ ; \quad (3)$$

which establishes a relation between the constants of the equations, and is an *equation of condition*, (Art. 39). Hence we infer that, if the relation between the eight constants of equations (1) gives equation (3), the lines will intersect, but otherwise they will not intersect. Lines in space which are not parallel may, therefore, have an infinite number of positions and not intersect. If equation (3) is true, either of equations (2) will give the ordinate z of the intersection, and this value in (1) gives for the other coördinates

$$x = \frac{ma' - m'a}{m - m'}\ ;\qquad y = \frac{nb' - n'b}{n - n'}.$$

If $m = m'$, equation (3) shows that $n = n'$, and we have

$$x = \infty\ ;\qquad y = \infty\ ;\qquad z = \infty\ ;$$

and the lines are parallel; and the equations $m = m'$ and $n = n'$ are *the equations of condition of the parallelism of two right lines in space.*

197. Equation to a line which passes through two points in space.—Let (x', y', z') be one point, and (x'', y'', z''), the other. Then proceeding as in Article 40, we have

$$x - x' = \frac{x'' - x'}{z'' - z'}(z - z')\ ;\qquad y - y' = \frac{y'' - y'}{z'' - z'}(z - z')\ ; \quad (1)$$

which are the required equations. If the line passes through one point only, the equations become

$$x - x' = m(z - z'),\qquad y - y' = n(z - z')\ ; \quad (2)$$

in which m and n are indeterminate.

198. Inclination of a line to the three Axes.—Draw a line through the origin parallel to the given line; it will make the same angles with the axes as those of the given line. Let the equations of the line be

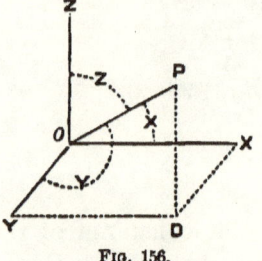
Fig. 156.

$$x = mz, \quad y = nz. \quad (1)$$

Let P be any point (x, y, z) of the line, and

$$X = POX, \quad Y = POY, \quad Z = POZ, \quad r = OP.$$

Then $x = r \cos X; \quad y = r \cos Y; \quad z = r \cos Z. \quad (2)$

Squaring these, and substituting in the equation

$$r^2 = x^2 + y^2 + z^2,$$

gives $\quad r^2 = r^2 \cos^2 X + r^2 \cos^2 Y + r^2 \cos^2 Z;$

$$\therefore \cos^2 X + \cos^2 Y + \cos^2 Z = 1; \quad (3)$$

by means of which, if two angles be given, the third may be found.

Substituting the values of x and y from equation (1) in the value of r^2 gives

$$r^2 = m^2 z^2 + n^2 z^2 + z^2$$

$$\therefore z^2 = \frac{r^2}{1 + n^2 + m^2} = r^2 \cos^2 Z \text{ (from (2))};$$

$$\therefore \cos Z = \pm \frac{1}{\sqrt{1 + n^2 + m^2}}. \quad (4)$$

Similarly

$$\cos X = \pm \frac{m}{\sqrt{1 + n^2 + m^2}}; \quad \cos Y \pm \frac{n}{\sqrt{1 + n^2 + m^2}}; \quad (5)$$

by means of which the required angles may be found when the equations to the lines are given.

199. Angle between two lines. — The angle between

two lines in space, whether the lines intersect or not, is the same as the angle between two lines drawn through the origin parallel respectively to the given lines. Let AP be parallel to one line, and AQ parallel to the other, then will the angle PAQ be the required angle. Since the angle will be independent of the length of the lines, take each equal to unity, and let $PQ = d$, and $PAQ = \theta$. Then, from Trigonometry, we have

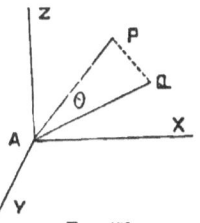

Fig. 157.

$$\cos \theta = \frac{AP^2 + AQ^2 - d^2}{2AP \cdot AQ} = \frac{2 - d^2}{2}. \quad (1)$$

Let the points P and Q be respectively (x', y', z') and (x'', y'', z''); then

$$d^2 = (x'' - x')^2 + (y'' - y')^2 + (z'' - z')^2. \quad (2)$$

But from equations (2) of Article 198, since $r = 1$, we have

$$\begin{matrix} x' = \cos X', & y' = \cos Y', & z' = \cos Z', \\ x'' = \cos X'', & y'' = \cos Y'', & z'' = \cos Z''. \end{matrix} \quad (3)$$

These values substituted in equation (2) give

$$d^2 = (\cos^2 X' + \cos^2 Y' + \cos^2 Z') + (\cos^2 X'' + \cos^2 Y'' + \cos^2 Z'')$$
$$- 2(\cos X' \cos X'' + \cos Y' \cos Y'' + \cos Z' \cos Z''). \quad (4)$$

Reducing this by means of equation (3) of Article 198, substituting in (1) above, and reducing gives

$$\cos \theta = \cos X' \cos X'' + \cos Y' \cos Y'' + \cos Z' \cos Z'', \quad (5)$$

which is one form of the required value.

Substituting in this equation the values corresponding to equations (4) and (5) of the preceding Article, we have

$$\cos \theta = \pm \frac{mm' + nn' + 1}{\sqrt{[(m^2 + n^2 + 1)(m'^2 + n'^2 + 1)]}}. \quad (6)$$

If the lines are mutually perpendicular, $\cos \theta = 0$, and we have

$$mm' + nn' + 1 = 0,$$

which is *the equation of condition of mutually perpendicular lines in space.*

If the lines are parallel, $\cos \theta = 1$, and squaring both members, transposing and reducing, gives

$$(m - m')^2 + (n - n')^2 + (mn' - m'n)^2 = 0;$$

and since each term is a square, they must separately be equal to zero, or

$$m = m', \qquad n = n', \qquad mn' = m'n,$$

the third of which results from the other two. This result agrees with that found in Article 196.

The Plane.

200. A Plane may be generated by a right line moving along another right line and remaining constantly parallel to its first position. Let ABC be a plane supposed to be generated by the line DE moving along the line BC and being constantly parallel to AB. The equation of every point of the moving line in every position will be the equation of the plane. *The trace of a plane* is its intersection with one of the coördinate planes.

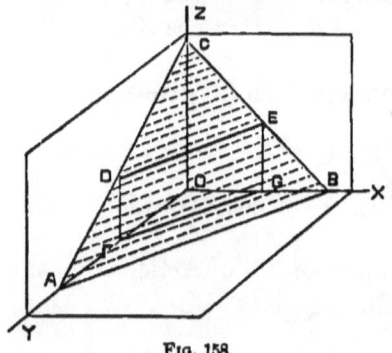

Fig. 158.

201. Equation to a Plane in terms of its Intercepts.—Let a, b, c, be the intercepts on the axes x, y, z, respectively. The equation of the line AB will be, (Eq. (5), Art. 29),

$$\frac{x}{a} + \frac{y}{b} = 1.$$

Let FG be the projection on the plane xy of the generating line DE, and let the intercepts of this projection be $a' = OG$ and $b' = OF$, then will its equation be

$$\frac{x}{a'} + \frac{y}{b'} = 1.$$

But from the similar triangles BOA and GOF, we have

$$\frac{a}{a'} = \frac{b}{b'}, \text{ or } \frac{1}{b'} = \frac{a}{a'b},$$

which, substituted in the preceding equation, gives

$$\frac{x}{a} + \frac{y}{b} = \frac{a'}{a}.$$

But the ordinate of the line DE is z, and the similar triangles BOC, BGE, give

$$\frac{c}{a} = \frac{z}{a-a'}; \quad \therefore \frac{a'}{a} = 1 - \frac{z}{c};$$

and the preceding equation becomes

$$\frac{x}{a} + \frac{y}{b} + \frac{z}{c} = 1,$$

which is the required equation.

202. Every Equation of the First Degree between three variables may represent a plane.—The equation will be in the form

$$Ax + By + Cz + D = 0,$$

which becomes, by dividing through by D,

$$\frac{A}{D}x + \frac{B}{D}y + \frac{C}{D}z + 1 = 0.$$

But whatever be the values of A, B, C, D, the coefficients of x, y, z, may be represented by $-\frac{1}{a}$, $-\frac{1}{b}$, $-\frac{1}{c}$, and the equation be thus reduced to that of the last equation of the preceding Article, which equation is that of a plane.

203. Discussion of the Equation $Ax + By + Cz + D = 0$. Let $z = 0$, then the equation becomes $Ax + By + D = 0$, which is the equation of the trace of the plane on the coördinate plane xy. Similarly, $Ax + Cz + D = 0$ is the equation of the trace on xz, and $By + Cz + D = 0$ is the trace on yz.

If $z = 0$, and $y = 0$, then $x = -\dfrac{D}{A}$, which is the abscissa OB of the point where the plane cuts the axis of x. Similarly, $y = -\dfrac{D}{B}$ gives the point A where it cuts the axis of y; and $z = -\dfrac{D}{C}$ gives the point C where it cuts the axis of z.

If the plane does not cut the axis of z, then the intercept on that axis will be infinite, or $\dfrac{D}{C} = \infty$; $\therefore C = 0$, and z *indeterminate*, we have

$$Ax + By + 0 \cdot z + D = 0$$

for the equation of a plane parallel to the axis of z and oblique to both the other axes.

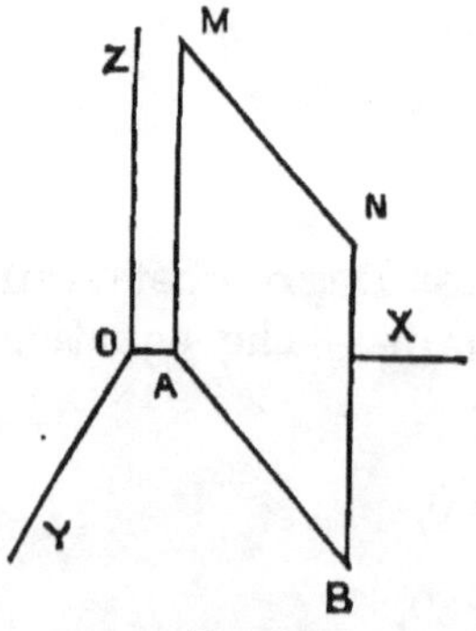

If the plane is also parallel to the axis of y, we have

$$Ax + 0 \cdot y + 0 \cdot z + D = 0.$$

It cannot be parallel to the three axes at the same time; for the sum of three zeros cannot equal a finite quantity.

Fig. 159.

If the plane passes through the origin, we have for that point $x = 0$, $y = 0$, $z = 0$; $\therefore D = 0$, that is, *the absolute term will be zero*.

If the plane coincides with the coördinate plane zy, it will pass through the origin, hence $D = 0$, and $x = 0$, and the equation becomes

$$0 \cdot y + 0 \cdot z = 0,$$

which is the equation of the coördinate plane yz. Similarly, for the other coördinate planes.

204. Equation to a Plane in terms of the perpendicular from the origin upon the plane and its direction-cosines.—Let p be the perpendicular, and X, Y, Z, the angles which it makes with the axes x, y, z, respectively. These angles will be the *direction-angles*, and $\cos X$, etc., the *direction-cosines*. Also, let a, b, c, be the intercepts of the plane on the axes. Then,

$$a \cos X = p, \quad b \cos Y = p, \quad c \cos Z = p. \quad (1)$$

Substituting the values of a, b, c, found from these equations, in the last one of Article 201, gives

$$x \cos X + y \cos Y + z \cos Z = p, \quad (2)$$

which is the required equation.

205. To reduce $Ax + By + Cz + D = 0$ to the form last found.—Dividing this equation by D and the former by p give

$$\frac{A}{D}x + \frac{B}{D}y + \frac{C}{D}z + 1 = 0,$$

$$-\frac{\cos X}{p}x - \frac{\cos Y}{p}y - \frac{\cos Z}{p}z + 1 = 0;$$

$$\therefore \cos X = -A\frac{p}{D}; \quad \cos Y = -B\frac{p}{D}; \quad \cos Z = -C\frac{p}{D}.$$

Squaring and adding, gives

$$\frac{p}{D} = \pm \frac{1}{\sqrt{A^2 + B^2 + C^2}}.$$

Therefore, if the coefficients of the given equation be divided by $\sqrt{A^2 + B^2 + C^2}$, the resulting coefficients will be respectively the values of $\cos X$, $\cos Y$, $\cos Z$, and p. The sign of the radical must be such as to make p always positive, which sign will determine the essential signs of $\cos X$, $\cos Y$, $\cos Z$.

206. Intersection of Two Planes.—Let the equations to the planes be

$$Ax + By + Cz + D = 0;$$

and
$$A'x + B'y + C'z + D' = 0.$$

Eliminating z gives

$$(AC' - A'C)x + (BC' - B'C)y + DC' - D'C = 0; \quad (1)$$

which is the equation of the projection of the line of intersection on the coördinate plane xy. Similarly, the equation of the projection on the plane xz, will be

$$(AB' - A'B)x + (B'C - BC')z + B'D - BD' = 0; \quad (2)$$

and on yz,

$$(A'B - AB')y + (A'C - AC')z + A'D - AD' = 0. \quad (3)$$

If the planes are parallel to each other, $x = \infty$, $y = \infty$, $z = \infty$, and the coefficients of x, y, z must be zero;

$$\therefore AB' = A'B, \quad AC' = A'C, \quad BC' = B'C; \quad (4)$$

which are the *equations of condition of the parallelism of two planes*.

207. Equation to a Plane passing through three Points.—Let the points be $(x', y', z'), (x'', y'', z''), (x''', y''', z''')$. Substitute these successively for x, y, z, in the equation

$$Ax + By + Cz + D = 0,$$

and find the values of A, B, C, in terms of x', x'', etc., and resubstitute the values of A, B, C, in the equation. In this way we may find

$$\begin{aligned}[y'(z''-z''') + y''(z'''-z') + y'''(z'-z'')]x \\ + [z'(x''-x''') + z''(x'''-x') + z'''(x'-x'')]y \\ + [x'(y''-y''') + x''(y'''-y') + x'''(y'-y'')]z \end{aligned} = \begin{cases} (y''z''' - y'''z'')x' \\ + (y'''z' - y'z''')x'' \\ + (y'z'' - y''z')x''', \end{cases}$$

for the required equation.

208. Inclination of a Plane to the Coördinate Planes.—The angle required is the angle between two perpendiculars from the origin to the respective planes. The angle between the given plane and the plane xy will be the angle between the axis of z and a perpendicular from the origin to the given plane. Letting Z be this angle, we have, (Art. 205),

$$\cos Z = \pm \frac{C}{\sqrt{(A^2 + B^2 + C^2)}}.$$

Similarly, for the inclination to the planes zy and zx respectively, we have

$$\cos X = \pm \frac{A}{\sqrt{(A^2 + B^2 + C^2)}}; \quad \cos Y = \pm \frac{B}{\sqrt{(A^2 + B^2 + C^2)}}.$$

209. Angle between two Planes.—Let X', Y', Z', and X'', Y'', Z'', be the angles between the coördinate axes and the respective perpendiculars from the origin upon the planes, and φ the angles between the planes. Then, from Article 199, we have

$$\cos \varphi = \cos X' \cos X'' + \cos Y' \cos Y'' + \cos Z' \cos Z''; \quad (1)$$

in which substitute the values of $\cos X'$, etc., (Art. 208), and we have

$$\cos \varphi = \pm \frac{AA' + BB' + CC'}{\sqrt{[(A^2 + B^2 + C^2)(A'^2 + B'^2 + C'^2)]}}, \quad (2)$$

which is the required equation.

If the two planes are perpendicular to each other, $\cos \varphi = 0$, and we have

$$AA' + BB' + CC' = 0, \quad (3)$$

which is *the equation of condition for the mutual perpendicularity of two planes.*

[If the two planes are parallel, $\cos \varphi = 1$ and we may find

$$AB' = A'B, \quad AC' = A'C, \quad BC' = B'C, \quad (4)$$

as found in Article 206.

If the second plane is parallel to the plane xy, the angle between them will be the same as between the first plane and the plane xy; and we will have, (Art. 203), $A' = 0$, $B' = 0$, and equation (2) becomes

$$\left. \begin{aligned} \cos \varphi_{xy} &= \frac{C}{\sqrt{(A^2+B^2+C^2)}}. \\ \text{Similarly,} \quad & \\ \cos \varphi_{xz} &= \frac{B}{\sqrt{(A^2+B^2+C^2)}}; \quad \cos \varphi_{yz} = \frac{A}{\sqrt{(A^2+B^2+C^2)}}, \end{aligned} \right\} \quad (5)$$

as found in the preceding Article.]

210. Equation to a Plane parallel to $Ax + By + Cz + D = 0.$—It will be of the form $A'x + B'y + C'z + D' = 0$. Substitute the values of A' and C', deduced from equation (4) of Article 206, and we have

$$Ax + By + Cz + \frac{B}{B'}D' = 0;$$

or, *the equations of parallel planes differ only in their absolute terms.*

211. Equations to a Plane perpendicular to $Ax + By + Cz + D = 0.$—The equation will be of the form of $A'x + B'y + C'z + D' = 0$. Substitute the value of A' from equation (3) (Art. 209), and we have

$$-\frac{BB' + CC'}{A}x + B'y + C'z + D' = 0,$$

for the required equation. But since this equation contains the undetermined constants B' C' D', there may be an infinite number of planes passed perpendicular to the given one.

Line and Plane.

212. To find the Point where a Line pierces a Plane.—Let $Ax + By + Cz + D = 0$ be the plane, and $x = mz + a, y = nz + b$, the line. Substitute the values of x and y from the equations of the line in the equation to the plane and find z. This gives

$$z = -\frac{aA + bB + D}{mA + nB + C},$$

for the z-coördinate.

In a similar manner the other coördinates may be found.

If the line is parallel to the plane, then $z = \infty$, and we have
$$mA + nB + C = 0$$
for *the equation of condition of parallelism of a line and plane.*

If the line coincides with the plane, it will be indeterminate and the numerator will also be zero.

213. Conditions existing between a Plane and a right line perpendicular to the Plane.—Let ABC be a plane and DP a line perpendicular to it. Every plane passed through DP will be perpendicular to the given plane since it contains a line perpendicular to that plane. Let any plane be passed through it and revolved about the line until it is perpendicular to the plane zx; the trace TS of the plane in this position will be the projection of the line on the plane zx. But since the projecting plane will be perpendicular both to the given plane and the plane zx, its trace TS, will be perpendicular to the trace AB of the given plane, (*Elementary Geometry*). Similarly, VQ will be perpendicular to CB, and UR to AC. Hence,—*If a line be perpendicular to a plane, its projections will be perpendicular respectively to its traces.*

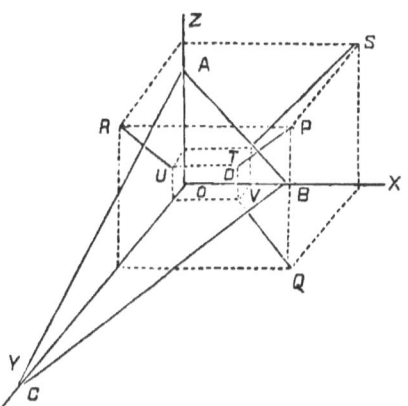

Fig. 160.

Let
$$Ax + By + Cz + D = 0 \qquad (1)$$
be the plane, and
$$x = mz + a, \qquad y = nz + b, \qquad (2)$$
the line perpendicular to the plane. Making $y = 0$ in the equation of the plane, we have
$$Ax + Cz + D = 0,$$

or
$$x = -\frac{C}{A}z - \frac{D}{A},$$

for the trace of the plane on xz. But the condition of perpendicularity requires, (Art. 44),

$$1 + m\left(-\frac{C}{A}\right) = 0;$$

$$\therefore A = mC. \qquad (3)$$

Similarly, we would find

$$B = nC. \qquad (4)$$

214. Equations to a Line perpendicular to a given Plane.—Let $Ax + By + Cz + D = 0$ be the plane. The equations of the line will be of the form

$$x = mz + a, \qquad y = nz + b,$$

in which substitute the values of m and n from the preceding Article, and we have

$$x = \frac{A}{C}z + a, \qquad y = \frac{B}{C}z + b; \qquad (5)$$

which are the required equations. Since the arbitrary constants a and b remain, there may be an infinite number of lines perpendicular to a given plane.

215. Equations to a Line perpendicular to a given Plane and passing through a given Point.—Let (x', y', z') be the point. The equations will be of the form

$$x = \frac{A}{C}z + a, \qquad y = \frac{B}{C}z + b;$$

and since the line contains the given point, we have

$$x' = \frac{A}{C}z' + a; \qquad y' = \frac{B}{C}z' + b;$$

and we find
$$x - x' = \frac{A}{C}(z - z'), \qquad y - y' = \frac{B}{C}(z - z'); \qquad (6)$$

which are the required equations.

To find where this perpendicular pierces the plane, eliminate x and y between equations (6) and (1) and find z. Similarly, find x and y. The length of the perpendicular will be
$$L = \frac{Ax' + By' + Cz' + D}{\sqrt{(A^2 + B^2 + C^2)}}.$$

216. *Equations to a Plane passing through a point and perpendicular to a given line.*—Let the point be (x', y', z'), and the line
$$x = mz + a, \qquad y = nz + b.$$

The equation of the plane will be of the form
$$Ax + By + Cz + D = 0,$$
and for the point, we will have
$$Ax' + By' + Cz' + D = 0.$$

Subtracting, we have
$$A(x - x') + B(y - y') + C(z - z') = 0.$$

But, from Article 213, we have
$$A = mC, \qquad B = nC;$$

which substituted above gives
$$m(x - x') + n(y - y') + z - z' = 0,$$

which is the required equation.

217. To find the Angle between a Line and Plane. It will be the angle between the line and its projection on the plane, and this equals the complement of the angle between the line and a perpendicular to the plane. From the

origin draw a line parallel to the given one, and let its equations be
$$x = mz, \quad y = nz.$$
Draw another line from the origin, perpendicular to the plane $Ax + By + Cz + D = 0$, and let its equations be
$$x = m'z, \quad y = n'z,$$
then, Article 213, equations (3) and (4),
$$m' = \frac{A}{C}, \quad n' = \frac{B}{C}.$$
Let U be the required angle, then from equation (6) Article 199, we have
$$\cos \varphi = \sin U = \frac{mA + nB + C}{\sqrt{[(1 + m^2 + n^2)(A^2 + B^2 + C^2)]}}.$$

If the line is parallel to the plane, we have $\sin U = 0$, therefore,
$$mA + nB + C = 0,$$
as before found, (Art. 212).

Transformation of Coördinates.

218. *The sum of the projections of any number of broken lines joining two points upon the right line joining those points, equals the length of the right line.*

If the projections all fall between the points, the proposition is evidently true. If any of the projections fall upon the prolongation of the line, there will be positive and negative projections (the signs being determined by the sign of the cosine of the inclination), the algebraic sum of which will equal the length of the line.

219. *To transform from a given rectangular system to a system of oblique planes.*

Let x, y, z be the rectangular axes, x', y', z', the oblique axes; X', X'', X''', the angles between x and x', y', z', respectively; Y', Y'', Y''', between y and the same axes; Z', Z'', Z''', between z and the same axes.

The projection of the coördinates x', y', z' of any point upon the axis of x, will equal the x-abscissa of the point.

For, the three oblique coördinates projected on the radius-vector of the point, will equal the radius-vector, and the projection of the radius-vector on the axis of x will equal the x-*abscissa* of the point. Similarly for the other axes. Hence we have

$$x = x' \cos X' + y' \cos X'' + z' \cos X''',$$
$$y = x' \cos Y' + y' \cos Y'' + z' \cos Y''', \quad (1)$$
$$z = x' \cos Z' + y' \cos Z'' + z' \cos Z'''.$$

If the coördinates of the new origin be a, b, c, we would have

$$x = a + x' \cos X' + y' \cos X'' + z' \cos X''',$$
$$y = b + x' \cos Y' + y' \cos Y'' + z' \cos Y''', \quad (2)$$
$$z = c + x' \cos Z' + y' \cos Z'' + z' \cos Z'''.$$

The angles are subject to the condition

$$\cos^2 X' + \cos^2 Y' + \cos^2 Z' = 1,$$
$$\cos^2 X'' + \cos^2 Y'' + \cos^2 Z'' = 1, \quad (3)$$
$$\cos^2 X''' + \cos^2 Y''' + \cos^2 Z''' = 1.$$

If the angles between the oblique axes are sought, we have (Art. 199, Eq. (5)),

$$\cos(x'y') = \cos X' \cos X'' + \cos Y' \cos Y'' + \cos Z' \cos Z'',$$
$$\cos(y'z') = \cos X'' \cos X''' + \cos Y'' \cos Y''' + \cos Z'' \cos Z''', (4)$$
$$\cos(z'x') = \cos X' \cos X''' + \cos Y' \cos Y''' + \cos Z' \cos Z'''.$$

Equations (3) and (4) will make known the angles between the oblique and the corresponding rectangular axis.

220. Polar Coördinates.—Let O be the pole; P any point; $\rho = OP =$ the radius-vector of the point; OX the initial line; XOY the initial plane; OD the projection of the radius-vector; $\theta = DOX =$ one variable angle,—corresponding to the azimuth angle in as-

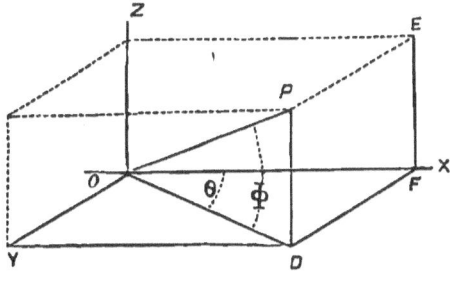

Fig. 161.

tronomy; $\varphi = POD$, the other variable angle,—corresponding to the angle of elevation in geodesy, or of declination in astronomy. Draw DF perpendicular to OX, and we have

$$OF = x = \rho \cos \varphi \cos \theta,$$
$$DF = y = \rho \cos \varphi \sin \theta,$$
$$PD = z = \rho \sin \varphi;$$

by means of which, rectangular coördinates may be changed to polar.

221. *To find formulas for passing from a polar system to a rectangular one.*

For this we find φ, θ, and ρ from the preceding equations. We have

$$x^2 + y^2 + z^2 = \rho^2$$

$$\sin \varphi = \frac{z}{\rho}; \therefore \cos \varphi = \sqrt{1 - \frac{z^2}{\rho^2}}.$$

$$\cos \theta = \frac{x}{\rho \cos \varphi}; \text{ or } \sin \theta = \frac{y}{\rho \cos \varphi};$$

from which θ and φ may be found in terms of x, y, and z.

EXAMPLES.

1. What is the distance between two points whose coördinates are

$$x' = -2, \; y' = 1, \; z' = 0; \text{ and } x'' = 0, \; y'' = -5, \; z'' = 4 \; ?$$

2. The equations of the projections of a straight line on the coördinate planes zx, yz, are

$$x = z + 1, \qquad y = \frac{1}{2}z - 2,$$

required its equation on the plane yx.

Ans. $2y = x - 5$.

3. Required the equations of the three projections of a straight line which passes through the two points whose coördinates are

$$x' = 2, \; y' = 1, \; z' = 0, \text{ and } x'' = -3, \; y'' = 0, \; z'' = -1.$$

Ans. $x = 5z + 2, \; y = z + 1, \; 5y = x + 3$.

4. Required the angle included between two lines whose equations are

$$\left.\begin{array}{l}x = 3z + 5 \\ y = 5z + 3\end{array}\right\} \text{ of the 1st, and } \left.\begin{array}{l}x = z + 1 \\ y = 2z\end{array}\right\} \text{ of the 2d.}$$

Ans. $14° 58'$.

5. Required the angles which a straight line makes with the coördinate axes, its equations being

$$x = -2z + 1, \qquad y = z + 3.$$

Ans. $\begin{cases} 144° 44' \text{ with } X, \\ 65° 54' \text{ with } Y, \\ 65° 54' \text{ with } Z. \end{cases}$

6. Having given the equations of two straight lines,

$$\left.\begin{array}{l}x = 2z + 1 \\ y = 2z + 2\end{array}\right\} \text{ of the 1st, and } \left.\begin{array}{l}x = z + 5 \\ y = 4z + \beta'\end{array}\right\} \text{ of the 2d,}$$

required the value of β' so that the lines shall intersect each other, and to find the coördinates of the point of intersection.

Ans. $\beta' = -6$, $x' = 9$, $y' = 10$, $z' = 4$.

7. To find the equations of a line that shall pass through a point, of which the coördinates are $x' = -2$, $y' = 3$, $z' = 5$, and be perpendicular to the plane, of which the equation is

$$2x + 8y - z - 4 = 0.$$

Ans. $\begin{cases} x = -2z + 8, \\ y = -8z + 43. \end{cases}$

8. To find the equation of a plane which shall pass through the three points, whose coördinates are

$$x' = 1, \quad y' = -2, \quad z' = 2; \quad x'' = 0, \quad y'' = 4, \quad z'' = -5;$$
$$x''' = -2, \quad y''' = 1, \quad z''' = 0.$$

Ans. $9x + 19y + 15z - 1 = 0$.

9. To find the equations of the intersection of two planes, of which the equations are

$$3x + 8y - 10z + 6 = 0, \text{ of the 1st,}$$

and
$$4x - 8y + z + 1 = 0, \text{ of the 2d.}$$

Ans. $\begin{cases} 7x - 9z + 7 = 0, \\ 56y - 43z + 21 = 0. \end{cases}$

10. To find the traces of a plane whose equation is

$$x - 9y + 11z - 12 = 0.$$

11. To find the length of a line drawn from a point whose coördinates are $x' = 2$, $y' = -3$, $z' = 0$, and perpendicular to a plane whose equation is
$$8x + 9y - z + 2 = 0.$$

Ans. $\dfrac{9}{\sqrt{146}}$.

12. Find the equation of a plane passing through the point $x' = 1$, $y' = -3$, $z' = 4$ and perpendicular to the line $x = 3z - 4$, $y = -2z + 5$.

13. Required the angle between the two planes

$$5x - 7y + 3z + 1 = 0, \text{ and } 2x + y - 3z = 0.$$

Ans. 100° 8′.

14. Required the angle which the plane $5x - 7y + 3z + 1 = 0$, makes with the coördinate planes.

Ans. $\begin{cases} 70° \ 46' \text{ with } XY. \\ 140° \ 12' \text{ with } ZX. \\ 56° \ 43' \text{ with } YZ. \end{cases}$

15. Find the equation of a right line which passes through the point (a, b, c) and is perpendicular to each of two right lines whose direction-cosines are l, m, n; l', m', n'.

Ans. $\dfrac{x-a}{mn'-m'n} = \dfrac{y-b}{nl'-n'l} = \dfrac{z-c}{lm'-l'm}$.

16. Find the equations of a line in space in terms of the direction-cosines of its angles with the three axes.

17. Find the equation to a plane passing through the points $(2, 3, 4)$, $(3, 4, 5)$, and perpendicular to the plane $x + 4y + 2z = 1$.

CHAPTER VII.

CURVED SURFACES.

222. Definitions.—A line may be generated by the movement of a point; a surface by the movement of a line; and a volume by the movement of a surface.

The generatrix is the moving element, whether it be a point, line, or surface.

The directrix is a fixed element, about which the generatrix moves.

A surface of revolution is a surface generated by the revolution of a line about an axis.

A curved surface is one from which a curve may be cut by a plane.

A single curved surface or *surface of single curvature* is one which may be generated by a right line having its consecutive positions in one plane; as a cylinder, or a cone.

A surface of double curvature is one which can be generated by a curve only; or one from which a right line cannot be cut; as a sphere, ellipsoid, paraboloid, etc.

A warped surface is one which may be generated by a right line no two consecutive positions of which are in one plane; as the hyperbolic paraboloid, conoids, etc.

An hyperbolic paraboloid is a surface from which parabolas may be cut by a certain system of parallel planes, and hyperbolas by another system of parallel planes, (Art. 236).

A conoid may be generated by a line moving around a curve, while some point of the line moves to and fro along a right line, the generatrix remaining parallel to a plane. The plane is called *the plane-directer*, and is generally perpendicular to the right-line directer.

A line of single curvature is one which changes its direction at every point, and all of whose points are in one plane; as a circle, ellipse, spiral, etc.

A line of double curvature is one which changes its direction at every point, and all of whose points cannot lie in one plane; as the thread of a screw; a thread wound spirally around a cone, or sphere, etc.

Of Cylindrical Surfaces.

223. A Cylinder may be generated by a right line moving around a fixed curve and remaining parallel to a fixed line. The fixed curve is the directrix, and the moving line, the generatrix.

If the generatrix is perpendicular to the plane of the directrix, the cylinder is called *right*. An oblique cylinder is one in which the generatrix is inclined to the plane of the directrix. *A circular cylinder* is one in which any right section is a circle. *An elliptic cylinder* is one in which any right section is an ellipse. Generally a cylinder takes its name from the character of its right sections, or by calling it an oblique cylinder with a given base, thus we may say an oblique cylinder with a circular base, or an oblique cylinder with an elliptical base, etc.

224. Equations of a Right Cylinder.—1°. *Let the base be a circle.* Take the origin at the centre of the circle, the plane xy in the plane of the circle, and the axis z perpendicular to that plane; then will the generatrix be parallel to the axis of z. Let P be any point in the surface, and CEP any section parallel to the base. Let fall the perpendicular PD upon CE, the radius CE being parallel to the axis of x; then will the coördinates of P be $z = CO$, $x = CD$, $y = DP$.

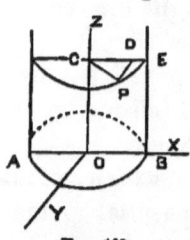

Fig. 162.

Let $r = CE = CP =$ the radius of the base; then

$$r^2 = CD^2 + DP^2;$$

or $$r^2 = x^2 + y^2,$$

and z *indeterminate;* hence the equation of the surface will be

$$x^2 + y^2 + 0 \cdot z^2 = r^2.$$

2°. *Let the base be an ellipse.*—Proceeding as before, we find

$$b^2 x^2 + a^2 y^2 + 0 \cdot z^2 = a^2 b^2,$$

for the equation of the surface.

3°. *Let the base be a parabola.*—Take the origin at the vertex, the plane of the curve being in the plane xy; then we find

$$y^2 + 0 \cdot z = 2px.$$

225. Equation of an Oblique Cylinder.—Let the directrix be in the plane xy and represented by

$$F(x, y) = 0, \qquad (1)$$

which is read, *function xy equal zero.* It implies that x and y are dependent upon each other, and may be made to represent the coördinates of any plane curve. For instance,

for the circle, $F(x, y) = x^2 + y^2 - r^2 = 0$;
for the ellipse, $F(x, y) = b^2 x^2 + a^2 y^2 - a^2 b^2 = 0$;
for the parabola, $F(x, y) = y^2 - 2px = 0$;
for the right line, $F(x, y) = y - mx - b = 0$.

For the oblique cylinder, the equation of the generatrix will be, (Art. 195),

$$x = mz + a, \quad y = nz + b, \qquad (2)$$

from which we have

$$a = x - mz, \quad b = y - nz; \qquad (3)$$

in which a and b are the coördinates of the point where the generatrix pierces the plane xy. But to generate the required cylinder, the point (a, b) must move around on the directrix, and its coördinates become the coördinates of the directrix; hence they take the place of x and y in the equation of the directrix, and *the functional equation of the surface becomes*

$$F(a, b) = F(x - mz, y - nz) = 0. \qquad (4)$$

In this equation, x, y, z, are the coördinates of the surface and a, b, the values of x, y, when $z = 0$. The particular value of $F(a, b)$ will be determined from the character of the base.

APPLICATIONS.—1°. *Let the base be a circle.* Then will the equation of the base be

$$F(a, b) = a^2 + b^2 - r^2 = 0.$$

Substituting in this equation the values of a and b from equations (3), give

$$(x - mz)^2 + (y - nz)^2 = r^2;$$

which is the equation of an oblique cylinder having a circular base.

If the generatrix be parallel to the axis of z, then $m = 0$, and $n = 0$, and we have

$$x^2 + y^2 + 0 \cdot z^2 = r^2,$$

as previously found for the equation of a right circular cylinder.

2°. *Let the base be a parabola.* Take the origin at the vertex, and we have

$$F(a, b) = a^2 - 2pb = 0;$$

and the equation of the surface becomes

$$(x - mz)^2 - 2p(y - nz) = 0;$$

which is the equation of the surface of an oblique cylinder having a parabola for its base.

3°. *Let the base be an ellipse.* We will have

$$b^2(x - mz)^2 + a^2(y - nz)^2 = a^2 b^2$$

for the required equation.

4°. *Let the directrix be a right line.* We will have, (Art. 28),

$$F(a, b) = b - m_1 a - b_1 = 0;$$

in which m_1 and b_1 are used so as to distinguish them from the letters in the preceding equations.

Substituting the values of a and b from equation (3), we have,

$$(y - nz) - m_1(x - mz) - b_1 = 0,$$

which is the equation of a plane, (Art. 202).

Of Conical Surfaces.

226. A Conical Surface may be generated by a right line constantly passing through a fixed point and moving around a fixed curve.

The moving line is the *generatrix*, the fixed curve the *directrix*, and the fixed point the *vertex* (or *apex*) of the cone.

The line will generate two parts of the surface, one part being on one side of the fixed point, and the other part on the opposite side; each of these parts is called a *nappe* of the cone.

A cone is described from the character of its base and inclination of its axis; thus, we may have a right or oblique cone with any given base. *A right cone* is one in which the line joining the apex with the centre of the base is perpendicular to the plane of the base. If the base has not a centre, the cone cannot properly be called *right*.

227. Equations to the surface of a Right Cone.—
1°. *Let the base be a circle.* Take the origin at the vertex, and the plane xy parallel to the plane of the circle, then will the axis of z coincide with the axis of the cone. Let P be any point in the surface, and pass a plane through it parallel to the plane xy; its section will be a circle. The coördinates of P will be

$$x = CD, \quad y = DP, \quad z = CO.$$

Let $h = AO =$ the altitude of the cone, $AB = r =$ the radius of the base, and $v = AOB =$ the angle between an ele-

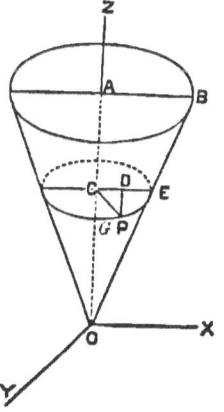

Fig. 163.

ment and the axis; then

$$CP = CE = z \tan v.$$

But
$$CD^2 + DP^2 = CP^2;$$

or
$$x^2 + y^2 = z^2 \tan^2 v,$$

which is the equation of a right cone having a circular base.

2°. *Let the base be an ellipse.* The horizontal section CE will be similar to the base. Let $AB = a$ be the semi-major axis, b the semi-minor axis, and we have

$$OA : OC :: AB : CE;$$

or
$$h : z :: a : CE = \frac{a}{h}z = a' \text{ (say)}.$$

Similarly,
$$CG = \frac{b}{h}z = b' \text{ (say)}.$$

For the point P of the ellipse, we have

$$a'^2 y^2 + b'^2 x^2 = a'^2 b'^2.$$

Substituting the values of a' and b', we have

$$a^2 y^2 + b^2 x^2 = \frac{a^2 b^2}{h^2} z^2;$$

which is the equation of a right cone having an elliptical base.

228. Equation to the surface of an Oblique Cone.— Take the plane xy in the plane of the directrix. The equation of the directrix will be $F(x, y) = 0$. The equations of the right line which forms the generatrix will be of the form

$$x = mz + a, \quad y = nz + b; \qquad (1)$$

but since this line must pass through a fixed point (x', y', z'), its equations become

$$x - x' = m(z - z'), \quad y - y' = n(z - z'), \qquad (2)$$

or
$$x = mz + (x' - mz'), \quad y = nz + (y' - nz'), \qquad (3)$$

in which the absolute terms $(x' - mz')$, $(y' - nz')$, are the coördinates of the points where the generatrix pierces the plane xy; their values in terms of the general variables are

$$\begin{aligned}(x' - mz') &= x - mz \\ (y' - nz') &= y - nz\end{aligned}\Bigg\}. \qquad (4)$$

But m and n will constantly vary, and should be expressed in terms of the general variables. Substituting their values from equations (2), and representing the left members of (4) respectively by M_x and N_y, gives

$$\left.\begin{aligned}M_x &= (x' - mz') = \frac{x'z - xz'}{z - z'} \\ N_y &= (y' - nz') = \frac{y'z - yz'}{z - z'}\end{aligned}\right\}, \qquad (5)$$

and the general equation of the surface becomes

$$F(M_x, N_y) = F\left(\frac{x'z - xz'}{z - z'}, \frac{y'z - yz'}{z - z'}\right) = 0. \qquad (6)$$

APPLICATIONS.—1°. *Let the base be an ellipse.* The coördinates of the base being M_x and N_y, its equation will be

$$a^2 N_y^2 + b^2 M_x^2 = a^2 b^2;$$

in which substituting the values of M_x and N_y from equation (5) gives

$$a^2 \left(\frac{y'z - yz'}{z - z'}\right)^2 + b^2 \left(\frac{x'z - xz'}{z - z'}\right)^2 = a^2 b^2 \qquad (7)$$

for the general equation of an oblique cone having an elliptical base.

2°. *Let the cone be right.* The vertex will be vertically over the origin, and we have

$$x' = 0, \quad y' = 0, \quad z' = h,$$

and the preceding equation becomes

$$a^2\left(\frac{-hy}{z-h}\right)^2 + b^2\left(\frac{-hx}{z-h}\right)^2 = a^2b^2,$$

or
$$a^2y^2 + b^2x^2 = \frac{(z-h)^2}{h^2}a^2b^2; \qquad (8)$$

which will become the same as the last equation of the preceding Article if the origin be transferred to the vertex.

3°. *Let the base be a circle and the cone right.* Then will $a = b$ in the preceding equation, and we have

$$h^2(x^2 + y^2) = (z-h)^2 a^2 \qquad (9)$$

for the equation of the surface.

4°. *Let the directrix be a right line.* The equation of the directrix will be

$$F(M_x, N_y) = N_y - m_1 M_x - b_1 = 0.$$

Substitute the values of N_y and M_x from equation (5) and we have

$$y'z - z'y - m_1(x'z - z'x) = b_1(z - z'),$$

which is *the equation of a plane*, (Art. 202).

Surfaces of Revolution.

229. The Sphere is a surface every point of which is equally distant from a point within called the centre. It may be generated by the revolution of a semicircle about a diameter. Let r be the radius of the sphere. The origin being at the centre, the distance of any point P from the centre will be, (Art. 193), $\sqrt{x^2 + y^2 + z^2}$, but this equals the radius; hence the required equation is

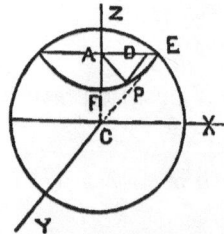

Fig. 164.

$$x^2 + y^2 + z^2 = r^2.$$

If the coördinates of the centre be a, b, c, the equation will be

$$(a - x)^2 + (b - y)^2 + (c - z)^2 = r^2.$$

230. General Equation.—When a surface is generated by the revolution of a plane curve, let the axis of z coincide with the axis of revolution, and the plane of the generatrix be in the plane zx. Every point of the curve will describe a circle having its centre on the axis of z. The length of the radius of this circle will generally depend upon the ordinate z, which may be expressed in the form

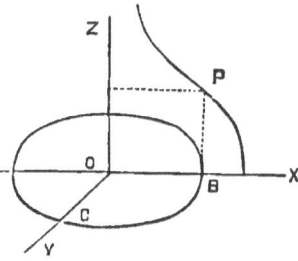

Fig. 165.

$$\rho = f(z), \qquad (1)$$

from which ρ may be found when the equation of the generatrix is known. But we also know, (Art. 57), that

$$\rho = \sqrt{x^2 + y^2}; \qquad (2)$$
$$\therefore x^2 + y^2 = (f(z))^2, \qquad (3)$$

which may be called the *functional equation of any surface of revolution*.

To apply this method to the sphere, the equation of the generatrix will be

$$z^2 + \rho^2 = r^2;$$
$$\therefore \rho^2 = (f(z))^2 = r^2 - z^2,$$

and equation (3) becomes

$$x^2 + y^2 + z^2 = r^2, \qquad (4)$$

as before found.

Of Ellipsoids.

231. The Prolate Ellipsoid may be generated by the revolution of a semi-ellipse about its major axis. Let the semi-major axis of the generating curve be b, the semi-minor, a. Let P be any point of the generating curve, ρ, the radius of the circle which it describes; then will the equation of the generatrix be

$$a^2 z^2 + b^2 \rho^2 = a^2 b^2;$$
$$\therefore \rho = f(z) = \frac{a}{b}\sqrt{b^2 - z^2}.$$

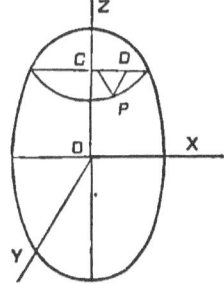

Fig. 166.

Hence equation (3) of the preceding Article, gives

$$\frac{x^2}{a^2} + \frac{y^2}{a^2} + \frac{z^2}{b^2} = 1, \qquad (5)$$

which is the equation of the surface of the prolate ellipsoid.

232. The Oblate Ellipsoid may be generated by the revolution of an ellipse about its minor axis. Let the axis of z be the axis of revolution. Then proceeding as before, we have

$$\frac{x^2}{b^2} + \frac{y^2}{b^2} + \frac{z^2}{a^2} = 1. \qquad (6)$$

233. The Ellipsoid of the most general character is a surface such that all sections of it by a plane are ellipses. It may be conceived to be generated by an ellipse revolving about one axis while the other axis increases or diminishes in such a way as to generate an ellipse during the revolution, and at the same time the generating curve be an ellipse in all positions.

Let the semi axes of the ellipsoid be a, b, c; then the equation of the surface will be

$$\frac{x^2}{a^2} + \frac{y^2}{b^2} + \frac{z^2}{c^2} = 1. \qquad (7)$$

Of Paraboloids.

234. The Paraboloid of Revolution may be generated by the revolution of the parabola about its axis. To find the equation of the surface, let the axis of x be the axis of revolution, and P any point in the surface. If the plane of the generating curve be in the plane zx, its equation will be $z^2 = 2px$. Let PD be perpendicular to EB; then

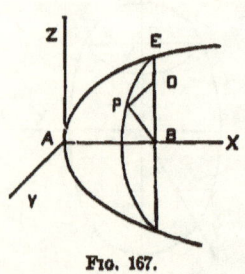

Fig. 167.

$$EB = PB = \sqrt{2px}.$$

But
$$PB^2 = BD^2 + DP^2 = z^2 + y^2;$$
$$\therefore z^2 + y^2 = 2px;$$
or
$$\frac{z^2}{2p} + \frac{y^2}{2p} - \frac{x}{1} = 0,$$

which is the equation of the surface of a paraboloid of revolution when revolved about the axis of x.

235. Elliptical Paraboloid.—Let the generatrix be a parabola and the directrix an ellipse, the parameter of the parabola being supposed to vary constantly, so that the generating arc may constantly pass through a point on the ellipse. The surface thus generated is an *elliptical paraboloid*.

Take the centre of the ellipse on the axis of x, and its plane parallel to the plane zy. Let P be any point on the surface whose coördinates are x, y, z, the origin being at the vertex of the parabola, and $2p$ the parameter of the parabola POC. Draw PD perpendicular to CD, then we have

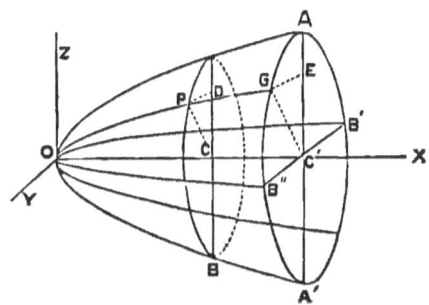

Fig. 168.

$$PC^2 = 2px,$$
or
$$z^2 + y^2 = 2px, \qquad (1)$$

which would be the required equation, *provided* that p varied according to the required law. To find p in terms of general variables: Let $b_1 = AC'$, $a_1 = B'C'$, $h = OC'$, G the point where the arc OP intersects the ellipse, and whose coördinates are x_1, y_1, z_1, and $2p_1$ the parameter of the parabola $B''OB'$. Then

$$a_1^2 = 2p_1 h; \quad GC'^2 = 2ph = y_1^2 + z_1^2; \qquad (2)$$

$$\therefore 2p = \frac{y_1^2 + z_1^2}{h} = \frac{y_1^2 + z_1^2}{a_1^2} 2p_1. \qquad (3)$$

The equation of the ellipse gives

$$a_1^2 z_1^2 + b_1^2 y_1^2 = a_1^2 b_1^2 \; ; \; \therefore \; a_1^2 = \frac{a_1^2 z_1^2 + b_1^2 y_1^2}{b_1^2}, \qquad (4)$$

which, substituted in the preceding equation, gives

$$2p = \frac{z_1^2 + y_1^2}{a_1^2 z_1^2 + b_1^2 y_1^2} 2 p_1 b_1^2. \qquad (5)$$

Since this equation does not contain x, it will be true for all corresponding values of y and z. Hence, dropping the subscripts to z and y and substituting in equation (1), gives

$$a_1^2 z^2 + b_1^2 y^2 = 2 p_1 b_1^2 x, \qquad (6)$$

for the required equation. Letting $2 p_1 b_1^2 = d_1^2$, and dividing through by $a_1^2 b_1^2 d_1^2$, it may be put under the form

$$\frac{z^2}{a^2} + \frac{y^2}{b^2} - \frac{x}{d} = 0, \qquad (7)$$

in which a^2, b^2, d, are used for the new denominators of z^2, y^2, x, respectively.

236. The Hyperbolic Paraboloid may be generated by the movement of a parabola whose parameter so varies that its arc shall follow the opposite branches of an hyperbola, the vertex remaining at the same point. Let the origin be at the vertex of the parabola, the plane zy be parallel to the plane of the hyperbola, and the axis of x pass through the centre of the hyperbola. If the equation to the hyperbola be $a_1^2 z_1^2 - b_1^2 y_1^2 = a_1^2 b_1^2$, we would find, in the same manner as shown in the pre-

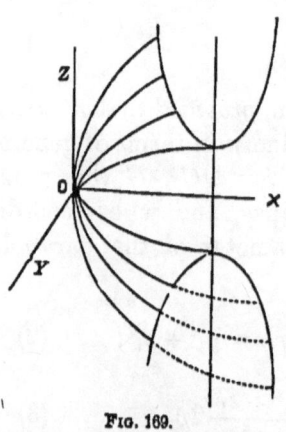

Fig. 169.

ceding Article, that the equation of the surface would be

$$\frac{x^2}{a^2} - \frac{y^2}{b^2} - \frac{x}{d} = 0;$$

but if the equation of the hyperbola be $a_1^2 z_1^2 - b_1^2 y_1^2 = -a_1^2 b_1^2$ (which is the conjugate of the preceding one), then we would find

$$-\frac{z^2}{a^2} + \frac{y^2}{b^2} - \frac{x}{d} = 0$$

for the required equation.

237. Problem.—*Required the equation of the surface generated by a right line moving parallel to a plane and along any other two right lines.*

Let one of the lines, as AB, lie in the plane xy, and the other in the plane xz. Take the origin on the line OC. Let the equation of the line OC be

$$z = nx,$$

of AB, $\quad y = mx + b,$

and let the generating line CE be constantly parallel to the plane zy, then will its projection, EF, on the plane xy be parallel to the axis of y. Let P be any point in the surface whose coördinates are $x = OF$, $y = FD$, $z = DP$. Then will

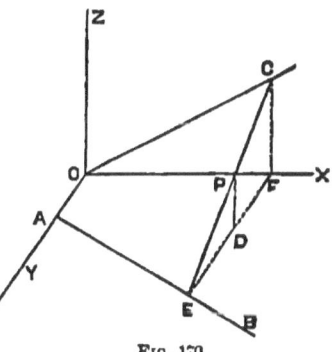

Fig. 170.

$$EF = mx + b,$$
and $\qquad CF = nx.$

We have $\quad ED : DP :: EF : FC;$

or $\qquad mx + b - y : z :: mx + b : nx;$

$$\therefore nmx^2 + bnx - nxy - mzx - bz = 0,$$

which is the required equation, and is an equation of the second degree between three variables. The character of the surface is not readily seen from this equation, but by a transformation of coördinates it may be reduced to the form

$$Mz^2 - Ny^2 - Qx = 0;$$

and hence it is an hyperbolic paraboloid, (Art. 236). It is a warped surface, (Art. 222). The same surface would be generated by the line OC moving on OA and CE in such a manner as to divide those lines proportionally.

Of Hyperboloids.

238. If the hyperbola EA revolves about its conjugate axis NO, the surface generated will be continuous, and is called **an hyperboloid of revolution of one nappe.**

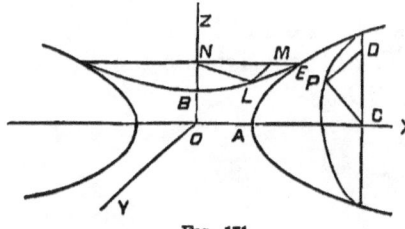

Fig. 171.

The equation of the hyperbola is

$$NE^2 = \frac{a^2 z^2 + a^2 b^2}{b^2}.$$

But $\qquad NE^2 = NL^2 = NM^2 + ML^2 = x^2 + y^2;$

$$\therefore x^2 + y^2 = \frac{a^2 z^2 + a^2 b^2}{b^2};$$

or, dividing by a^2, we have

$$\frac{x^2}{a^2} + \frac{y^2}{a^2} - \frac{z^2}{b^2} = 1; \qquad (1)$$

which is the equation of the hyperboloid of revolution of one nappe.

239. If the hyperbola revolves about the axis of x, two surfaces will be generated, one by the right branch of the curve, and the other by the left. This is called *an hyperbo-*

loid of two nappes. Its equation is

$$\frac{x^2}{a^2} - \frac{y^2}{b^2} - \frac{z^2}{b^2} = 1. \qquad (2)$$

In these equations the negative signs apply to those axes which do not intersect the surface.

240. Elliptical Hyperboloid.—If the arc of the hyperbola be made to follow an elliptical directrix, the surface generated will be an elliptical hyperboloid. Its equation will be of the form

$$\frac{x^2}{a^2} + \frac{y^2}{b^2} - \frac{z^2}{c^2} = 1, \qquad (3)$$

when the surface is of *one nappe;* and

$$\frac{x^2}{a^2} - \frac{y^2}{b^2} - \frac{z^2}{c^2} = 1, \qquad (4)$$

when of two *nappes.*

Observe that the character of these surfaces may be determined by finding the character of the curve of intersection of the coördinate planes with the respective surfaces. Thus, in equation (1) if $z = 0$, we have $x^2 + y^2 = a^2$, which is the equation of a circle. If $x = 0$, we have $b^2y^2 - a^2z^2 = a^2b^2$, which is the curve of intersection of the plane yz with the surface, and is an hyperbola.

In equation (4), if $z = 0$, we have the equation of an hyperbola, which is the intersection of the plane xy with the surface. Similarly, if $y = 0$, we have the equation of the intersection of the surface by the plane xz, which is also an hyperbola. If $x = 0$, the curve is imaginary, or the plane yz does not cut the surface. The surface represented by equation (3) is cut by all the coördinate planes.

241. Problem.—*To find the surface generated by any line revolving about another line.* Take the axis of z as the directrix, and the axis of x perpendicular both to the directrix and the generatrix, and passing through the point where the

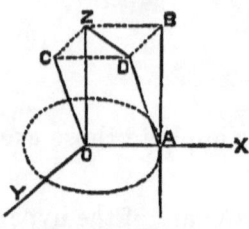

Fig. 172.

plane xz cuts the given line. Then will the projection of the generatrix on the plane xz be parallel to the axis of z, and the projection on zy will pass through the origin. The line will make a constant angle, COZ, with the axis of z, and the point A will describe the arc of a circle. Let $\tan COZ = l$, and $OA = d$; then will the square of the distance of any point D from the axis of z be

$$CZ^2 + CD^2,$$

or, $\qquad l^2 z^2 + d^2;$

hence we have $\qquad x^2 + y^2 = l^2 z^2 + d^2, \qquad (1)$

or, dividing by d^2 it may be written in the form

$$\frac{x^2}{a^2} + \frac{y^2}{a^2} - \frac{z^2}{b^2} = 1, \qquad (2)$$

which is the equation of an hyperboloid of revolution of one nappe, (Art. 238). Hence, an hyperboloid of one nappe is a warped surface.

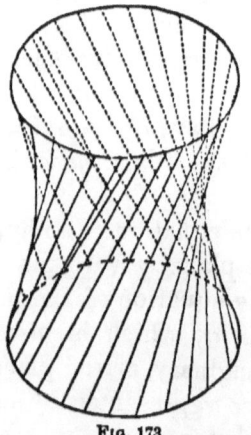

Fig. 173.

[This problem may be solved in a more general way. Let the equations of the generatrix be

$$x = mz + a, \quad y = nz + b \qquad (3)$$

in which a and b are the coördinates of the points where the generatrix pierces the plane xy. The locus of this point will be a circle, hence we have

$$a^2 + b^2 = r^2 = (x - mz)^2 + (y - nz)^2. \qquad (4)$$

In this equation m and n are variables, but they must be subjected to the condition that the angle which the generatrix makes with the axis of z is constant; hence, (Art. 198),

$$m^2 + n^2 = a\ constant = d^2,\ (\text{say}), \qquad (5)$$

and equation (4) becomes

$$r^2 = x^2 + y^2 - 2(mz \cdot x + nz \cdot y) + d^2z^2. \qquad (6)$$

Substituting the values of mz and nz from equations (3), we find

$$x^2 + y^2 = 2(ax + by) - r^2 + d^2z^2 ; \qquad (7)$$

in which substitute from (3) the values

$$ax = amz + a^2, \; by = bnz + b^2, \qquad (8)$$

and we find $\quad x^2 + y^2 = 2(am + bn)z + r^2 + d^2z^2. \qquad (9)$

Here are four arbitrary constants a, b, m, n, between which two conditions have been fixed; we may therefore make an arbitrary assumption and leave them still indeterminate. Assuming that the origin of coördinates is fixed by the condition

$$am + bn = 0, \qquad (10)$$

we finally have $\quad x^2 + y^2 = r^2 + d^2z^2, \qquad (11)$

which is of the same form as equation (1).

If $z = 0$, we have

$$x^2 + y^2 = r^2,$$

which is the equation of the curve of intersection of the plane xy with the surface, which curve is a circle.

If $x = 0$, we have

$$y^2 - d^2z^2 = r^2 ;$$

which is the equation of an hyperbola, and is the equation of the curve of intersection of the plane yz with the surface. Similarly, the intersection of the plane xz with the surface is an equal hyperbola. The condition of equation (10) places the origin so that the surface will be symmetrical in reference to the axes. All horizontal sections are circles, and that part of the surface whose section is least is called the *gorge*. In equation (11) the origin is at the centre of the gorge. In equation (9) the origin may be anywhere on the axis of z.

If the directrix were an ellipse the surface generated by the line would be an *elliptical hyperboloid of one nappe;* and similarly for other directrices.]

Of Intersections.

242. Problem.—*To find the intersection of a plane and sphere.*

The equation of the sphere is

$$x^2 + y^2 + z^2 = r^2. \qquad (1)$$

Let the cutting plane be parallel to the plane xy, then, in the equation of the plane

$$Ax + By + Cz + D = 0, \qquad (2)$$

will A and B be zero, and we will have

$$z = -\frac{D}{C},$$

which, substituted in the equation of the sphere, gives

$$x^2 + y^2 = r^2 - \frac{D^2}{C^2},$$

which is the equation of a circle. It is real when $\frac{D^2}{C^2}$ is less than r^2, and imaginary if it is greater that r^2. It will be greatest when $D = 0$; but when D is zero the plane passes through the centre of the sphere, as shown by equation (2), since, in that case, equation (2) has no absolute term, and the origin is at the centre.

In a similar manner we find that the intersection of the sphere by any plane parallel to the coördinate planes is a circle.

Generally, substitute the value of z from equation (2) in equation (1), and we have

$$(A^2 + C^2)x^2 + (B^2 + C^2)y^2 + 2ABxy + 2ADx + 2BDy = C^2 r^2 - D^2;$$

which is the projection of the curve on the plane xy; and is an ellipse, (Art. 178). But in order to determine the character of the curve, we must find its equation in its own plane, that is in the plane the equation of which is given by equation (2). This process will be explained in Article 249.

243. *To find the intersection of a plane with an hyperbolic paraboloid.*

The equation of the hyperbolic paraboloid is, (Art. 236),

$$\frac{z^2}{a^2} - \frac{y^2}{b^2} - \frac{x}{d} = 0. \qquad (1)$$

Let the plane be parallel to the plane zy, then will its equation be

$$Ax + D = 0, \quad y \text{ and } z \text{ indeterminate};$$

$$\therefore x = -\frac{D}{A};$$

and this value reduces the preceding equation to the following;

$$\frac{z^2}{a^2} - \frac{y^2}{b^2} = \frac{D}{dA},$$

which is the equation to an hyperbola, (Art. 78), hence, *All sections of an hyperbolic paraboloid made by a plane perpendicular to the axis of the parabolic sections, are hyperbolas.*

Let the plane be parallel to the plane xz, then will the equation of the plane be

$$By + D = 0, \quad x \text{ and } z \text{ indeterminate};$$

$$\therefore y = -\frac{D}{B},$$

and this value in the equation of the surface gives

$$\frac{z^2}{a^2} = \frac{x}{d} + \frac{D^2}{b^2 B^2},$$

which is the equation of a parabola, (Art. 88).

Let the cutting plane be parallel to xy, and the equation of the plane be

$$z = g, \quad x \text{ and } y \text{ indeterminate}.$$

This value in equation (1) gives

$$\frac{y^2}{b^2} = -\frac{x}{d} + \frac{g^2}{a^2},$$

which is imaginary if $\dfrac{x}{d} > \dfrac{g^2}{a^2}$, and a real parabola if $\dfrac{x}{d} < \dfrac{g^2}{a^2}$.

244. Surfaces of the Second Order are those whose equations are of the second degree. It will be observed that all the curved surfaces discussed in this chapter are of the *second order*.

245. The intersection of a surface of the second order by a Plane is a Conic Section.—For the equation to the curve of intersection is found by eliminating one of the variables between the equations of the plane and surface. But the equation of the surface is of the second degree, and of the plane, of the first degree; hence, according to the principles of algebra, the resulting equation will be of the second degree, and hence will be the equation of a conic, (Art. 177). The curve thus found is the projection of the required curve on one of the coördinate planes, but by transforming the coördinates so as to represent the equation of the curve in its own plane, the degree of the equation is not changed, (Arts. 187a and 187b).

246. Intersection of two surfaces of the second order.—Proceeding as before to eliminate the variables from the equations to the surfaces, the resulting equations will be the equations of the *projections* of the curves of intersection on the respective planes. If the intersection be a *plane curve*, its equation may be found in its own plane. *Generally*, however, the curve of intersection of surfaces of the second order will be a curve of double curvature, (Art. 222), in which case its character can be determined only by considering its three projections.

247. *Intersection of a sphere and ellipsoid of revolution.*
Let the equations be

$$x^2 + y^2 + z^2 = r^2, \qquad ax^2 + ay^2 + bz^2 = d.$$

Eliminating z gives

$$(a - b)x^2 + (a - b)y^2 = d - br^2,$$

which is the equation of a circle whose radius is

$$\sqrt{\frac{d - br^2}{a - b}}.$$

Eliminating x gives

$$z\sqrt{a-b} = \sqrt{ar^2 - d},$$

which gives a point on the axis of z. If the intersection of the two surfaces be not a point, the preceding equation will be the equation of a line and may be written

$$0 \cdot y + \sqrt{a-b} \cdot z = \sqrt{ar^2 - d},$$

which is the equation to the trace of a plane on yz parallel to the plane xy. The line of intersection is therefore a plane curve, and is a circle.

248. *Intersection of a sphere and hyperbolic paraboloid.*
Let the equations be

$$x^2 + y^2 + z^2 = r^2, \quad ax^2 - by^2 - cz = d.$$

Eliminating x gives

$$(a+b)y^2 + az^2 + cz = ar^2 - d,$$

which is the equation of an ellipse, the centre being on the axis of z at a distance from the origin equal to $-\dfrac{c}{2a}$. It is the projection of the curve on the plane yz.

Eliminating y gives

$$(a+b)x^2 + bz^2 - cz = br^2 + d,$$

which is also the equation of an ellipse and is the projection of the curve on xz.

Eliminating z gives an equation of the form

$$Ax^4 + Bx^2y^2 + Cy^4 + Dx^2 + Ey^2 + F = 0,$$

which is the equation to the projection of the curve on the plane xy. This is a curve of the fourth order, and, not being plane, has no special name.

249. Intersection of a Plane and Cone. — Take a right cone having a circular base; its equation will be, (Art. 227),

$$x^2 + y^2 = z^2 \tan^2 v, \qquad (1)$$

the origin being at the apex. Transfer the origin to O, a point on the axis of z, the distance ZO being c; then will the equation become

$$x^2 + y^2 = (z - c)^2 \tan^2 v. \qquad (2)$$

Let the secant plane embrace the axis of y, then will it be perpendicular to the plane xz, and BO will be its trace on that plane. Let the angle between the secant plane and the axis of z be $u = BOZ$. Then will the equation of the secant plane be (Art. 203),

$$0 \cdot y + x = z \tan u. \qquad (3)$$

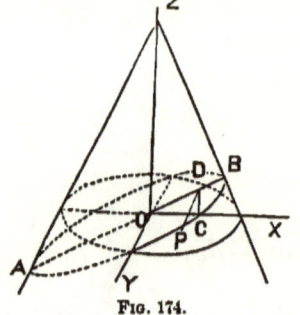

Fig. 174.

Eliminating z between equations (2) and (3), we have

$$y^2 \cot^2 v + (\cot^2 v - \cot^2 u)x^2 + 2cx \cot u = c^2, \qquad (4)$$

for the equation of the projection on the plane xy of the curve of intersection. To find the equation of the curve in its own plane, let P be any point of the curve whose coördinates are

$$OD = x' = x \operatorname{cosec} u, \quad DP = y' = y. \qquad (5)$$

Substitute the values of x and y from these equations in the preceding equation, and dropping the accents we have

$$y^2 \cot^2 v + \sin^2 u (\cot^2 v - \cot^2 u) x^2 + 2cx \cos u = c^2.$$

This is an equation of the second degree between two variables; and, by comparing it with the general form given in Article 177, gives

$$A = \sin^2 u\,(\cot^2 v - \cot^2 u)\,;\quad B = \cot^2 v\,;\quad H = 0\,;$$
$$G = c\cos u\,;\quad F = 0\,;\ \text{and}\ C = -c^2.$$

Hence, according to Article 178, the curve of intersection will be (since H is zero)

an ellipse if . . . $(\cot^2 v - \cot^2 u)$ *is positive;*

a parabola if . . $(\cot^2 v - \cot^2 u) = 0$;

an hyperbola if $(\cot^2 v - \cot^2 u)$ *is negative.*

This expression will be positive when $\cot u < \cot v$, or $u > v$; that is,

When the secant plane makes a greater angle with the axis of the cone than the elements do, the intersection is an ellipse, (Art. 185).

The expression will be zero when $u = v$; hence,

When the secant plane is parallel to an element of the cone, the curve of intersection will be a parabola.

The expression will be negative when $u < v$; hence,

When the secant plane makes a less angle with the axis of the cone than the elements do, the curve of intersection will be an hyperbola.

If the secant plane is perpendicular to the axis of the cone, u will be $90°$. The coefficient of x^2 is

$$\sin^2 u \cot^2 v - \cos^2 u,$$

or for $u = 90°$, it becomes

$$\cot^2 v,$$

and the equation of the curve becomes

$$y^2 + x^2 = c^2 \tan^2 v,$$

which is the equation of a circle.

250. *To find the intersection of a plane with a right cone having an elliptical base.*

The equation of the curve will be of the second degree, and hence the curve will be a conic.

251. *To find the intersection of a plane with a right cone having an hyperbola for its base.*

The equation of the curve of intersection will be of the second degree, but if the intersection be an ellipse it will not be re-entrant.

Similarly, the intersection of a plane with a cone having a parabola for a base will be a conic.

252. *To find the intersection of a right cone having an elliptical base with an ellipsoid.*

Let the equations be,

of the cone, $\qquad ax^2 + by^2 = cz^2;$

of the ellipsoid, $\quad a_1x^2 + b_1y^2 + c_1z^2 = d,$

Eliminating z gives

$$(a_1c + ac_1)x^2 + (b_1c + bc_1)y^2 = cd,$$

which is the equation of an ellipse. In a similar manner find the equations of the curve on the other planes. It may not be a plane curve.

Discussion of the General Equation of the Second Degree having Three Variables.

253. A full discussion is not here attempted; but some of the steps are indicated by which a more complete discussion may be made. The general equation is of the form:

$$Ax^2 + 2Hxy + By^2 + 2Kyz + Ez^2 + 2Lzx + 2Gx + 2Fy + 2Dz + C = 0. \quad (1)$$

This equation may be divided through by C (or any other coefficient) and be equally general, after which there will be nine arbitrary constants. This equation is called THE QUADRIC. Hence, in general, *a quadric may be passed through nine points not in the same plane.*

254. The extent and character of a surface may be determined by intersecting it with planes and determining the extent and character of the intersections. Intersecting a

Quadric by a plane will give an equation of the second degree; hence *every plane section of a Quadric is a Conic.*

If the Quadric be cut by two planes parallel to xy, whose equations are $z = k$ and $z = k_1$, the curves of intersection will be found by eliminating z from equation (1). The two resulting equations will contain the same values of A, H, and B; hence *parallel sections of a Quadric are similar Conics.*

255. Transform the coördinates by changing the direction of the axes, the origin remaining the same. For this purpose use the equations, (Art. 219),

$$x = x' \cos X' + y' \cos X'' + z' \cos X''',$$
$$y = x' \cos Y' + y' \cos Y'' + z' \cos Y''',$$
$$z = x' \cos Z' + y' \cos Z'' + z' \cos Z'''.$$

These values in equation (1) reduce it to the form

$$A'x^2 + 2H'xy + B'y^2 + 2K'yz + E'z^2 + 2L'zx + 2G'x + 2F'y + 2D'z + 1 = 0, \quad (2)$$

in which the coefficients are functions of the angles. There being nine arbitrary constants, we may make nine equations of condition among these coefficients. Let the new axes be at right angles to each other; this gives the following six equations, (Art. 219),

$$\cos X' \cos X'' + \cos Y' \cos Y'' + \cos Z' \cos Z'' = 0,$$
$$\cos X' \cos X''' + \cos Y' \cos Y''' + \cos Z' \cos Z''' = 0,$$
$$\cos X'' \cos X''' + \cos Y'' \cos Y''' + \cos Z'' \cos Z''' = 0;$$

$$\cos^2 X' + \cos^2 Y' + \cos^2 Z' = 1,$$
$$\cos^2 X'' + \cos^2 Y'' + \cos^2 Z'' = 1,$$
$$\cos^2 X''' + \cos^2 Y''' + \cos^2 Z''' = 1.$$

We may make three more equations of condition; hence we may have

$$H' = 0; \quad K' = 0; \quad L' = 0;$$

and the equation reduces, after dropping the accents, to the form

$$Ax^2 + By^2 + Ez^2 + 2Gx + 2Fy + 2Dz + C = 0. \quad (3)$$

256. *To transform to parallel axes.*

Make $\quad x = a + x', \quad y = b + y', \quad z = c + z',$

in equation (3). There being three new arbitrary constants, a, b, c, such values may be assigned to them as will make the coefficients of the first powers of x, y, z, each equal to zero. These values will be

$$a = -\frac{G}{A}, \quad b = -\frac{F}{B}, \quad c = -\frac{D}{E};$$

which will be real and finite unless A, B, or E is zero; that is, they will be real when the equation contains the second powers of all three variables. When such is the case the equation becomes

$$Ax^2 + By^2 + Ez^2 + C = 0. \qquad (4)$$

In this equation, if $-x$ be substituted for x, $-y$ for y, and $-z$ for z, neither the form nor value of the equation will be changed; hence, every line drawn through the origin and terminated by the surface is bisected at the origin. The origin, therefore, is at the centre of the surface. Such quadrics are called CENTRAL QUADRICS.

257. THE CENTRAL QUADRIC.—When the equation of the central quadric is reduced to the form

$$Ax^2 + By^2 + Ez^2 + C = 0; \qquad (5)$$

the axes are called *Principal Axes*, and the sections made by the coördinate planes, *Principal Planes*. It has been shown that this transformation is always possible for central quadrics; hence, conversely, *Every central quadric has at least one set of three conjugate planes and three diameters which are mutually perpendicular.*

If A, B, and E are positive, and C negative, we have

$$Ax^2 + By^2 + Ez^2 = C;$$

which is the general equation of *the ellipsoid*, (Art. 233).

If $A = B$, it is the equation of the *ellipsoid of revolution*.
If $A = B = E$, it is the equation of *the sphere*.
If $C = 0$, then for a real locus $x = 0$, $y = 0$, $z = 0$, which are the equations of *a point*. If x, y, z, are not zero, it is the equation to an *imaginary ellipsoid*.

If C is positive, the surface is *imaginary*.

Again, if two of the coefficients A, B, E, are positive and one negative, let A and B be positive and E and C negative; then we have

$$Ax^2 + By^2 - Ez^2 = C;$$

which is the equation of an *elliptical hyperboloid*, (Art. 240).

If $A = B$, it is the equation of an *hyperboloid of revolution* of one nappe.

If $A = B = E$, it is the equation of the *equilateral hyperboloid of revolution* of one nappe.

If $C = 0$, it is the equation of the surface of *a right cone*, (Art. 227).

If C be positive, the equation becomes

$$- Ax^2 - By^2 + Ez^2 = C,$$

which is the equation of *the hyperboloid of two nappes*. If $A = B$, it is an hyperboloid of revolution, and if $A = B = E$, it is an equilateral hyperboloid of revolution of two nappes.

258. Suppose that one of the coefficients A, B, E, as A, is zero in equation (3). Again transform the origin, and determine the values of the new constants by making the coefficients of z and y, and the absolute term, separately equal to zero, and we will find

$$By^2 + Ez^2 + 2Gx = 0;$$

which is the equation of a *paraboloid*. It is a NON-CENTRAL QUADRIC.

If B and E are positive and G negative, it will be an *elliptic paraboloid*, (Art. 235). If $B = E$ and G negative, it is a paraboloid of revolution, (Art. 234).

If B is negative and E positive, or the reverse, it is an *hyperbolic paraboloid*, (Art. 236).

259. The preceding transformation is impossible if all the terms containing x are zero;—that is, if $A = 0$, and $G = 0$. These reduce equation (3) to

$$By^2 + Ez^2 + 2Fy + 2Dz + C = 0,$$

which is the equation of a cylindrical surface whose axis is parallel to the axis of x and whose base is a circle, ellipse, or hyperbola, depending upon the relative values and signs of B and E.

260. If $A = 0$ and $B = 0$, equation (3) becomes

$$Ez^2 + 2Gx + 2Fy + 2Dz + C = 0.$$

Intersecting this surface by planes parallel to xy, the equation of one of which will be $z = k$, we have

$$2Gx + 2Fy = K \text{ (say)},$$

which is the equation of a straight line. Hence the elements are parallel straight lines, parallel to the plane xy. Making $x = 0$, we have

$$Ez^2 + 2Fy + 2Dz + C = 0,$$

which is a parabola, and is the intersection of the plane zy with the surface. Similarly, making $y = 0$, we have

$$Ez^2 + 2Gx + 2Dz + C = 0,$$

which is also the equation of a parabola, and is the intersection of the plane zx with the surface. The surface, therefore, is a cylinder having a parabolic base.

Of Tangent Planes.

261. *To find the equation of a plane tangent to a central quadric.*

The general equation of the surface is

$$Ax^2 + By^2 + Ez^2 + C = 0. \tag{1}$$

Let the point of tangency be (x', y', z'), and we have the equation of condition

$$Ax'^2 + By'^2 + Ez'^2 + C = 0, \qquad (2)$$

which subtracted from the preceding equation gives

$$A(x^2 - x'^2) + B(y^2 - y'^2) + E(z^2 - z'^2) = 0. \qquad (3)$$

The equation of a plane passing through the point of tangency is

$$a(x - x') + b(y - y') + c(z - z') = 0, \qquad (4)$$

in which such values must be substituted for a, b, c, as will make the plane tangent to the surface.

If two lines be passed through the point tangent to the surface, the plane of these lines will be the tangent plane required. Let a plane be passed through the tangent point parallel to xz; it will cut a section from the quadric and a secant to the section from the cutting plane. The equation of the plane will be

$$y = y',$$

which substituted in equations (3) and (4) will give

$$A(x + x')(x - x') + E(z + z')(z - z') = 0,$$

and $\qquad a(x - x') + c(z - z') = 0.$

The value of $(x - x')$ from the last equation substituted in the preceding gives

$$a = \frac{A}{E} \cdot \frac{x + x'}{z + z'} c.$$

Let the secant turn about (x', z') until the point (x, z) coincides with (x', z'), then will the secant become a tangent, and we have

$$a = \frac{Ax'}{Ez'} c.$$

Similarly, we may find

$$b = \frac{By'}{Ez'} c;$$

and these values substituted in equation (4), and the result reduced by means of equation (2), give

$$Axx' + Byy' + Ezz' + C = 0,$$

for the required equation.

262. *To find the equation of a plane tangent to a non-central quadric.*

The equation of the surface is

$$By^2 + Ez^2 + 2Gx = 0.$$

Proceeding as before, we find

$$Byy' + Ezz' + G(x + x') = 0,$$

for the required equation.

Of Normals to Quadrics.

263. *A normal is a line perpendicular to a tangent plane at the point of contact.* Hence, the equations of the normal for central quadric are, (Art. 214),

$$x - x' = \frac{Ax'}{Ez'}(z - z'), \qquad y - y' = \frac{By'}{Ez'}(z - z'),$$

and for paraboloids, are

$$x - x' = \frac{G}{Ez'}(z - z'), \qquad y - y' = \frac{By'}{Ez'}(z - z').$$

[For a further discussion of Quadrics, see Salmon's *Geometry of Three Dimensions*.]

EXAMPLES.

1. Determine the class of surfaces to which the following quadrics belong:

$$7x^2 + 6y^2 + 5z^2 - 4yz - 4xy = 6,$$

$$3x^2 - 4y^2 + 2z^2 - xy + 2yz - 3xz = 12,$$

$$3x^2 - 2y^2 - 3z + 4x = 5,$$

$$2x^2 + 3y^2 + 4z^2 = 0.$$

2. Tangent planes at the extremities of any diameter are parallel.

3. In any central quadric the sum of the squares of three conjugate diameters is constant.

4. The locus of the intersection of three planes tangent to an ellipsoid which are mutually perpendicular, is the sphere

$$x^2 + y^2 + z^2 = a^2 + b^2 + c^2.$$

CHAPTER VIII.

LOCI OF HIGHER ORDERS.

264. Definitions.—*Higher plane curves* include all loci whose equations cannot be reduced to the first or second degree. In earlier works upon this subject they were divided into the two classes, *Algebraic* and *Transcendental;* which classification would be proper if bilinear coördinates only were used; but in Modern Geometry there are numerous systems of coördinates, and the same curve may be expressed algebraically in one system, or transcendentally in another. Thus, for example, the circle is expressed algebraically by the equation

$$x^2 + y^2 = R^2,$$

and transcendentally by the equation

$$\sin \varepsilon = a \cos [\sqrt{-1} (\log r + b)],$$

in which r denotes the radius vector of the curve, ε the angle between r and the tangent at the point, a and b constants. (See *Math. Monthly*, 1858, pp. 11 and 58.)

When curves are classed as transcendental, it is implied that their equations involve trigonometrical, circular, logarithmic, or exponential quantities. Algebraic curves are such as may be expressed by algebraic quantities. They are classed according to the degree of the equation; thus, an equation of the second degree represents a curve of the second order; of the third degree, the third order; and of the nth degree, the nth order.

The number of higher plane curves is unlimited. It is known that there are at least eighty species of lines of the third order, and more than 5000 species of the fourth order. Only a few of the higher curves, and those most noted, will here be considered.

Of Spirals.

265. A Spiral is a curve which may be generated by a point moving uniformly around a fixed point and whose distance from the fixed point varies according to an assigned law.

The fixed point O is called *the pole* of the spiral.

A spire, or whorl, is a portion of the spiral generated during one revolution of the generating point. Thus $LMNE$ is one-half of a spire.

The measuring circle is one described with a radius unity, having the pole as the centre. Thus, if OB be a unit radius, then will the circle BCD, etc., be the measuring circle.

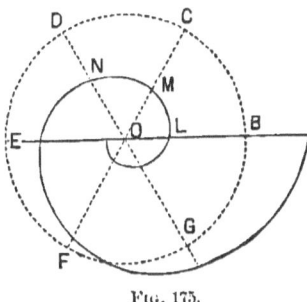

Fig. 175.

The radius vector is the distance from the pole to any point of the curve, as OL, OM, etc.

266. The Spiral of Archimedes is a curve which may be generated by a point moving *uniformly* along the radius vector while that radius has an *uniform* rotary motion.

To construct it, divide the measuring circle BCD, etc., into equal parts, $BC = CD = DE$, etc., and draw radial lines OB, OC, etc., and let OX be the polar axis (or initial line). Assume OJ (or find it from given conditions), and make $OK = 2OJ$, $OL = 3OJ$, etc., then will J, K, L, etc., be points in the spiral.

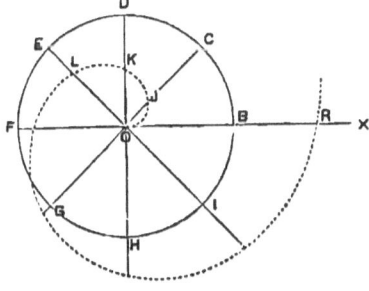

Fig. 176.

To find the equation to the locus, let θ be the variable angle, measured by the arc BCD, etc., r the radius vector, a the ratio of the radius vector to the variable angle; then, according to the definition, we have

$$r = a\theta, \qquad (1)$$

which is the required equation.

If the radius vector OR at the end of the first spire be taken as unity, we have

$$1 = a \cdot 2\pi; \;\therefore\; a = \frac{1}{2\pi},$$

and equation (1) becomes

$$r = \frac{\theta}{2\pi}. \qquad (2)$$

If $\theta = 0$, $r = 0$, hence the curve passes through the pole. If θ be negative, r will be negative, and the locus will be the same as for θ positive. r increases directly with θ from zero to infinity.

For one whorl $\qquad r_1 = 1$,
for two whorls $\qquad r_2 = 2$;
$\qquad\qquad \therefore r_2 - r_1 = 1 = r_1$;

and generally, the radial distance between any two consecutive whorls is constant, and equals the radius at the end of the first whorl.

EXAMPLE.—A string is wound spirally around a cone, extending from the apex to the base, dividing the slant height into n equal parts, the radius of the base being R, and the slant height l. The cone is placed on a plane and rolled in such a way as to unwind the string, the string remaining on the plane as it is unwound; required the equation of the curve of the string.

$$Ans. \; r = \frac{l^2}{2\pi n R}\theta.$$

267. The Reciprocal or Hyperbolic Spiral is a curve in which the radius vector varies inversely as the measuring arc.

Its equation is

$$r = \frac{a}{\theta}. \qquad (1)$$

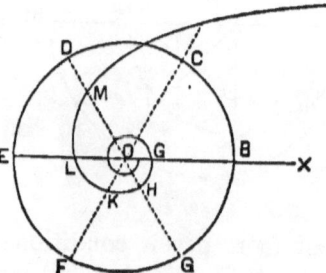

Fig. 177.

If the radius at the end of one whorl is unity, we have

$$1 = \frac{a}{2\pi}; \therefore a = 2\pi,$$

and the equation becomes

$$r = \frac{2\pi}{\theta}. \quad (2)$$

If $\theta = 0, r = \infty$. As θ increases, r diminishes, and when $\theta = 2\pi$, $r = 1$, so that the radius vector passes from 1 to ∞ in one whorl. If $r = 0, \theta = \infty$, hence there are an infinite number of whorls between the pole and the measuring circle.

The two preceding spirals are special cases of the curve

$$r = a\theta^n,$$

in which n may be $+1$ or -1.

268. The Logarithmic or Equiangular Spiral is defined by the equation

$$a\theta = \log r, \text{ or } r = e^{a\theta};$$

from which it follows that the logarithm of the radius vector is proportional to the measuring arc.

If $r = 1 = OB, \theta = 0$; hence if $r = 1$ be initial, the measuring arc will begin at B on the polar axis.

If $r = 0, \theta = -\infty$; hence, between the pole and $r = 1$, there will be an infinite number of whorls.

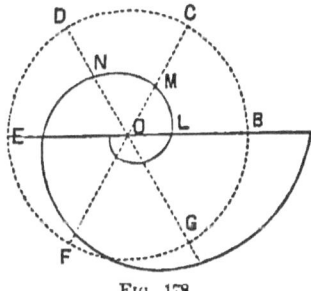

Fig. 178.

The curve may also cross the polar axes an infinite number of times for values of r greater than unity; for r will be real for values of $\theta = 2\pi, 4\pi, 6\pi$, etc.

To construct the curve make the measuring arcs $BC = CD = DE$, etc., and lay off OL, OM, ON, etc., in geometrical progression; that is, $\dfrac{OM}{OL} = \dfrac{ON}{OM}$, etc.

This curve is called equiangular because the tangent at

any point makes a constant angle with the radius vector at that point.*

269. **The Involute of a Circle is the locus of any point of a line as it rolls upon a circle.**

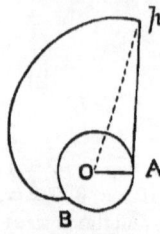

Fig. 179.

Thus, let p be any point of the line Ap, then if Ap be rolled upon the curve, P will describe the arc pB, and if it continues to roll in that direction, it will describe a second branch of the curve. The arc pB is the involute of the circle whose centre is O and radius OA. The same curve may be described by the end of a string as it is unwound from the circle. In a similar manner the involute of any other curve may be described.

To find an equation to the involute of the circle, let $r = OA$, $\rho = Op = $ the radius vector, $Ap = \text{arc } AB = r\theta$; then since Ap is tangent to the circle, the right triangle pAO gives

$$\rho^2 = r^2 + (r\theta)^2 = r^2(1 + \theta^2),$$

which is the required equation in one form; but the system to which it is referred is peculiar. To find the equation referred to rectangular coördinates, take the origin at the centre O, the axis of x passing through the point B and positive from O towards B, and θ positive from B. Then we may find from the figure that

$$x = r \cos \theta + r\theta \sin \theta,$$
$$y = r \sin \theta - r\theta \cos \theta;$$

---

* [Differentiating the equation to the curve, gives

$$\frac{dr}{d\theta} = ae^{a\theta} = ar;$$

$$\therefore \frac{dr}{rd\theta} = a = a \text{ constant}.$$

But the first member is the tangent of the angle between the normal and the radius vector; hence *the curve cuts the radius vector at a constant angle.*

EXAMPLE.—An equiangular spiral whose equation is $a\theta = \log r$, rolls on a straight line; required the locus of the pole.

Ans. A right line cutting the given line at an angle whose tangent is a.]

and if θ could be eliminated from these three equations we should find the rectangular equation to the locus; but if eliminated the result would contain circular functions, and, therefore, the equation would be transcendental.

(*Rem.*—The involute is often used in the construction of teeth in gearing. The length of the involute of the circle is given by the expression $Bp = \frac{1}{2}r\theta^2$; and the area $ABp = \frac{1}{6}r^2\theta^3$.)

270. The lituus is a curve defined by the equation

$$\rho^2\theta = a^2.$$

If $\rho = 0$, $\theta = \infty$;

if $\theta = 0$, $\rho = \infty$.

The equation may be written

$$\rho = \pm \frac{a}{\sqrt{\theta}};$$

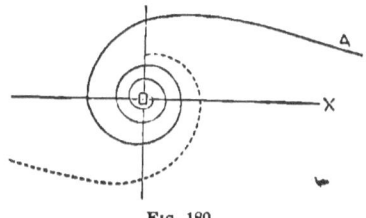

Fig. 180.

hence there are two equal values of ρ corresponding to each value of θ, one of which is positive and the other negative.

If $\rho = 1$ when $\theta = 2\pi$, then

$$1 = \frac{a}{\sqrt{2\pi}}; \therefore a = \sqrt{2\pi},$$

and the equation becomes

$$\rho^2\theta = 2\pi.$$

It is found by higher analysis that the initial line OX is an asymptote to the curve.

271. Parabolic Spiral.—If the axis of a parabola whose equation is $y^2 = 2px$, be wrapped around the circumference of a circle, and the corresponding ordinates of the parabola be laid off on the radius prolonged, the locus will be a parabolic spiral. Let $\rho = CB$, $r = CA$ =the radius of the circle, then will the ordinate y be

$$y = \rho - r,$$

and the corresponding abscissa,

$$x = r\theta,$$

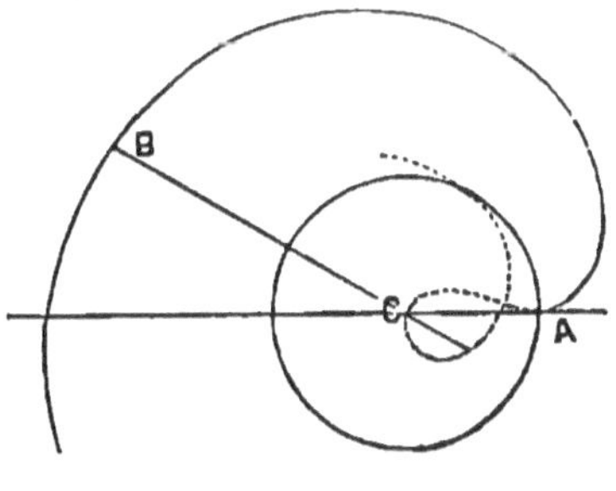

Fig. 181.

and these in the equation of the parabola give

$$(\rho - r)^2 = 2rp\theta,$$

which is the required equation. Let $\rho = nr$, then the equation becomes

$$(n-1)^2 r = 2p\theta.$$

If $n = 0, \rho = 0$, and $\theta = \dfrac{r}{2p}$, which being real shows that the locus passes through the centre of the circle. As n increases θ decreases until $n = 1$, for which value $\theta = 0$, and $\rho = r$. This is the initial point of the curve, from which point θ increases as ρ increases. If n (or r) be negative, θ increases, and when $\rho = -r$ we have $n = -1$, and $\theta = \dfrac{2r}{p}$, at which point the locus again crosses the circumference, and θ will increase as r increases negatively.

Trigonometrical Curves.

272. Trigonometrical curves are such as involve a trigonometrical expression in the equations of the curve.

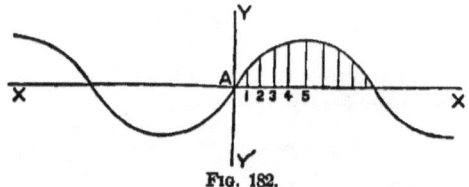

Fig. 182.

273. The Sinusoid is a curve whose equation is of the form

$$y = a \sin \frac{x}{b}, \qquad (1)$$

in which a and b are constants. If a and b are each unity, the equation becomes

$$y = \sin x, \qquad (2)$$

which is called the equation of sines, and the corresponding curve is called *the curve of sines*.

To construct the sinusoid, let

$$x = \tfrac{1}{10} b\pi = A1, \text{ then } y = a \sin \tfrac{1}{10} \pi;$$
$$x = \tfrac{2}{10} b\pi = A2, \qquad y = a \sin \tfrac{2}{10} \pi;$$
$$x = \tfrac{3}{10} b\pi = A3, \qquad y = a \sin \tfrac{3}{10} \pi;$$
$$\text{etc.} \qquad\qquad \text{etc.}$$

Erecting ordinates at the points 1, 2, 3, etc., equal to the corresponding values of y, gives points in the curve.

274. Remark.—The sinusoid is used in Physics to express certain laws of motion. It expresses the law of movement of the vibration of perfectly elastic solids ; of the vibratory movement of a particle acted upon by a force which varies directly as the distance from the origin ; approximately, the vibratory movement of a pendulum ; and exactly the law of vibration of the so-called mathematical pendulum.

275. The curve of Tangents is expressed by the equation

$$y = \tan x.$$

If $x = \tfrac{1}{2}\pi$, $y = \infty$,

$x = 0$, π, 2π, etc., $y = 0$.

The curve begins at the origin A. The ordinate DC, whose abscissa is $\tfrac{1}{2}\pi$, is an asymptote to the curve. EF is another asymptote, etc. The curve has an infinite number of branches.

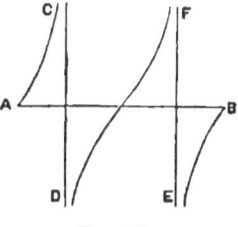

Fig. 183.

276. The curve of Secants is expressed by the equation

$$y = \sec x.$$

If $x = 0$, $y = 1 = A1$.

If $x = \tfrac{1}{2}\pi$, $y = \infty$, and DC is an asymptote. Similarly, FE, at a distance $\tfrac{3}{2}\pi$ from the origin is an asymptote. If $x = \pi$, then $y = -1$; if $x = 2\pi$, then $y = 1 = B1'$. This curve also has an infinite number of branches.

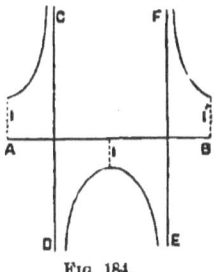

Fig. 184.

277. In a similar manner curves may be found for each of the remaining five trigonometrical functions. Circular functions give corresponding curves.

Logarithmic Curves.

278. A Logarithmic Curve is one in which the abscissa is the logarithm of the ordinate, or the ordinate is the logarithm of the abscissa. Its equation is

$$x = \log y.$$

If a be the base of the system of logarithms, its equation may be written

$$y = a^x.$$

If $x=0$, we find $y=1 = AB$;
$x=1=A1,$ " " $y=a^1=1b$;
$x=2=A2,$ " " $y=a^2=2c$;
etc.   etc.
$x=-1=-1A,$ " " $y=a^{-1}=-1d$;
$x=-\infty,$ " " $y=0.$

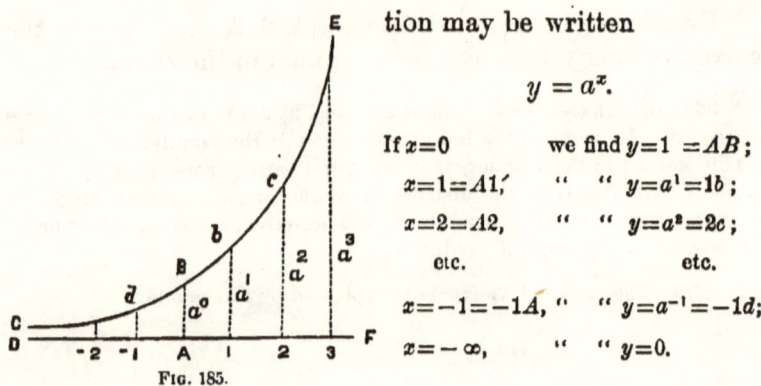

Fig. 185.

The ordinate at the origin will always be unity of the scale on which the locus is constructed, but the other ordinates will depend upon the value of the base of the system. The curve extends indefinitely in both directions from the origin, and the axis of x is an asymptote to the curve on the negative side of the origin.

If the base is unity, the locus will be parallel to the axis of x, and the ordinates cannot express a *series* of numbers; hence the base of a system of logarithms cannot be unity.

Of Parabolas.

279. All curves expressed by the equation

$$y = mx^n$$

are called *parabolas*. The values of n and m may be fractional or entire, and positive or negative; but we shall here consider m as positive.

If $x = 0$, $y = 0$, hence all these curves pass through the origin.

280. The Common Parabola.—Making $n = \frac{1}{2}$, we have

$$y^2 = m^2 x,$$

which is the equation of the common parabola, m^2 being the parameter, (Art. 86).

281. The Cubical Parabola.—Making $n = 3$, we have

$$y = mx^3,$$

which is called the equation to the *cubical parabola*. It extends indefinitely to the right of the origin and above the axis of x, and to the left of the origin and below the axis of x; and is convex to that axis.

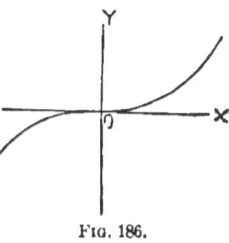
Fig. 186.

If $n = \tfrac{1}{3}$ we have

$$y^3 = m^3 x,$$

which is the equation of a cubical parabola convex to the axis of y.

282. The Semi-Cubical Parabola.—Making $n = \tfrac{3}{2}$, we have

$$y = mx^{\tfrac{3}{2}},$$

which is called the equation to the *semi-cubical parabola*. It extends indefinitely to the right of the origin and above and below the axis of x. It is convex to that axis, and symmetrical in reference to it.

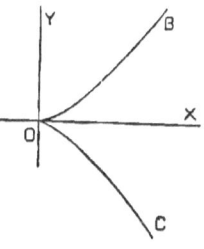
Fig. 187.

If $n = \tfrac{2}{3}$, we have

$$y^{\tfrac{3}{2}} = m^{\tfrac{3}{2}} x,$$

which is a semi-cubical parabola, convex to the axis of y.

283. The Biquadratic Parabola is given by the equation

$$y = mx^4, \text{ or } y^4 = m^4 x,$$

and is deduced from the general equation, (Art. 279), by making $n = 4$, or $n = \tfrac{1}{4}$. These curves, in their general shape, resemble the common parabola, the former being *concave* to the axis of y, and the latter concave to the axis of x.

Of Trochoids.

284. A Trochoid is the locus of a point in a circle rolling upon a line. The generating point may be within the circle, on its circumference, or entirely without; and the directrix may be a right line or a curve.

285. The Cycloid is a trochoid in which the path is described by a point in the circumference of a circle rolling on a straight line. Thus, if the circle CP rolls on the straight line AX, the point P will describe the arc of a cycloid $APBX$.

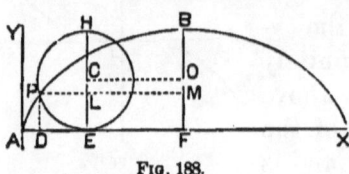

Fig. 188.

To find its equation, take the origin at A, C the centre of the circle, $r = CE$, P any point in the curve, and we have
$$x = AD, \quad y = PD = LE.$$

But $\qquad AE = \text{arc } PE = r \text{ vers}^{-1} \dfrac{y}{r}.$

Also $AD = AE - DE$, and $DE = PL = \sqrt{HL \cdot LE}$;

$$\therefore x = r \text{ vers}^{-1} \frac{y}{r} - DE,$$

$$= r \text{ vers}^{-1} \frac{y}{r} - \sqrt{2ry - y^2}, \qquad (1)$$

which is the required equation.

If $y = 0$, $x = 0$, $\pm 2\pi$, $\pm 4\pi$, etc.; hence the curve has an infinite number of branches above the axis of x.

If $y = 2r$, $x = \pi r = AF = \tfrac{1}{2} AX$.

If the origin be at B, $BM = y$, and $MP = x$, the equation becomes, by changing x to $\pi r - x$, and y to $2r - y$, (Art. 49),

$$x = r \text{ vers}^{-1} \frac{y}{r} + \sqrt{2ry - y^2}. \qquad (2)$$

Draw a radius PC, and let $PCE = \varphi$, = the angle described by the radius of the circle, while the point P describes the arc AP, then, the origin being at A, we find

$$\left. \begin{array}{c} AE = \text{arc } PE = r\varphi \\ x = r(\varphi - \sin \varphi) \\ y = r(1 - \cos \varphi) \end{array} \right\}, \qquad (3)$$

which are equations of the curve in terms of φ, and are often convenient in solving problems.

REMARKS.—The curve of quickest descent of a body from one point to another down a smooth surface, is a cycloid. The involute of a cycloid is an equal cycloid in another position. (This is proved by higher analysis.) The cycloidal pendulum, in which the pendulum describes the arc of a cycloid, is of historical interest, but is not considered of much practical value.

286. The Prolate Cycloid is the path described by a point *within* the circle rolling on a straight line.

Let P be the generating point at a distance from the centre $= PC = b$, O the origin,

$r = CR$, $\varphi = PCR$, $x = OD$,
$y = DP$;

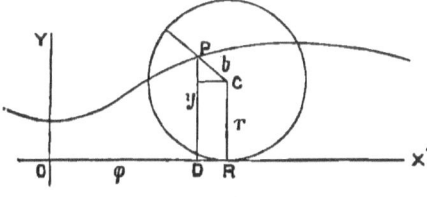

FIG. 189.

then
$$x = r\varphi - b \sin \varphi,$$
$$y = r - b \cos \varphi, \tag{4}$$

are the equations of the locus, in which $b < r$.

287. The Curtate Cycloid is the path described by a point *without* the rolling circle. The equations are the same as in the preceding Article, excepting that we must make $b > r$. If $b = r$, we have the cycloid.

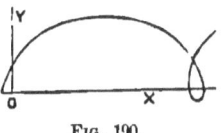

FIG. 190.

288. The Hypercycloid (or Epicycloid) is the path traced by a point in the circumference of the generating circle as it rolls on the convex side of another circle.

Let R be the radius of the directrix; r the radius of the generatrix, $\theta = COA$, $\varphi = BCP$, P the point in the curve AP.

Then arc $BA = $ arc BP, or
$$R\theta = r\varphi\,;\ \therefore\ \varphi = \frac{R}{r}\theta.$$

The inclination of CP to the axis of x is $\theta + \varphi$, or substituting the value of φ, it becomes $\dfrac{r+R}{r}\theta$. The

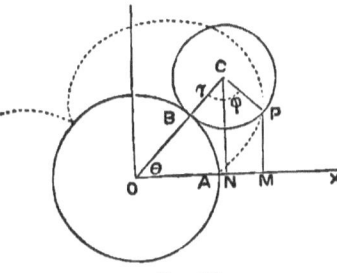

FIG. 191.

projection of CO on the axis of x will be $(R+r)\cos\theta$; and of CP, $r\cos\dfrac{r+R}{r}\theta$, hence the abscissa of P will be

$$x = (R + r) \cos\theta - r \cos\frac{R+r}{r}\theta,$$

and similarly, $\quad y = (R + r) \sin\theta - r \sin\frac{R+r}{r}\theta,$

which are the equations of the curve in terms of the angular movement of the line of centres. If R and r are incommensurable, the curve will have an infinite number of branches, but if they are commensurable, the curve will repeat itself. When they are commensurable θ may be eliminated, and a single algebraic equation found for the locus.

259. The Hypertrochoid is the path traced by *any* point on the radius of the generating circle as it rolls on the convex arc of a fixed circle.

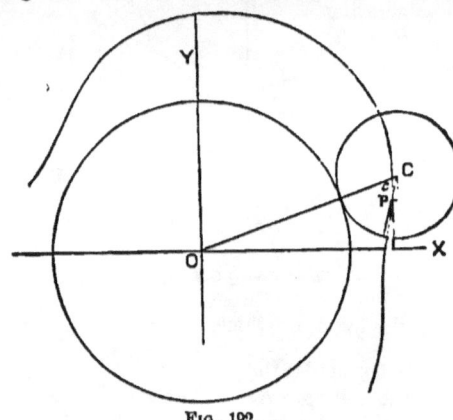

Fig. 192.

Let b be the distance of the generating point from the centre of the generating curve, and the other notation as in the preceding Article; then we find

$$\left.\begin{array}{l} x = (R + r) \cos\theta - b \cos\dfrac{R+r}{r}\theta, \\[6pt] y = (R + r) \sin\theta - b \sin\dfrac{R+r}{r}\theta, \end{array}\right\} \quad (6)$$

for the equations of the curve.

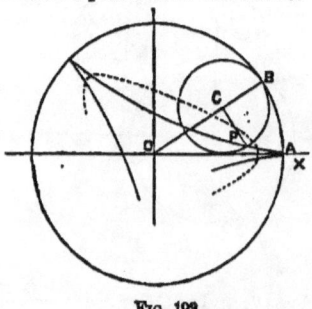

Fig. 193.

290. The Hypotrochoid is the path traced by *any point* in the radius of the generating circle rolling on the concave arc of another curve.

Let $R = OA =$ the radius of the generating circle, $r = CB =$ the radius of the moving circle, $b = CP =$ the distance of the generating point P from the centre C (P being on the dotted line), $\theta = BOA$, $\varphi = BCP =$ the angle through which the generating circle has revolved from the initial point; then will $\varphi =$

θ + supplement of the *inclination of CP* to the axis of x. Also

$$\text{arc } BA = \text{arc } BP \; ; \; \therefore \; \varphi = \frac{R}{r}\theta \; ;$$

$$\textit{inclination of } CP = \varphi - \theta = \frac{R-r}{r}\theta \; ;$$

and the equations to the curve become

$$\left.\begin{array}{l} x = (R-r)\cos\theta + b\cos\dfrac{R-r}{r}\theta , \\[1ex] y = (R-r)\sin\theta - b\sin\dfrac{R-r}{r}\theta . \end{array}\right\} \quad (7)$$

291. The Hypocycloid is the path traced by a point on the circumference of the generating circle rolling on the concave arc of the fixed circle.

Hence the equations of the curve are formed by making $b = r$ in equations (7). They are

$$\left.\begin{array}{l} x = (R-r)\cos\theta + r\cos\dfrac{R-r}{r}\theta , \\[1ex] y = (R-r)\sin\theta - r\sin\dfrac{R-r}{r}\theta . \end{array}\right\} \quad (8)$$

The curve is represented by the full line passing through A.

292. Four Cusped Hypocycloid.—Let $r = \frac{1}{4}R$, then will the curve consist of four branches, and form four cusps. Making $r = \frac{1}{4}R$ in equations (8), we have

$$x = 3r\cos\theta + r\cos 3\theta,$$

$$y = 3r\sin\theta - r\sin 3\theta. \quad (9)$$

From Trigonometry we have

$$\cos 3\theta = 4\cos^3\theta - 3\cos\theta,$$

$$\sin 3\theta = 3\sin\theta - 4\sin^3\theta,$$

which substituted give

$$x = R\cos^3\theta,$$

$$y = R\sin^3\theta \; ;$$

$$\therefore \; x^{\frac{2}{3}} = R^{\frac{2}{3}}\cos^2\theta \; ; \quad y^{\frac{2}{3}} = R^{\frac{2}{3}}\sin^2\theta.$$

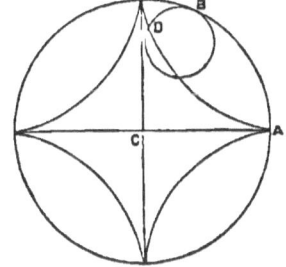

Fig. 194.

Adding, observing that $\sin^2\theta + \cos^2\theta = 1$, we have

$$x^{\frac{2}{3}} + y^{\frac{2}{3}} = R^{\frac{2}{3}}, \quad (10)$$

which is the equation to the curve.

REMARKS.—*Hyper* is from the Greek and means over or above ; and *Hypo*, under. The hypercycloid and hypocycloid are often used in the theoretical construction of the teeth of gear wheels. In the hypercycloid, if the radius of the fixed circle be infinite, the curve becomes a cycloid. If the radius of the generating circle be infinite, the hypercycloid becomes an involute. If the diameter of the generating circle equals the radius of the fixed circle, the hypocycloid becomes a straight line and will be the diameter of the fixed circle, and *any trochoid* will be an ellipse. If the radius of the generating circle equals the diameter of the fixed circle, and the generating circle be conceived to roll within the fixed one, the centre of the generating circle will describe the circumference of the fixed one, and any point on the radius will describe a curve called the Limaçon (see next Article), and any point on the circumference a *Cardioid*. If the diameter of the generating circle equals that of the fixed one, and rolls outside the fixed one, any point on the radius of the generating circle will describe a *Limaçon*, and any point on the circumference a *Cardioid*.

Several interesting problems involve the four cusp hypocycloid:

If the back of a chimney be vertical and the floor be horizontal, and the edge of the front piece be x feet from the back, and y feet from the floor; then the length of the longest rod that can be run squarely up the chimney will be the value of R in Equation (10), that is, $R = (x^{\frac{2}{3}} + y^{\frac{2}{3}})^{\frac{3}{2}}$.

If a smooth bar rests on a curve and against a vertical wall, the bar will be in equilibrium in all positions if the curve be a certain hypocycloid.

The length of the tangent of a four cusp hypocycloid limited by the axes is constant, and equals the radius of the directrix.

293. The Limaçon may be defined as in the preceding Remarks. But it was originally defined as follows: If a secant be drawn through a fixed point on a circle, and equal distances be laid off both ways on this secant from the other point where the secant cuts the circle, the locus is a Limaçon.

Taking the pole at the fixed point, r the radius of the circle, the initial line passing through the centre, and b the constant distance, then will the polar equation be

$$\rho = 2r \cos \theta \pm b,$$

and the rectangular equation

$$(x^2 + y^2)^2 - (4rx + b^2)(x^2 + y^2) + 4r^2x^2 = 0.$$

294. The Cardioid is a particular case of the Limaçon, in which $b = 2r$. The polar equation is

$$\rho = 2r (\cos \theta \pm 1),$$

and the rectangular equation

$$(x^2 + y^2)^2 - 4rx(x^2 + y^2) - 4r^2y^2 = 0.$$

The Conchoid of Nicomedes.

295. If a line OP be drawn through a fixed point O across a fixed line CX at F, and a constant distance FP be laid off on the line both ways from the point F, the locus of the point P is called the *Conchoid of Nicomedes*. The fixed point O is the *pole*, XCX the *directrix*, and BC the *parameter* of the curve.

296. To find the equation of the Conchoid, let P be any point, $BOP = \varphi$, $OC = a$, $CB = b = FP$, and $\rho = OP =$ the radius vector; then for the polar equation we have,

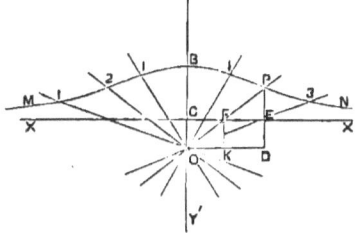

Fig. 195.

$$OP = OF + FP,$$

or $\quad\quad \rho = a \sec \varphi \pm b.\quad\quad (1)$

For the *rectangular* equation, let $x = OD$, and $y = DP$, then

$$\sec \varphi = \frac{\rho}{y}, \quad x^2 + y^2 = \rho^2,$$

which substituted in (1) give

$$(x^2 + y^2)(y - a)^2 = b^2 y^2. \quad\quad (2)$$

If the origin be transferred to C, we will have (writing $y + a$ for y)

$$x^2 y^2 = (a + y)^2 (b^2 - y^2), \quad\quad (3)$$

which is the required equation.

These equations give both branches of the curve. The branch nearest the pole is called the *inferior branch*, and the more remote portion the *superior branch*.

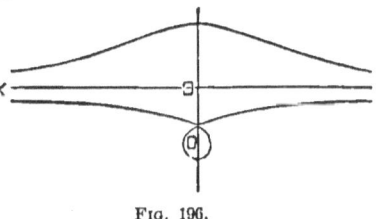

Fig. 196.

If $b > a$ there will be a loop inclosing the pole.

If $a = b$, there will be a cusp at the pole.

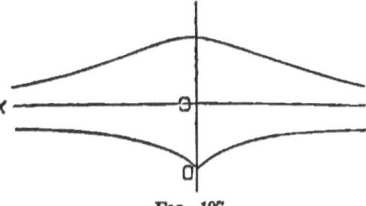

Fig. 197.

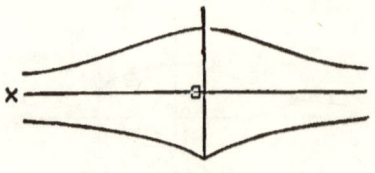

Fig. 198.

If $b > a$ the lower branch will be more or less rounded when it crosses the axis of y.

297. To Trisect an Angle.—Let EOC be the angle.

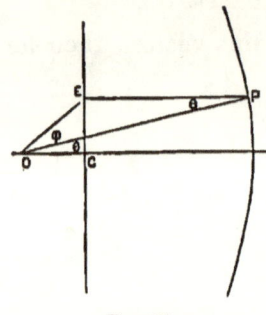

Fig. 199.

Take the directrix EC of the Conchoid perpendicular to OC at any point. Construct a conchoid having its pole at O, and parameter b equal to $2OE$. Draw EP parallel to OC, P being the point where it intersects the conchoid, and draw OP; then will $POC = \frac{1}{3}EOC$. For, if $\theta = EPO = POC$, then $EP = b \cos \theta = 2OE \cos \theta$; and the triangle EOP gives

$$\frac{EP}{EO} = \frac{\sin \varphi}{\sin \theta} = \frac{b \cos \theta}{\frac{1}{2}b} = 2 \cos \theta;$$

$$\therefore \sin \varphi = 2 \sin \theta \cos \theta = \sin 2\theta$$

or
$$\varphi = 2\theta;$$

$$\therefore \theta = \tfrac{1}{2}\varphi = \tfrac{1}{3}EOC,$$

which was to be proved. Bisecting EOP by well-known methods, the angle O becomes trisected.

298. Remark.—Among the noted problems of the ancient mathematicians were the Trisection of an Angle and the Duplication of the Cube. The geometrical construction of these problems is beyond the reach of elementary geometry, since it involves curves of a higher order than the circle. Both these problems involve the solution of a cubic equation, and both may be made to depend upon the construction of two mean proportionals between two given straight lines. This has been accomplished in various ways by means of higher plane curves, many of which were invented for this purpose. The Conchoid was invented by one Nicomedes, a Greek mathematician, for trisecting an angle. The Cissoid of Diocles was invented by the Greek geometer Diocles, for the purpose of solving these problems.

The Cissoid.

299. Let O be the centre of a circle, AB a diameter. Erect a pair of ordinates DE, $D'E'$, equidistant from the centre, and from A draw a secant AE through the extremity E of one ordinate; its intersection P' with the other ordinate determines a point in the required locus. Similarly the intersection of AE' and DE at P determines another point. The locus thus constructed is called the *Cissoid of Diocles*.

300. Rectangular Equation of the Cissoid.—Let P' be any point of the curve, $AD' = x$, $D'P' = y$, $AO = r$; then

$$AD : DE :: AD' : D'P',$$

or $\quad 2r - x : \sqrt{(2r-x)x} :: x : y;$

$$\therefore y^2 = \frac{x^3}{2r - x}, \qquad (1)$$

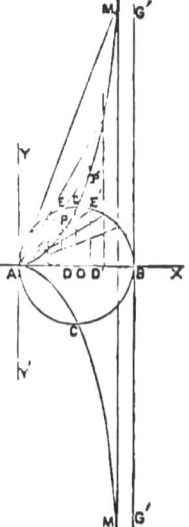

Fig. 200.

which is the required equation.

301. Polar Equation to the Cissoid.—Take the origin at A, $AP' = \rho$, $P'AX = \theta$; then

$$y = \rho \sin \theta, \quad x = \rho \cos \theta,$$

which substituted in equation (1) will give

$$\rho = 2r \sin \theta \tan \theta, \qquad (2)$$

for the required equation. The curve has two infinite branches and is symmetrical in reference to the axis of x.

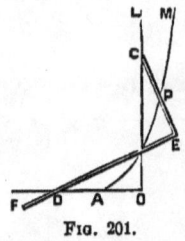

Fig. 201.

[Sir Isaac Newton gave the following mechanical construction for this curve. Let O be the centre of the circle, $AO = AD$ its radius, and OC a line perpendicular to OA. Take a rectangular ruler DEC whose leg EC equals OD. Let the end C slide along OL while EF constantly passes through the point D, then will P the middle point of EC describe a cissoid.

The locus of the vertex of a common parabola rolling upon an equal parabola is a cissoid.]

302. *To insert two mean proportionals between two given lines.*

Let a and b be the given quantities. With $a = AC$ as a

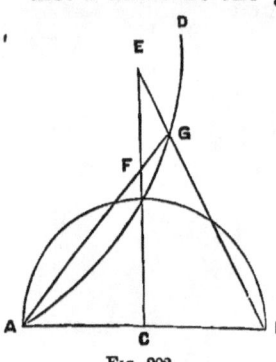

Fig. 202.

radius describe a semi-circumference, and construct the corresponding cissoid AGD. At the centre of the circle, erect an ordinate $CE = b$ and join E and B, noting the point G where it crosses the cissoid. Draw GA and note the point F where it crosses CE; then will CF be one of the mean proportionals.

Let fall a perpendicular GH upon the diameter AB (not shown in the figure), then will $x = AH$, and $y = GH$, and the similar triangles ACF and AHG, and BHG and BCE, give

$$\frac{a}{x} = \frac{CF}{y}, \text{ and } \frac{2a-x}{a} = \frac{y}{b},$$

which combined with the equation of the curve will give

$$CF = \sqrt[3]{a^2 b}.$$

By exchanging the quantities a and b in the construction, we would find the value of $\sqrt[3]{ab^2}$, and the required proportion will be

$$a : \sqrt[3]{a^2 b} :: \sqrt[3]{ab^2} : b.$$

303. Duplication of the Cube.—*To find the edge of a cube whose volume shall be double that of a given cube.* Let a be the length of one edge of the given cube, and making $b = 2a$ in the preceding Article, we have

$$CF = a \sqrt[3]{2},$$

which is the required length.

To find a cube whose volume is n times that of a given cube, make $b = na$, and we find

$$CF = a\sqrt[3]{n}$$

for the length of one edge.

Quadratrix of Dinostratus.

304. If an ordinate DP moves uniformly along the diameter of the circle AB, while the radius rotates uniformly from B to A, both beginning at the same time at B, their intersection P will be the locus of the *Quadratrix*.

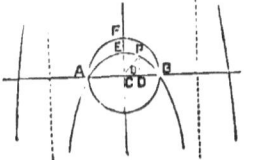

Fig. 203.

305. Equation of the Quadratrix.—Let $r = CB =$ the radius, $\theta = PCB$, $x = CD$, $y = DP$. Then from the definition we have

$$\frac{\frac{1}{2}\pi}{r} = \frac{\theta}{r-x}, \quad y = x \tan \theta;$$

$$\therefore y = x \tan (r-x)\frac{\pi}{2r},$$

which is the equation of the curve. If $x = 0$, $y = 0 \times \infty$; which, by the principles of vanishing fractions, is found to be $\dfrac{2r}{\pi}$; hence

$$\frac{2r}{CE} = \pi.$$

[REMARK.—This curve was invented by Dinostratus for the purpose of finding the area of the circle (hence the name *Quadratrix*), and also for dividing an angle into any number of equal parts. Thus, to trisect the angle PCD, trisect DB and erect ordinates at the points of division; then will the radial lines from C to the points of intersection of the ordinates with the quadratrix trisect the angle.

If the law of construction for BEA be continued outside the circle the curve will become an asymptote to the dotted lines, and another branch, shown by the full lines outside the dotted ones, may be described, and so on indefinitely.]

Witch of Agnisi.

306. To construct the Witch of Agnisi, let AB be the diameter of a circle perpendicular to AX, draw a tangent to the circle at B (not shown in the figure); it will be parallel to AX, and through any point F of the circle draw a secant AF and prolong it to meet the tangent through B; project the intersection thus found on the ordinate EF prolonged, the point P thus found will be a point of the locus. To find its equation let $x = AD$, $y = DP$, then we will find

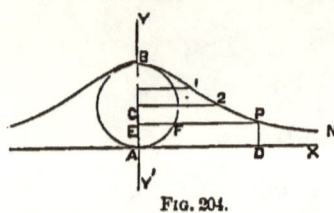

Fig. 204.

$$x^2 y = 4r^2(2r - y).$$

Ovals.

307. An Oval is a reëntrant curve in which the distances of any point from two fixed points have a constant relation. The fixed points are called *the foci*. The term is also applied to figures made with arcs of circles, which resemble the ellipse in form.

308. The Cartesian Oval is one in which a fixed multiple of one radius vector of any point differs from the other by a constant quantity.

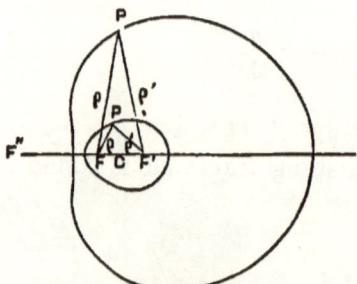

Fig. 205.

To find its equation, let P be any point in the locus, F and F'' the foci, ρ and ρ' the radii vectors, k the fixed multiple, and d the constant difference; then, from the definition, we have

$$\rho - k\rho' = \pm d. \qquad (1)$$

Let the distance between the foci be $FF' = c$; take the pole at F and the variable angle $PFF'' = \varphi$, then from the figure we have

$$\rho'^2 = \rho^2 + c^2 - 2c\rho \cos \varphi. \qquad (2)$$

Eliminating ρ' gives
$$(k^2 - 1)\rho^2 - 2(ck^2 \cos\varphi \mp d)\rho + c^2k^2 - d^2 = 0, \qquad (3)$$
which is of the form
$$\rho^2 + 2(A + b\cos\varphi)\rho + C = 0, \qquad (4)$$
and is the required equation. Equation (3) shows that there are two ovals answering the required condition, as shown in the figure.

REMARKS.—This oval was first investigated by Descartes, hence its name. It has been shown that this curve has a third focus F' outside of both the ovals. (See *Reprint* of Solutions from the *Educational Times*, Vol. XXV., p. 68.) If $k = -1$ and d is positive, the locus becomes a single curve and is an ellipse.

If $k = 1$ and d, positive, it is an hyperbola.

If $d = ck$, we have, (Eqs. (3) and (4)),
$$\rho + 2(A + B\cos\varphi) = 0,$$
which is the equation of the limaçon, (Art. 293).

If $k = 1$ and $d = c$ it becomes the equation of the cardioid, Art. 294.

[*Mechanical Construction.*—The following mode of constructing the Cartesian is given by Prof. Hammond, of Bath, England. Wind a string around two concentric wheels and let it pass around smooth pins A and B, and be joined at P. A pencil point at P will trace a Cartesian when the wheels C and D are turned on their common axis. To prove this, differentiate equation (1), and thus find $d\rho = kd\rho'$, which gives the relation between the rates of change of the radii. But from the figure we see that the rate of increase of BP is to the rate of decrease of AP, as the diameter of D, is to the diameter of C. The last ratio being constant, may be represented by $-k$; which shows that the locus is a Cartesian, the foci being at A and B. If the circles C and D, are equal, or $k = -1$, the locus is an ellipse. If the circles are of the same size, or both threads are wound the same way about D, in which case $k = 1$, the locus will be an hyperbola. (*Am. Jour. Math.* 1878, p. 283.)]

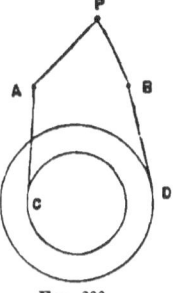

Fig. 206.

PROB.—Prove that the locus of the triple foci of a series of Cartesian Ovals passing through five points is an equilateral hyperbola. (Reprint *Ed. Times*, Lond., Vol. XVII., p. 24, 1877.)

309. **The Cassian Oval** is the locus of a point the product of whose distances from two fixed points is constant.

Let P be any point, F and F' the fixed points called *foci*, O the middle point of FF', $x = OD, y = DP, OF = c$, m^2 the constant product of the radii.

We have from the definition $\rho\rho' = m^2$, and from the figure

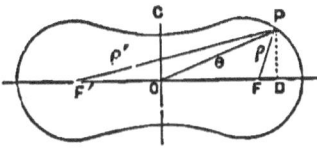

Fig. 207.

$$\rho^2 = (x-c)^2 + y^2 \; ; \; \rho'^2 = (x+c)^2 + y^2.$$

These equations give

$$(x^2 + y^2 + c^2)^2 - 4c^2x^2 = m^4, \qquad (5)$$

which is the rectangular equation to the curve.

Changing to *polar coördinates*, O being the pole, $r = OP$ the radius vector, we have

$$r^4 + 2c^2(1 - 2\cos^2\theta)r^2 = m^4 - c^4, \qquad (6)$$

which is the required equation.

310. Lemniscata of Bernoulli.—This is a special case of the Cassian in which $m = c$, and hence the rectangular equation is

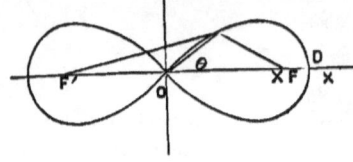

Fig. 208.

$$(x^2 + y^2)^2 - 2c^2(x^2 - y^2) = 0. \qquad (7)$$

The polar equation is

$$r^2 = 2c^2\cos 2\theta. \qquad (8)$$

If $\theta = 0$, we have $r = c\sqrt{2}$, which call a, and the equation will become

$$r^2 = a^2\cos 2\theta, \qquad (9)$$

which is the more usual form. The curve crosses itself at the origin.

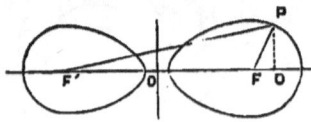

Fig. 209.

If $m < c$ the Cassian does not cut the axis of y, and the locus divides itself into two distinct ovals.

311. A Catenary is the curve assumed by a perfectly flexible chain when suspended at its ends. If the origin be at the centre of the curve, x horizontal and y vertical, the equation is

Fig. 210.

$$y = \frac{1}{4a}(e^{ax} - e^{-ax})^2,$$

in which e is the base of the Napierian system of logarithms and a the ratio of the weight per unit of length to twice the tension at the lowest point. (See the Author's *Analyt. Mech.*, p. 134.)

[The following are some of the properties of the catenary:

The directrix is a line parallel to the axis of x, and below the vertex a distance equal to $\frac{1}{2a}$.

If a common parabola be rolled on a fixed line, the locus of the focus will be a catenary;—also the envelope of the directrix will be a catenary symmetrical with the former in reference to the fixed line. (Reprint, *Ed. Times*, Vol. XXV., p. 93.)

The radius of curvature at any point of the catenary equals (in length) the normal limited by the directrix.

The tension at any point equals the weight of the chain whose length is the ordinate of the point from the directrix.

If an indefinite number of strings (without weight) be suspended from the catenary and terminated by a horizontal line, and the catenary be then drawn out to a horizontal line, the locus of the lower ends of the strings will be a parabola.

The centre of gravity of the catenary is lower than for any other curve of the same length terminated by the fixed points A and B.]

Miscellaneous.

312. Curves of Pursuit.—If a point moves along any path and another point is made to move directly towards it according to any law, the path of the latter is called *a curve of pursuit*.

PROBLEM.—A fox runs uniformly along the straight line AX, and when the fox is at A a dog starts at C and runs at an uniform rate towards the fox ; required the equation of the path described by the dog, and the distance run by each.

At any instant let the dog be at P and the fox at A', then will PA' be a tangent to the curve. This is a curve in which the length of arc CP bears a constant ratio to the distance AA'.

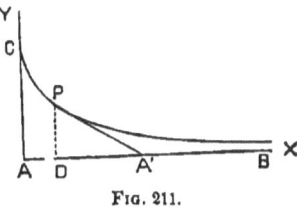

Fig. 211.

Let $AA' + CP = n$, and $CA = a$, $x = AD$, $y = DP$, then it may be found that (*Simpson's Fluxions*),

$$2x = \frac{y^{1+n}}{a^n(1+n)} - \frac{a^n y^{1-n}}{1-n} + 2\frac{an}{1-n^2},$$

which is the rectangular equation to the curve. When the dog overtakes the fox, we have

$$y = 0 \; ; \; \therefore \; x = \frac{an}{1-n^2},$$

which will be the distance run by the fox ; hence the distance run by the dog will be $\dfrac{a}{1-n^2}$.

There are numerous *curves of pursuit* depending upon the laws to which the moving bodies are subjected, and the path described by the leading body.

313. The Folium of Descartes is expressed by the equation

$$x^3 + y^3 - 3axy = 0.$$

It has an asymptote whose equation is

$$x + y + a = 0.$$

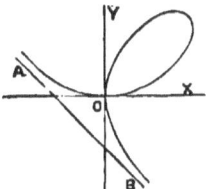

Fig. 212.

314. Discontinuous Curves.—The equations of some curves give real values for certain values of one of the variables and imaginary values for other values, and when the imaginary parts fall between real portions, the locus is called *discontinuous*.

315. The locus whose equation is

$$y = ax^2 \pm \sqrt{x \sin bx}$$

is an example of a discontinuous curve, in which one portion of the locus is represented by points only. Thus, for negative values of x, the radical is imaginary for all values of x except when $\sin(-bx)$ is zero. When that is zero y is real and gives the isolated points A', B', C', etc., all of which are located on the curve whose equation is $y = ax^2$. All positive values of x, (bx being less than a multiple of $\tfrac{1}{2}\pi$) give real values for y, and the locus will be a series of ovals symmetrical in reference to the positive branch of the parabola $y = ax^2$.

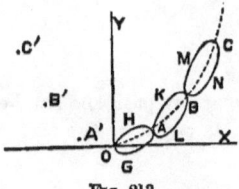

Fig. 213.

316. The locus $y = \sqrt{2 \sin x - 1} \cdot \sqrt{\cos x - 1}$, is another example. If $x > 0$ and $< 30°$ both radicals are imaginary and hence y will be real. For $x > 30°$ and $< 150°$, the first is real and the second imaginary, and hence y, imaginary; and so on throughout the circumference and multiples of the circumference.

317. If the locus be referred to polar coördinates, the same condition may exist, as may be seen from the equation

$$\rho = \sqrt{2 - 3 \sin 4\varphi}.$$

318. A Loxodromic Curve is one that cuts the meridians of a sphere at a constant angle. It is found, by higher analysis, that the equation of a loxodromic of 45° is

$$x = \log \tan (45° + \tfrac{1}{2}y),$$

in which y is the latitude and x the longitude, the origin being on the equator.

[REMARK.—If a ship should start at the equator and sail continually north-east at a finite rate, it would reach the north pole in a finite time. It would go around the pole an infinite number of times, but the length of the path would be finite. When it passed 360 degrees of longitude, its latitude would be 89° 53', or it would be within about 8 miles of the pole.]

319. The Logocyclic Curve is one whose polar equation is

$$\rho = a \sec \theta \,(1 \pm \sin \theta),$$

in which a is the value of ρ for $\theta = 0$. The rectangular equation is

$$(x^2 + y^2)(x - 2a) + a^2 x = 0.$$

The locus of the foci of all elliptic sections whose planes pass through a tangent to a circular cylinder parallel to the base is a logocyclic curve.

320. An Helix is a curve which cuts the rectilinear elements of a cylinder at a constant angle.

To find its equations, let the axis of z coincide with the axis of the cylinder, x and y, horizontal, θ, the variable angle measured from the axis of x, r, the radius of the base, and φ the angle between the helix and the rectilinear elements; then we find

$$x = r \cos \theta, \quad y = r \sin \theta, \quad z = r\,\theta \cot \varphi,$$

which are the required equations. If c be the slope of the helix; that is, the tangent of the angle which the helix at a unit's distance from the axis makes with the base, then

$$z = c\,\theta,$$

and the values of x and y remain the same.

321. A Conoid is a surface generated by a right line remaining constantly parallel to a plane and moving on two other lines one or both of which is curved. The plane is called a *plane-directer*. It is a warped surface.

322. Problem.—*To find the equation to a conoid in which one directrix is an ellipse and the other a right line, the right line being parallel to the major axis of the ellipse and perpendicular to the plane-directer.*

Let the axis of x coincide with the major axis of the ellipse, and the plane yz be the plane-directer passing through the principal vertex of the ellipse; BE the generating line, P, any point in the surface, $h = CB$, the altitude; $x = AC$, $y = CD$, and $z = DP$. The equation to the ellipse gives, (Art. 72),

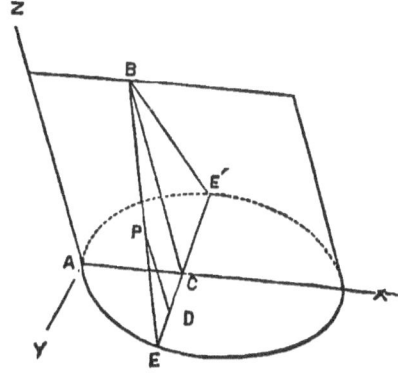

Fig. 214.

$$EC^2 = \frac{b^2}{a^2}(2ax - x^2),$$

and the similar triangles EDP and ECB give

$$ED : EC :: DP : CB,$$

or $$EC - y : EC :: z : h;$$

hence $$h^2 y^2 = (h - z)^2 EC^2;$$

$$\therefore a^2 h^2 y^2 = b^2 (h - z)^2 (2ax - x^2),$$

which is the required equation.

EXAMPLES.

1. Find the curve of intersection of a plane with the conoid.

2. Show that a plane section parallel to the curved directrix is a curve of the same class.

PART II.

QUATERNIONS.

$$\left(\sqrt{-1}\right)^{\frac{2\theta}{\pi}}$$

QUATERNIONS.

CHAPTER I.

ADDITION AND SUBTRACTION OF VECTORS.

Definitions.

323. Quaternions is a system of analytical geometry, invented by Sir William Rowan Hamilton about the year 1843. He gave to the system the above name because it involves, in its fundamental expressions, four arbitrary units (from the Latin word *quaternio*, meaning *a set of four*.)*

324. A Vector implies the transferrence of a point a given distance in a given direction. Thus, if a point be transferred from A to B, the length and direction of AB being known, any quantity which will represent this action is a vector. Therefore, in this system, *a vector is the representative of transferrence through a given distance in a given direction.* Geometrically it is represented by a right line whose direction is parallel to the transferrence, and whose length equals the distance through which the point has been carried. Analytically it is represented by some letter of the Greek alphabet, α, β, γ, etc.

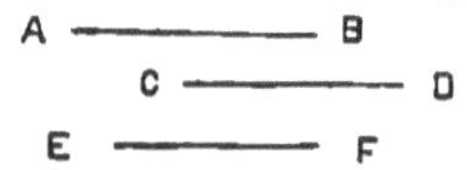

Fig. 215.

325. The Sign of a Vector.—If the transferrence be considered as positive in one direction, a transferrence in the opposite direction will be negative. Either direction may be assumed, arbitrarily, as positive. Thus, if AB be posi-

---

* Hamilton's Lectures, Preface, pp. (40), (62); also pp. 89, 109, 112, 128, 449.

tive BA will be negative, the letters being arranged in the order of the transferrence, AB signifying a transferrence *from A towards B.* This principle will be observed in this system.

326. Equal Vectors are such as are parallel and equal in length. Thus if AB, CD, and EF are parallel and equal in length, we write

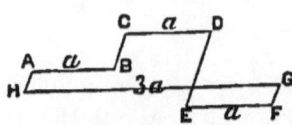

Fig. 216.

$$AB = CD = EF.$$

Equal vectors are added the same as similar quantities in algebra; hence we have, for this case,
$$AB + CD + EF = 3AB.$$

If HG is parallel to AB and equal in length to $3AB$, we have
$$AB + CD + EF = HG = 3AB,$$
or $\qquad AB + CD + EF + GH = 0.$

327. Parallel Vectors are Multiples of each other. This follows directly from the preceding Article. Since they are parallel, we have, in comparing one vector with another, only to compare their lengths. If α be a vector, then will $n\alpha$ be a parallel vector n times as long. In the triangle CDE, if DE is parallel to AB and n times as long, and if vector AB be α, then will vector DE be $n\alpha$.

Generally, we have

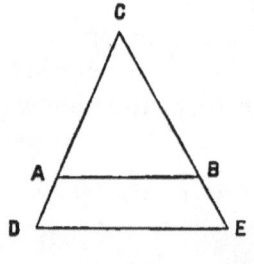

Fig. 217.

$$\alpha + l\alpha + m\alpha + \text{etc.} = (1 + l + m + \text{etc.})\,\alpha.$$

in which l, m, etc., may be positive or negative, entire or fractional.

Similarly, in Fig. 216, if $HA = \beta$, $BC = l\beta$, $DE = -n\beta$, etc., we have

$$\beta + l\beta - n\beta + \text{etc.} = (1 + l - n + \text{etc.})\beta.$$

Vectors not parallel must not be represented by the same letter.

328. A Unit Vector *is one whose length is unity*, its direction being given or assumed. The length of the unit will be the same for all the vectors in any particular problem. Unit vectors are generally represented by the same Greek letters as the entire vector, in which case they are especially designated as such; thus, let α, β, γ, etc., be unit vectors, but we will, at present, distinguish them by subscripts, thus, α_1, β_1, γ_1, etc., are unit vectors, of which α, β, γ, etc., are the entire corresponding vectors, and we may have

$$\alpha = l\alpha_1, \ \beta = x\beta_1, \text{ etc.},$$

and similarly for the others.

329. A Tensor* *is the numerical factor* by which a unit vector is multiplied to produce the real vector. Thus, l, m, n, etc., in the preceding expressions, are tensors. Tensors were represented by Hamilton, in general discussions, by the letter T; thus, $T\alpha$, $T\beta$, etc., and are read, 'tensor α, tensor β,' etc. This notation is still retained in many cases. Thus, we have

$$\alpha = T\alpha\,(\alpha_1), \qquad \beta = T\beta\,(\beta_1), \text{ etc.}$$

(REMARK.—The definition of the other terms, SCALAR and VERSOR, will be given in the second chapter. A Quaternion, strictly speaking, involves all four units, and hence the analysis of this chapter, involving, as it does, only two of the required units, is at best a restricted and partial case of the more general analysis. Still, the principles here developed are a necessary part of the subject.)

Vector Equations.

330. Let ABC be a triangle, the direction and length of AB being represented by vector α, of BC, by vector β, of AC, by vector γ. The transferrence of a point from A to B, followed by a transferrence from B to C, gives the same *result* as a transferrence directly from A to C. This is expressed in the *form* of an equation, thus

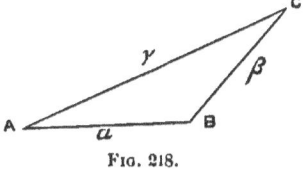

Fig. 218.

$$\alpha + \beta = \gamma, \qquad (1)$$

---

* Literally, *that which stretches.*

The symbols + and = have not the same meaning here as in algebra. The addition is not numerical, neither is the equality a numerical one; still, their meaning here is not *opposed* to that in algebra. They are used in an enlarged sense. The expression may be read 'a transferrence expressed by vector α, followed by a transferrence expressed by vector β, is equivalent to a transferrence expressed by vector γ.' In this sense the expression may be called a *vector-equation*, and read in the usual way, thus, 'α plus β equals γ.' If the vectors are parallel, they will be represented by multiples of the same vector, and the equation will express a numerical equality, (326).

331. Law of Signs.—A separate vector may be positive in either direction along the line, (325). When they are connected with each other the sign depends upon the order of the transferrence. Thus, if the transferrence be from A to B,

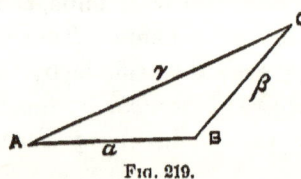
Fig. 219.

thence from B to C, thence from C to A, and all these directions be considered positive, we have

$$\alpha + \beta + \gamma = 0. \qquad (2)$$

But if we make AB, BC, positive, and also AC positive *from A*, then will CA be negative; in which case we have

$$\alpha + \beta - \gamma = 0, \qquad (3)$$

which is the same as transposing γ in equation (1) to the first member; hence, *The rule for the transposition of terms in a vector equation in regard to signs, is the same as for algebraic equations.*

If the transferrence be positive, and from B to A, and A to C, the result will be the same as from B to C, and we have

$$\alpha + \gamma = \beta.$$

Or, if the transferrence had continued from C towards B, BC being positive, we would have

$$\alpha + \gamma - \beta = 0.$$

The transferrence may begin at any angle of the triangle.

[OBS.—This freedom in regard to signs may, at first, seem to lead to uncertainty in regard to the result; but it is only necessary to observe that the result must be interpreted in accordance with the original assumptions.

If the right member consists of one term only, it may be considered as *a measure of the result*, while the left member may be considered as *the expression of an operation*.]

332. Unless otherwise stated the vectors *from* the initial point will be considered positive; *but the directions may be assumed arbitrarily.*

EXAMPLES.

If vectors $AB = \alpha$, $BC = \beta$, $CD = \gamma$, $AD = \delta$, excepting that the positive signs are not necessarily in the order of the letters here given; interpret the following vector equations; that is, give the order of the transferrence, the position of the point after transferrence, and the direction in which the vector is positive.

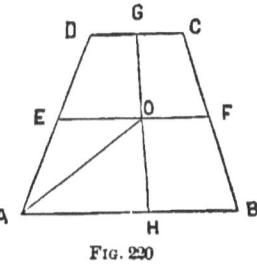

FIG. 220

$$\alpha + \beta = \gamma + \delta$$
$$\alpha = \gamma + \delta - \beta$$
$$\alpha + \beta + \gamma + \delta = 0$$
$$0 = \delta + \gamma + \beta + \alpha$$
$$\alpha + \beta + \gamma = \delta$$
$$AH + HO = AO$$
$$AO = AD + DG - OG.$$

333. The sign of a vector in one direction being fixed, the other vectors parallel and in the same direction should have the same sign. Thus if BD be positive from B towards D; then should DE and EA also be positive. This principle, if observed, will prevent confusion in regard to signs.

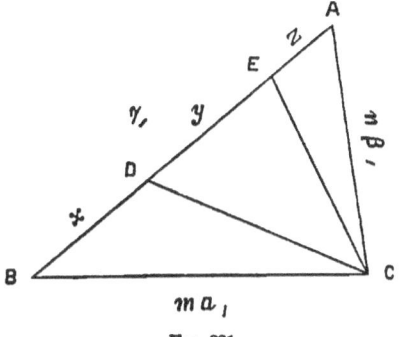

FIG. 221.

EXERCISE.—Let α_1, β_1, γ_1 be unit vectors, and $m\alpha_1$, $n\beta_1$, $l\gamma_1$, the sides of the triangle. Draw radial vectors CD, CE, from C; and let $x = BD$, $y = DE$, $z = EA$; it is required to show that $l\gamma_1 = m\alpha_1 + n\beta_1$ by passing from the triangle BCD to DCE, and thence to ECA.

We have, (330),

$$\text{for triangle } BCD \quad x\gamma_1 = m\alpha_1 + CD,$$
$$\text{`` \quad `` } \quad DEC \quad y\gamma_1 = -CD + CE,$$
$$\text{`` \quad `` } \quad ECA \quad z\gamma_1 = -CE + n\beta_1.$$

Adding we have

$$(x + y + z)\gamma_1 = m\alpha_1 + n\beta_1.$$

But $$x + y + z = l;$$

$$\therefore l\gamma_1 = m\alpha_1 + n\beta_1,$$

as required.

Let the reader deduce the same result by taking D as the initial point.

334. Proposition.—If $\Sigma\alpha + \Sigma\beta = 0$, then will $\Sigma\alpha = 0$, and $\Sigma\beta = 0$.

For the vectors being entirely independent of each other, either sum may be zero independently of the other.

We may also show it directly from Article 326; thus, no amount of transferrence along vectors α will affect β, hence if their sum is zero, each must separately be zero. This may be illustrated by the figure. We have

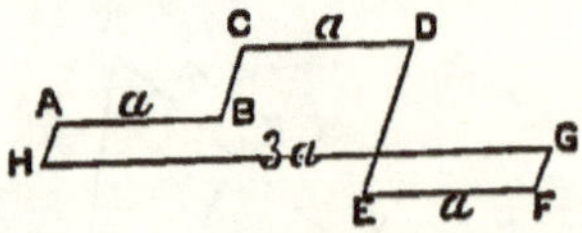

$$AB + CD + EF + GH = 0 = \Sigma\alpha,$$
$$HA + BC + DE + FG = 0 = \Sigma\beta,$$
$$\text{and } \Sigma\alpha + \Sigma\beta = 0,$$

Fig. 222.

from which we see that each may be independently equal to zero. The expressions developed become

$$(l + m + n + p + \text{etc.})\alpha = 0,$$
$$(a + b + c + d + \text{etc.})\beta = 0.$$

This principle is similar to that of indeterminate coefficients in Algebra.

APPLICATIONS.

1. *The corresponding sides of mutually equiangular triangles are proportional.*

Let α_1, β_1, γ_1, be unit vectors; vector $BC = m\alpha_1$, vector $CA = n\beta_1$, vector $BA = l\gamma_1$. (Observe that α is opposite A, β, opposite B, and γ, opposite C, corresponding to similar notation in trigonometry.) Then, since the sides of the triangle CDE are parallel to ABC, we will have

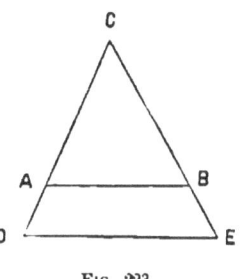

Fig. 223.

$$EC = a\alpha_1, \qquad CD = b\beta_1, \qquad ED = c\gamma_1.$$

From the triangle ABC we have, (330),

$$m\alpha_1 + n\beta_1 = l\gamma_1, \tag{1}$$

and from CDE
$$a\alpha_1 + b\beta_1 = c\gamma_1. \tag{2}$$

Multiplying the first equation by c; the second by l; and taking the difference, gives

$$(cm - la)\alpha_1 + (cn - lb)\beta_1 = 0.$$

Hence, according to Article 334, we have

$$cm - la = 0,$$
$$cn - lb = 0;$$
$$\therefore \frac{cm}{cn} = \frac{la}{lb};$$

and dropping the common factors c and l, and multiplying by α_1 and dividing by β_1 gives

$$\frac{m\alpha_1}{n\beta_1} = \frac{a\alpha_1}{b\beta_1};$$

and, substituting lines for the corresponding vectors, we have

$$\frac{CB}{CA} = \frac{CE}{CD};$$

or $$CB : CA :: CE : CD,$$

which was to be proved.

Similarly, eliminating α_1 from equations (1) and (2) will give
$$CA : AB :: CD : DE.$$

2. *If the opposite sides of a quadrilateral are parallel, they equal each other in length.*

The term *vector* being understood, it is generally omitted in speaking of a line. Thus, instead of saying 'vector $CD = \alpha$,' we simply say 'let $CD = \alpha$.'

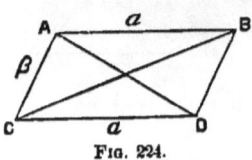

Fig. 224.

Let $CD = \alpha$, $CA = \beta$; then since AB is parallel to CD it will equal $m\alpha$; and similarly $DB = n\beta$.

Calling CD positive, BA will be negative. We have
$$\text{vector } CD + \text{vector } DB + \text{vector } BA + \text{vector } AC = 0,$$
or $$\alpha + n\beta - m\alpha - \beta = 0;$$
or $$(1-m)\alpha - (1-n)\beta = 0,$$

which, according to Article 334, gives
$$1 - m = 0, \quad 1 - n = 0;$$
$$\therefore m = 1, \text{ and } n = 1;$$

therefore $$AB = CD, \quad CA = DB,$$

which was to be proved.

3. *If the lengths of the opposite sides of a quadrilateral are equal, they will be parallel.*

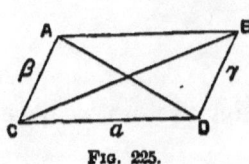

Fig. 225.

Let the unit vectors along CA, CD, DB, AB, be β_1, α_1, γ_1, δ_1, respectively; then will the sides as vectors be
$$CD = m\alpha_1, \quad CA = n\beta_1,$$
$$AB = m\delta_1, \quad DB = n\gamma_1,$$

and we have, (**330**),

$$m\alpha_1 + n\gamma_1 - m\delta_1 - n\beta_1 = 0,$$

or $$(\alpha_1 - \delta_1)m + (\gamma_1 - \beta_1)n = 0.$$

But since m and n are mutually independent, we have

$$\alpha_1 = \delta_1, \qquad \gamma_1 = \beta_1;$$

hence the opposite sides are parallel and the figure is a parallelogram.

4. *The diagonals of a parallelogram mutually bisect each other.*

Let
$$AB = DC = \alpha,$$
$$AD = BC = \beta;$$

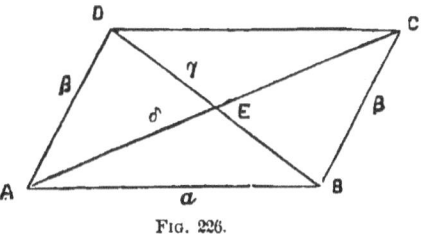

Fig. 226.

a unit vector along $AE = \delta_1$, and along $BD = \gamma_1$. Then, (**327**),

$$AE = x\delta_1, \qquad EC = y\delta_1,$$
$$BE = u\gamma_1, \qquad ED = v\gamma_1.$$

We have, (**330**),

$$AE = AB + BE,$$

or $$x\delta_1 = \alpha + u\gamma_1;$$

also $$y\delta_1 = v\gamma_1 + \alpha.$$

Subtracting, $(x - y)\delta_1 = (u - v)\gamma_1,$

or $(x - y)\delta_1 - (u - v)\gamma_1 = 0;$

$$\therefore \ (\mathbf{334}), x = y, \qquad u = v,$$

or $$AE = EC, \qquad BE = ED,$$

which was to be proved.

5. *The lines joining the middle points of the opposite sides of* ANY *quadrilateral mutually bisect each other.*

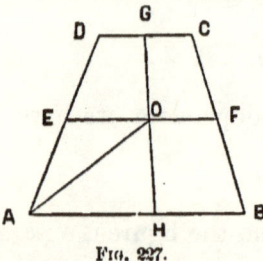

Fig. 227.

The proposition will be proved to be true whether the lines all lie in one plane or not. If they do not, the surface will be warped, (**222**), and is sometimes called a surface *gauche*, a French word signifying *twisted*.

Let three vectors, AB, AC, AD, co-initial at A, be drawn in any direction and of any length, and their extremities joined by the lines BC, CD, making a quadrilateral $ABCD$. (Vector AC is not shown in the figure.) The three vectors here given will always determine the figure, and by using them, instead of the four sides as vectors, the solution is simplified, and a more symmetrical expression found for the result.

Let $AB = \alpha$, $AC = \beta$, $AD = \gamma$, all positive *from* A; E, G, F, H, middle points of the respective sides, and O the middle point of GH; we are to prove that it is also the middle point of EF.

We have $\qquad AD + DC = AC,$

or $\qquad \gamma + DC = \beta;$

$\qquad \therefore DC = \beta - \gamma,$

and $\qquad DG = \tfrac{1}{2}(\beta - \gamma);$

also, $\qquad AH + HG = AD + DG,$

or $\qquad \tfrac{1}{2}\alpha + HG = \gamma + \tfrac{1}{2}(\beta - \gamma)$

$\qquad = \tfrac{1}{2}(\beta + \gamma);$

$\qquad \therefore HG = \tfrac{1}{2}(\beta + \gamma - \alpha),$

and $\qquad HO = \tfrac{1}{2}HG$

$\qquad = \tfrac{1}{4}(\beta + \gamma - \alpha);$

then, $\qquad AO = AH + HO$

$\qquad = \tfrac{1}{2}\alpha + \tfrac{1}{4}(\beta + \gamma - \alpha)$

$\qquad = \tfrac{1}{4}(\alpha + \beta + \gamma). \qquad (1)$

Let the middle point of EF be O', then
$$AO' = AE + EO'$$
$$= \tfrac{1}{2}\gamma + \tfrac{1}{2}EF. \quad (2)$$

To find EF, we have
$$AE + EF = AB + BF;$$
$$\therefore EF = \alpha + \tfrac{1}{2}BC - \tfrac{1}{2}\gamma,$$
and $$\tfrac{1}{2}EF = \tfrac{1}{2}\alpha - \tfrac{1}{4}\gamma + \tfrac{1}{4}BC. \quad (3)$$

To find BC, we have
$$AB + BC = AC,$$
or $$BC = \beta - \alpha,$$
and $$\tfrac{1}{4}BC = \tfrac{1}{4}(\beta - \alpha), \quad (4)$$

which in (3) gives
$$\tfrac{1}{2}EF = \tfrac{1}{4}(\alpha + \beta - \gamma),$$

and this in (2) gives
$$AO' = \tfrac{1}{4}(\alpha + \beta + \gamma), \quad (5)$$

which being identical with equation (1) shows that the points O and O' coincide; hence the lines EF and HG mutually bisect. The point O is called the mean point of the polygon.

It will be observed that the mode of solution consists, chiefly, in reaching the same point by two different routes.

335. The mean point of a polygon is a point to which the vector is the average of the vectors to all the angles of the polygon. The initial point may be chosen arbitrarily.

The mean point coincides with the centre of gravity of a system of equal particles, one particle being at each angle of the polygon.

6. *The mean point of a quadrilateral is at the middle point of the line joining the points of bisection of the diagonals.*

Using the notation of the preceding example, and letting O be the middle point of PQ, we have,

$$AO = AP + PO$$
$$= \tfrac{1}{2}\beta + \tfrac{1}{2}PQ$$
$$= \tfrac{1}{2}\beta + \tfrac{1}{2}(AB + BQ - AP)^*$$
$$= \tfrac{1}{2}\beta + \tfrac{1}{2}\alpha + \tfrac{1}{4}BD - \tfrac{1}{4}\beta$$
$$= \tfrac{1}{4}\beta + \tfrac{1}{2}\alpha + \tfrac{1}{4}BD$$
$$= \tfrac{1}{4}\beta + \tfrac{1}{2}\alpha + \tfrac{1}{4}(AD - AB)$$
$$= \tfrac{1}{4}\beta + \tfrac{1}{2}\alpha + \tfrac{1}{4}\gamma - \tfrac{1}{4}\alpha$$
$$= \tfrac{1}{4}(\alpha + \beta + \gamma);$$

hence the point O of this figure coincides with O of the preceding figure.

An interpretation of this result gives a geometrical method of finding the mean point, for $4AO$ is one side of a polygon of which the other sides are α, β, γ.

---

* A little practice will enable the student to make these substitutions without formally writing the equation, and determining the value of the unknown quantity. In this case the equation would be

$$AP + PQ = AB + BQ;$$
$$\therefore PQ = AB + BQ - AP,$$

which compared with the figure shows that the terms to be substituted will be the remaining sides of the polygon $APQB$, and that the signs of the vectors will be determined by passing backwards around the polygon. Thus, instead of passing around in the direction from P towards Q, go in the opposite direction. As AP was considered positive from A towards P, it will be negative from P towards A, and we have $-AP$. Similarly, $+AB$ and $+BQ$. The order of the sides (or terms) is immaterial, thus we may have $+BQ - AP + AB$. It would be well for the student to mark on the figure the *positive* directions of the vectors, as soon as the direction has been fixed, as shown in Fig. 232.

Construct a polygon whose successive sides are α, β, γ, and join AD'. Then we have, (330),

$$AD' = \alpha + \beta + \gamma,$$

and

$$AO = \tfrac{1}{4}(\alpha + \beta + \gamma);$$

$$\therefore AO = \tfrac{1}{4} AD'.$$

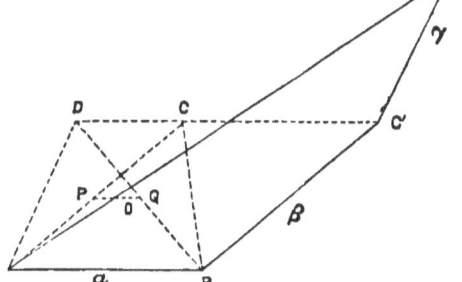

Fig. 229.

The order of the sides, as shown by the equation, is immaterial, hence the successive sides may be α, γ, β, or γ, α, β. It is not necessary for them to be all in one plane.

To find the mean point we simply add the vectors and divide by the number of angles of the polygon.

EXAMPLES.

1. Prove that the mean point of a triangle is in a line joining the vertex with the centre of the base, and at two-thirds its length from the vertex. (It will be at the same point as the centre of gravity of three equal particles, one at each angle of the triangle.)

2. Take the initial point on the diagonal of a square prolonged, and show that the mean point of the square is at the middle point of the diagonal of the square.

(OBS.—Let the adjacent sides of the square be α, β, and the vectors to the four angles be $\gamma, \delta, \varepsilon, \mu$, then if δ be the vector to the nearest angle, prove that AO (the mean vector) equals $\delta + \tfrac{1}{2}(\alpha + \beta)$.)

3. If the initial point be at the middle of the base of a triangle, prove that the vector to the mean point is one-third the vector to the opposite angle.

(OBS.—These examples should be solved by writing the equations, and not deduced from the value of AO given above.)

4. In a regular pyramid having a rectangle for the base, equal heavy particles are placed one at each corner of the base and vertex of the pyramid; show that the centre of gravity of the five particles will be on the altitude, and at $\tfrac{4}{5}$ its length from the vertex.

Medial Vectors.

336. A **medial vector** is one drawn from the common point of two given vectors to the middle of the line joining the extremities of the given vectors.

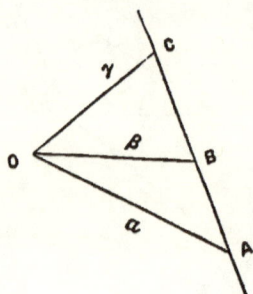

Fig. 230.

Thus if OA and OC are given vectors, and OB a line drawn from O to B, the middle of AC, then will OB be the medial vector.

337. Problem.—*To find an expression for a medial vector.*

Let α and γ be the given vectors, and β the medial, all positive from the initial point. We have

$$OB = OA + AB$$
or $\qquad \beta = \alpha + AB;$
also $\qquad OB + BC = OC,$
or $\qquad \beta = \gamma - BC;$
adding gives $\qquad \beta = \tfrac{1}{2}(\alpha + \gamma),$ since $AB = BC,\quad$ (1)

which is the required value.

The rules for signs apply to this expression the same as for other cases. If γ be positive *towards* the origin, we have

$$\beta = \tfrac{1}{2}(\alpha - \gamma); \qquad (2)$$

and if α and γ are both positive *towards* the origin, and β positive *from* it, we have

$$\beta = -\tfrac{1}{2}(\alpha + \gamma),$$
or $\qquad -\beta = \tfrac{1}{2}(\alpha + \gamma); \qquad (3)$

and if α and γ are both positive *from* the origin and β positive *towards* it, we have

$$-\beta = \tfrac{1}{2}(\alpha + \gamma)$$
or $\qquad \beta = -\tfrac{1}{2}(\alpha + \gamma), \qquad (4)$

which are the same as the two preceding equations.

APPLICATIONS.

1. *One-half the diagonal of a parallelogram whose adjacent sides are α and β is the medial of those sides.*

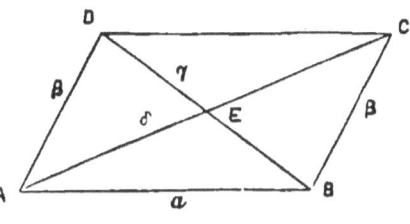

Fig. 231.

In the parallelogram $ABCD$ we are to show that one-half of AC is the medial of AB and AD.

Let $AD = \beta = BC$, and $AB = \alpha$; then

$$AC = \alpha + \beta;$$
$$\therefore AE = \tfrac{1}{2} AC,$$
$$= \tfrac{1}{2}(\alpha + \beta);$$

hence, according to equation (1), AE is the medial of AB and AD, and E is in the line BD. Similarly, $BE = \tfrac{1}{2} BD$ is the medial of BA and BC. This is another mode of proving that the diagonals of a parallelogram are mutually bisected.

2. *The medials of a triangle meet in a point and mutually trisect each other.*

(The medials are the lines from the angles to the middle of the opposite sides.)

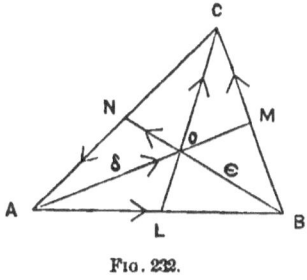

Fig. 232.

Let AM, BN, CL, be drawn from the angles to the middle of the opposite sides, and take the vectors all positive in a left-handed direction starting from A, in the triangles ABC, AOC, AON, as indicated by the arrow-heads. The lines AM and BN will intersect at some point, as O; draw from O the line OC, then if OC

and the medial LC coincide, the three medials will all meet in O.

Let $AB = \gamma$, $BC = \alpha$, $CA = \beta$, $AO = \delta$, $OM = m\delta$, $BO = \varepsilon$, $ON = n\varepsilon$.

We have
$$AB + BC + CA = 0,$$
or
$$\gamma + \alpha + \beta = 0; \tag{1}$$
also,
$$AB + BO - AO = 0,$$
or
$$\gamma + \varepsilon - \delta = 0; \tag{2}$$
also,
$$OM + MC + CN + NO = 0,$$
or
$$m\delta + \tfrac{1}{2}\alpha + \tfrac{1}{2}\beta - n\varepsilon = 0. \tag{3}$$

From (1) and (2),
$$\delta + \alpha + \beta - \varepsilon = 0; \tag{4}$$

from twice (3) subtract (4),
$$(2m-1)\delta - (2n-1)\varepsilon = 0;$$

therefore, (**334**), $2m = 1$, and $2n = 1$,

or $AO = 2OM$ and $BO = 2ON$,

or $AM = 3OM$ and $BN = 3ON$;

hence AM and BN trisect each other. (Equation (4) may be deduced directly from the quadrilateral $AOBCA$.)

Next we have
$$AO + OC + CA = 0,$$
or
$$\delta + OC + \beta = 0;$$
$$\therefore OC = -\beta - \delta$$
$$= -\beta - \tfrac{2}{3} AM$$

because AM is medial,
$$= -\beta - \tfrac{2}{3} \times \tfrac{1}{2}(\gamma - \beta)$$

from (1),
$$= -\beta - \tfrac{1}{3}(-\alpha - 2\beta)$$
$$= \tfrac{1}{3}(\alpha - \beta).$$

But because CL is medial, we have
$$-LC = \tfrac{1}{2}(\beta - \alpha),$$

or
$$LC = \tfrac{1}{2}(\alpha - \beta);$$
$$\therefore OC = \tfrac{2}{3} LC,$$

therefore, (327), CL and CO are parallel, and because they have the point C common they coincide; and the medial CL passes through O; and since $CO = \tfrac{2}{3} CL$ it is trisected at O.

(OBS.—In order to find the numerical values of the tensors we find an equation between two vectors and apply Article 334.)

EXAMPLES.

1. If α, β, γ, make a closed triangle, and δ is medial between α and γ, find its value in terms of α and β.

2. If $\beta = \tfrac{1}{2}(\alpha + \gamma)$ is the medial of α and γ, all being positive from the origin; if the vectors be prolonged, the negative vectors being equal to the positive ones, and the medial positive from the origin; show that the medial in one of the supplementary angles will be $\beta' = \tfrac{1}{2}(\alpha - \gamma)$, and in the other $\beta'' = \tfrac{1}{2}(-\alpha + \gamma)$.

3. Of three co-initial vectors α, β, γ, find the medial of the medials of α, β, and β, γ.

4. Given three co-initial vectors α, β, γ; if the extremity of the medial of α and β be joined by a straight line to the extremity of the medial of β and γ; show that it will be parallel to the line joining the extremities of α and γ.

Co-planar Vectors.

338. Any three co-planar vectors which give the relation
$$a\alpha + b\beta + c\gamma = 0,$$

may be represented by the sides of a triangle.

For a line may be drawn equal and parallel to $a\alpha$, and through either extremity of it a line, equal and parallel to $b\beta$. Join the extremities of these lines, and call the closing side $c'\gamma'$. Consider them all as positive as we pass around the triangle in the direction of $+ a\alpha$ and $+ b\beta$, and we have, (330),
$$a\alpha + b\beta + c'\gamma' = 0,$$

subtracting from the given equation gives
$$c\gamma - c'\gamma' = 0,$$

or
$$\gamma = \frac{c'}{c}\gamma';$$

hence, (327), the closing side will be parallel to the third vector, and the relation $c\gamma = c'\gamma'$ shows that they will be of the same length, which was to be proved.

339. Three non-planar vectors, cannot give the relation $a\alpha + b\beta + c\gamma = 0$, unless $a\alpha = 0$, $b\beta = 0$, $c\gamma = 0$.

For a closed triangle cannot be made from the three vectors; hence, each term must be separately equal to zero. It does not, however, follow that each vector is zero, for the coefficients, a, b, c, may be the algebraic sum of several numbers; as, $a = m + n + p +$ etc. $= 0$.

It follows from the above that if the sum of three vectors is zero, they must be co-planar.

340. If three co-initial vectors give the relations

$$a\alpha + b\beta + c\gamma = 0, \quad (1)$$

and

$$a + b + c = 0, \quad (2)$$

they will terminate in a right line.

Multiplying the second equation by γ and subtracting from the first, will give

$$a(\alpha - \gamma) + b(\beta - \gamma) = 0,$$

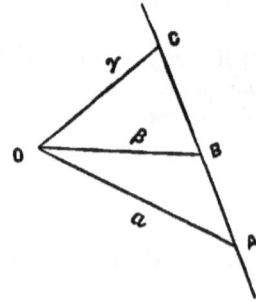

Fig. 233.

or
$$\alpha - \gamma = -\frac{b}{a}(\beta - \gamma). \quad (3)$$

But from the figure, we have

$$\alpha = \gamma + CA,$$

and
$$\beta = \gamma + CB;$$

$$\therefore \ \alpha - \gamma = CA,$$

$$\beta - \gamma = CB;$$

which in (3) gives

$$CA = -\frac{b}{a}CB;$$

hence, (**327**), CA and CB are parallel, and since they have a point C common, they coincide; hence they terminate in the same right line.

(Ons.—Since CA and CB are co-initial it would at first appear that they ought to have the same sign in the result. But an examination of equation (2) shows that one or two of the coefficients must be negative while the other two or one is positive. If a and b have contrary signs then we have $CA = \frac{b}{a}CB$. The first equation shows that the three vectors, $a\alpha$, $b\beta$, $c\gamma$, taken in succession, will make a closed triangle.)

APPLICATIONS.

1. *The extremities of two adjacent sides of a parallelogram and the middle point of the diagonal between those sides are in a right line.*

Let $AB = \alpha$, $AD = \beta = BC$, $AE = \frac{1}{2}AC = \delta$.

We have

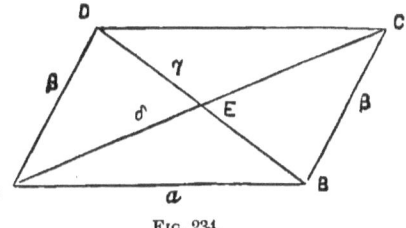

Fig. 234.

$$AC = AB + BC,$$

or $$2\delta = \alpha + \beta;$$

$$\therefore 2\delta - \alpha - \beta = 0.$$

But the coefficients of δ, α, β, give

$$2 - 1 - 1 = 0;$$

hence, (**340**), the extremities B, D, and E of the vectors are in a right line. This, then, is another mode of proving that the diagonal BD bisects AC. Similarly, AC bisects BD.

2. *The medials of a triangle meet in one point.*

This has already been solved, (**337**, 2), but we here present another demonstration. In Fig. 232, we will show that the extremities of the vectors AL, AO, AC, are in a right line. If they are in a right line, we must find values for a, b, c, which will satisfy the equations

$$aAL + bAO + cAC = 0, \qquad (1)$$
$$a + b + c = 0. \qquad (2)$$

Substituting the values $AL = \tfrac{1}{2}\gamma$, $AO = \tfrac{1}{3}(\gamma - \beta)$, $AC = -\beta$, gives

$$\tfrac{1}{2} a\gamma + \tfrac{1}{3} b(\gamma - \beta) - c\beta = 0,$$
or
$$(3a + 2b)\gamma - (2b + 6c)\beta = 0. \qquad (3)$$

But β and γ being independent we must have

$$3a + 2b = 0,$$
$$2b + 6c = 0;$$

from which we have

$$a = -\tfrac{2}{3} b, \text{ and } c = -\tfrac{1}{3} b; \qquad (4)$$

or they are indeterminate. Substituting these values in the second equation, gives

$$-\tfrac{2}{3} b + b - \tfrac{1}{3} b,$$

which reduces identically to zero; hence every value of b will reduce (1) and (2) to zero, and the points L, O, C, are in a right line. Making $b = -3$ in (4), gives $a = 2$, and $c = 1$, and (1) and (2) become

$$2AL - 3AO + AC = 0,$$
$$2 - 3 + 1 = 0.$$

3. *The altitudes of a triangle meet in a common point.*

(The altitudes are the perpendiculars from the angles upon the opposite sides.)

Let AE, BF, CP, be the respective perpendiculars, the lengths of the sides be $l = AB$, $m = BC$, $n = CA$, γ_1, α_1, β_1, unit vectors along the corresponding sides. Let the directions be positive as indicated in Fig. 232; then we have

$$\alpha = m\alpha_1, \quad \beta = n\beta_1, \quad \gamma = l\gamma_1.$$

Let $AO = p\delta_1$, $OC = q\varepsilon_1$, $BO = r\mu_1$. As lines we have

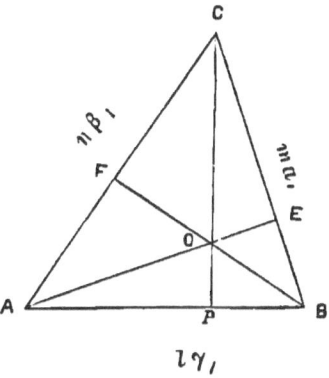

Fig. 235.

$$CP = n \sin A, \quad AP = n \cos A,$$
$$BF = l \sin A, \quad AF = l \cos A;$$

and as vectors

$$PC = n \sin A \cdot \varepsilon_1, \quad AP = n \cos A \cdot \gamma_1,$$
$$BF = l \sin A \cdot \mu_1, \quad FA = l \cos A \cdot \beta_1.$$

We have from $\triangle APC$,

$$AP + PC + CA = 0,$$

or $\qquad n \cos A \cdot \gamma_1 + n \sin A \cdot \varepsilon_1 + n\beta_1 = 0;$

$$\therefore \varepsilon_1 = -\frac{\beta_1 + \cos A \cdot \gamma_1}{\sin A};$$

and from $\triangle ABF$

$$\mu_1 = -\frac{\gamma_1 + \cos A \cdot \beta_1}{\sin A};$$

From $ABOCA$, we have

$$l\gamma_1 + r\mu_1 + q\varepsilon_1 + n\beta_1 = 0;$$

in which substituting the values of ε_1 and μ_1, collecting terms, and making the coefficients of γ_1 and β_1 separately equal to zero, we find

$$r = \frac{l - n \cos A}{\sin A},$$

$$q = \frac{n - l \cos A}{\sin A}.$$

The triangle ABO gives

$$p\delta_1 = l\gamma_1 + r\mu_1$$

$$= l\gamma_1 + \frac{(n \cos A - l)(\gamma_1 + \cos A \cdot \beta_1)}{\sin^2 A}$$

$$= \frac{\cos A}{\sin^2 A}[(n \cos A - l)\beta_1 + (n - l \cos A)\gamma_1]. \quad (1)$$

We now find that P, O, C, are in a right line, for if in the equation

$$aAP + bAO + cAC = 0, \qquad (2)$$

we substitute the value of AO from equation (1), and reduce, we find

$$c = \frac{\cos A (n \cos A - l)}{n \sin^2 A} b,$$

$$a = -\frac{l \cos A - n}{n \sin^2 A} b;$$

$$\therefore a + b + c = 0;$$

which was to be proved; therefore the altitudes of a triangle meet in a common point.

(It will be found best to get the value of vector AO in terms of the adjacent vectors, γ and β, and the included angle A, as has been done in the preceding solution.)

EXAMPLES.

1. Three co-initial vectors 2α, 3β, $c\gamma$, terminate in a right line; find c.

2. Three co-initial vectors 7α, -4β, 3α, are connected by the relation $2\beta = 3\beta - \alpha$; do they terminate in a right line ?

3. The perpendiculars to the sides of a triangle from the middle points meet in one point. (The point will be the centre of the circumscribed circle.)

Let FO''' be perpendicular to AB at its middle point, EO''' perpendicular to AC at its middle point; then if a line from O''' to the middle point of BC is perpendicular to that side, the problem is true. We have to prove that vector $O'''D$ is a multiple of the vector from A perpendicular to CB, the value of which is given by equation (1) in the third problem above.

4. To find vector AO''' drawn from the angle A to the centre of the circumscribed circle.

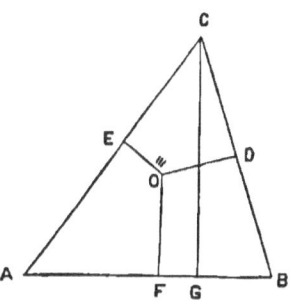

Fig. 236.

Solution.—Letting the signs of the vectors be represented as in Fig. 232, we have from the preceding figure,

$$\text{Vector } GC = -n \cos A \cdot \gamma_1 - n\beta_1,$$

$$\text{Vector } BE' = -l \cos A \cdot \beta_1 - l\gamma_1,$$

(where BE' is an altitude from B). The polygon $AFO'''EA$ gives

$$AF + FO''' + O'''E + EA = 0,$$

or $\frac{1}{2} l\gamma_1 - xn(\beta_1 + \cos A \cdot \gamma_1) - yl(\gamma_1 + \cos A \cdot \beta_1) + \frac{1}{2}n\beta_1 = 0.$

Expanding, collecting terms, and equating to zero, separately the coefficients of β_1 and γ_1, we find

$$x = \frac{n - l \cos A}{2n \sin^2 A};$$

$$\therefore FO''' = -\frac{(n - l \cos A)(n\beta_1 + n \cos A \cdot \gamma_1)}{2n \sin^2 A}.$$

Then $AO''' = AF + FO'''$

$$= \tfrac{1}{2} l\gamma_1 - \frac{(n - l \cos A)(n\beta_1 + n \cos A \cdot \gamma_1)}{2n \sin^2 A}$$

$$= \frac{1}{2 \sin^2 A}[(l \cos A - n)\beta_1 + (l - n \cos A)\gamma_1],$$

which is the quantity sought.

5. If the sides of two polygons are parallel and lie in the same direction, their sides are proportional.

6. If the corresponding vertices of two triangles are in lines radiating from a point, the corresponding sides prolonged will meet in a right line. (This is solved by Transversals, further on.)

Angle-Bisectors.

341. A line which bisects an angle of a triangle is called an *angle-bisector*.

342. Problem.—*To find the vector of an angle-bisector.*

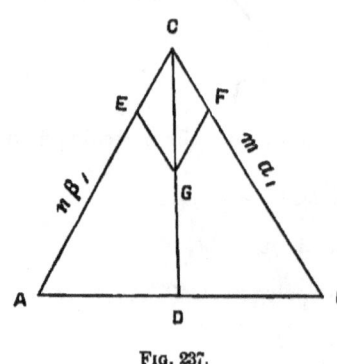

Fig. 237.

Let CB be α, CA, β, and CD, the vector which bisects their angle, be δ. Take $CF = \alpha_1$, a unit vector, $CE = \beta_1$, and complete the rhombus $CEGF$, then will $EG = \alpha_1$, and $FG = \beta_1$; and all being positive *from* C, we have

$$CG = CF + FG,$$
or $\quad CG = \alpha_1 + \beta_1.$

But the bisecting line CD may be of any length; hence we have

$$CD = xCG,$$
or $\quad \delta = x(\alpha_1 + \beta_1), \qquad (1)$

in which x is indeterminate. This is a general expression for the required vector. If, however, the vector CD is limited by the line which joins the extremities of two given vectors, its value becomes determinate. For we have

$$CA + AB - CB = 0,$$
or $\quad n\beta_1 + l\gamma_1 - m\alpha_1 = 0,$

$$\therefore \gamma_1 = \frac{m\alpha_1 - n\beta_1}{l}.$$

Also $\quad \delta = x(\alpha_1 + \beta_1) = CA + AD$
$$= n\beta_1 + y\gamma_1$$

Substituting γ_1, $\quad = n\beta_1 + \dfrac{y}{l}(m\alpha_1 - n\beta_1)$

reducing, $\quad = \dfrac{n(l-y)\beta_1 + my\alpha_1}{l}$;

transposing,
$$\left.\begin{array}{r}(lx - my)\alpha_1 \\ + (lx - n(l-y))\beta_1\end{array}\right\} = 0;$$

$\therefore$ (334), $\quad lx - my = 0$,
$$lx - n(l-y) = 0;$$

from which we find

$$x = \frac{mn}{m+n}, \qquad (2)$$

$$y = \frac{nl}{m+n}. \qquad (3)$$

Substituting the value of x in (1) gives

$$\delta = \frac{mn}{m+n}(\alpha_1 + \beta_1), \qquad (4)$$

which is the definite value of the vector which bisects the angle between $m\alpha_1$ and $n\beta_1$, and limited by the line joining their extremities.

APPLICATIONS.

1. *The medial to the base of an isosceles triangle is an angle-bisector.*

Since the triangle is isosceles, $m = n$. The medial to the base will be, (336),

$$\delta = \tfrac{1}{2}(m\alpha_1 + n\beta_1)$$

(since $m = n$), $\quad = \tfrac{1}{2}m(\alpha_1 + \beta_1).\qquad (5)$

The angle-bisector is given by equation (4) by making $m = n$, and hence is

$$\delta = \tfrac{1}{2} m (\alpha_1 + \beta_1);$$

which being the same as (5) shows that the medial bisects the vertical angle.

2. *The angle which bisects the vertical angle of an isosceles triangle, bisects the base.*

When CD bisects the angle C, the value of AD is given by equation (3), which gives, when $m = n$,

$$y = \tfrac{1}{2} l;$$

that is, $AD = \tfrac{1}{2} AB$, which was to be proved.

3. *Any angle-bisector of a triangle divides the opposite side into parts proportional to the adjacent sides.*

When CD bisects C, we have found, (Eq. (3)),

$$AD = y = \frac{nl}{m+n};$$

$$\therefore DB = l - y = \frac{ml}{m+n}.$$

Dividing, we have

$$\frac{AD}{DB} = \frac{n}{m} = \frac{CA}{CB},$$

which was to be proved.

4. *To find the conditions which will cause the diagonal of a parallelogram to bisect the angles through which it passes.*

We have $\alpha = m\alpha_1$, $\beta = m\beta_1$; and

$$\delta = \alpha + \beta$$
$$= m\alpha_1 + n\beta_1;$$

and if δ bisects the angle, we have

$$\delta = x(\alpha_1 + \beta_1);$$

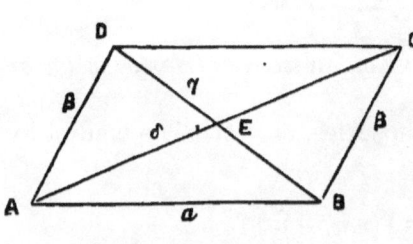

Fig. 238.

therefore we must have

$$x\alpha_1 + x\beta_1 = m\alpha_1 + n\beta_1,$$

or $\qquad (x-m)\alpha_1 + (x-n)\beta_1 = 0;$

$\therefore$ **(334)**, $x = m$, and $x = n$,

or $\qquad\qquad m = n,$

that is, the adjacent sides must be equal, and hence the figure must be a rhombus.

5. *The angle-bisectors of a triangle meet in a point.*

Let AM, BN, CL, bisect the respective angles A, B, C; then considering the vectors positive in the directions indicated in Fig. 232, we have

$$AM = x(\gamma_1 - \beta_1),$$
$$BN = y(\alpha_1 - \gamma_1),$$
$$-LC = z(\beta_1 - \alpha_1).$$

AM and BN intersect in O; draw CO; then if CO is a multiple of CL, the latter will pass through O. AO will be a multiple of AM, and BO of BN, hence

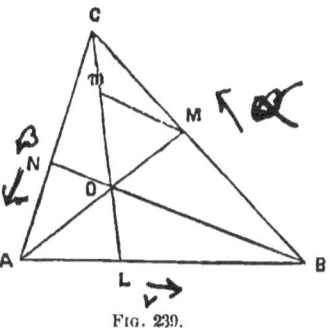

Fig. 239.

$$AO = u(\gamma_1 - \beta_1),$$
$$BO = v(\alpha_1 - \gamma_1).$$

We have from the triangle ABC,

$$l\gamma_1 + m\alpha_1 + n\beta_1 = 0;$$

$$\therefore \gamma_1 = -\frac{m\alpha_1 + n\beta_1}{l}. \qquad (1)$$

Also $\qquad AO + OC + CA = 0,$

or $\qquad u(\gamma_1 - \beta_1) + OC + n\beta_1 = 0;$

$$\therefore CO = n\beta_1 + u(\gamma_1 - \beta_1)$$
$$= n\beta_1 + u\left(\frac{-m\alpha_1 - n\beta_1}{l} - \beta_1\right)$$
$$= \frac{1}{l}[(ln - nu - lu)\beta_1 - mu\alpha_1]. \quad (2)$$

Also $\quad OC + CB + BO = 0,$

or $\quad OC - m\alpha_1 + v(\alpha_1 - \gamma_1) = 0;$

$\therefore CO = -m\alpha_1 + v(\alpha_1 - \gamma_1)$

$= -m\alpha_1 + v\left(\alpha_1 + \dfrac{m\alpha_1 + n\beta_1}{l}\right)$

$= \dfrac{1}{l}[(lv + mv - ml)\alpha_1 + nv\beta_1].$ (3)

Making these values of CO equal, and resolving according to Article 334, we have

$$lv + mv - ml + mu = 0,$$
$$nv - ln + nu + lu = 0;$$

from which we find

$$u = n\dfrac{l}{l + m + n}$$

$$v = m\dfrac{l}{l + m + n}.$$

Substituting the value of u in equation (2) gives

$$CO = \dfrac{mn}{l + m + n}(\beta_1 - \alpha_1)$$

$$= \dfrac{mn}{z(l + m + n)} CL;$$

hence, (327), CO and CL are parallel and, having the point C common, they coincide.

The point O is the centre of the inscribed circle.

EXAMPLES.

1. Prove that the line which bisects the external angle of a triangle divides the opposite side (prolonged) into segments proportional to the sides about the angle bisected.

2. In Fig. 239, if Mm be drawn parallel to BN, show that it bisects CO.

3. If a line be drawn from L parallel to BO it will bisect AO. Let n be the point of bisection, show that mn will be parallel to AC.

4. The centre of the circumscribed circle, the common point of the

altitudes, and the common point of the medials of a triangle, are in a right line.

SOLUTION.—We have the following values for the vectors from A to the several points:

Fig. 236, $\quad AO''' = \dfrac{1}{2\sin^2 A}[(l\cos A - n)\beta_1 + (l - n\cos A)\gamma_1]$

Fig. 235, $\quad AO = \dfrac{\cos A}{\sin^2 A}[(n\cos A - l)\beta_1 + (n - l\cos A)\gamma_1]$

Fig. 229, $\quad AO' = \tfrac{1}{3}(l\gamma_1 - n\beta_1)$,

Substituting y for $(l\cos A - n)$, and z for $(l - n\cos A)$; then, according to Article 236, we must have

$$\dfrac{a}{2\sin^2 A}(y\beta_1 + z\gamma_1) + \dfrac{b\cos A}{\sin^2 A}(-z\beta_1 - y\gamma_1) + \dfrac{c}{3}(l\gamma_1 - n\beta_1) = 0,$$

which reduced gives $\quad a = 2b$, and $c = -3b$;

$$\therefore 2AO''' + AO - 3AO'' = 0.$$
$$2 + 1 - 3 = 0.$$

Hence the three points O, O'', O''', lie in one straight line.

Transversals.

343. A Transversal is a right line which cuts a system of lines.

The transversal of a triangle is a line which cuts the three sides, the sides being considered as prolonged indefinitely. The transversal may cut all the sides on the prolonged parts.

A segment of a side is the distance from the point where it is cut by the transversal to an angle of the triangle, measured along the side.

APPLICATIONS.

1. *A transversal divides the sides into segments such that the product of three non-conterminous segments equals the product of the other three segments.*

Let $BN = lAN$, $BM = mMC$, $CL = nLA$, $MC = \alpha$, $LA = \beta$, $AN = \gamma$.

We have from the triangle ABC,

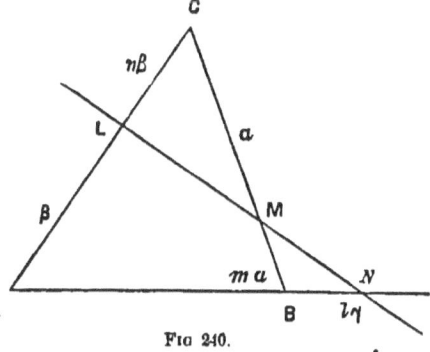

Fig. 240.

$$(1-l)\gamma + (1+m)\alpha + (1+n)\beta = 0; \qquad (1)$$

triangle MCL, $\quad\quad \alpha + n\beta - ML = 0$; $\quad\quad\quad$ (2)

triangle ANL, $\quad\quad NL + \beta + \gamma = 0$; $\quad\quad\quad$ (3)

also $\quad\quad\quad\quad\quad\quad NL = xML$. $\quad\quad\quad\quad$ (4)

Eliminating between equations (1), (2), (3), and (4) gives, (334),

$$l = mn;$$

$$\therefore \frac{BN}{AN} = \frac{BM}{MC} \cdot \frac{CL}{LA},$$

or $\quad\quad BN \cdot MC \cdot LA = AN \cdot BM \cdot CL,$

which was to be proved.

344. Conversely.—*If three points be taken on the sides of a triangle, one being taken on a side prolonged, or on all three prolonged, dividing the sides into parts such that the product of three non-conterminous parts equals the product of the other three, the points will be in a right line.*

345. *The three angle-transversals of a triangle through a common point, divide the sides so that the product of three non-conterminous segments equals the product of the other three segments.*

(The point may be within or without the triangle.)

Let
$$AO = x\delta, \ OC = y\varepsilon, \ BO = z\mu,$$
and
$$LB = lAB, \ MC = mBC, \ AN = nAC.$$

From the triangle AOC we will get

$$x\delta + y\varepsilon + \beta = 0;$$

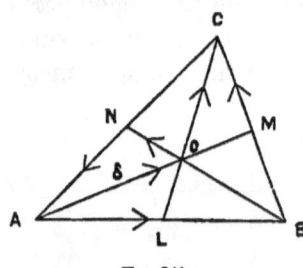

Fig. 241.

but since $\quad \delta = \gamma + (1-m)\alpha, \ \varepsilon = \alpha + l\gamma, \ \beta = -\alpha - \gamma,$

we have $\quad x(\gamma + (1-m)\alpha) + y(\alpha + l\gamma) - \alpha - \gamma = 0,$

or $\quad\quad x\gamma + x(1-m)\alpha + y\alpha + ly\gamma - \alpha - \gamma = 0;$

$\quad\quad \therefore x(1-m) + y - 1 = 0 = x + ly - 1,$

which gives $\quad\quad mx = (1-l)y.$ $\quad\quad\quad\quad$ (1)

Similarly from the triangle BOC we will obtain

$$ly = (1-n)z, \quad\quad\quad\quad (2)$$

and from $BOA \quad nz = (1 - m)x.$ \hfill (3)

Multiplying these three equations together and dividing through by xyz, gives
$$lmn = (1-l)(1-m)(1-n).$$
Substituting the proper values and reducing we finally have
$$LB \cdot MC \cdot NA = AL \cdot BM \cdot CN,$$
which was to be proved.

346. Conversely.—*If a point be taken on each of the sides of a triangle, or on one side and two prolonged, such that the product of three non-conterminous segments equals the product of the other three, then will the angle-transversals to those points pass through a common point.*

347. The coördinates used in quaternions are the lines of the figure, and such auxiliary lines as are necessary in order to solve the problem. There is not a set of fixed lines used as axes, as in the Cartesian system.

348. Remark.—Several of the solutions given in this chapter, are much longer than by other well-known methods; and the reader may be inclined to say that if this feature is characteristic of the system, it must be chiefly valuable for its novelty rather than for its use. But the merits of any system of analysis should not be judged from its ability to solve elementary problems, whether the solutions be long or short. The most powerful analysis generally appears to a disadvantage when applied to such problems. Were the processes of the Calculus exhibited to the student in solving such problems only as: The shortest distance between two points is a straight line; The evolute of a circle is a point; The shortest distance between two points on the surface of a sphere is the arc of a great circle; the student might infer that it was a cumbersome and tedious process of proving what could very easily be learned by very simple means, instead of the most powerful system of analysis ever devised. One merit, at least, is apparent in the preceding solutions; it furnishes an *uniform method* for the solution of this class of problems.

Solutions by Means of Transversals.

349. By discussing a general property, all the particulars contained in it may be deduced. We here present a solution of several of the problems already given, by deducing them from the proposition contained in Article 343.

1. *To prove the proposition contained in Article* 345.

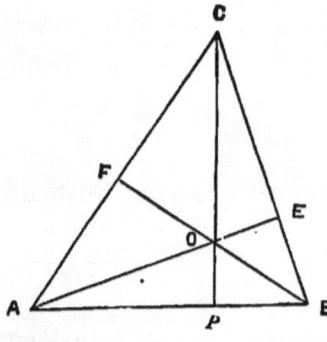

Let O be the point through which the angle-transversals are drawn. Let BF be a transversal to the triangle ACP, then we have, (343),

$$PB \cdot OC \cdot FA = AB \cdot PO \cdot CF;$$

and triangle CPB, transversal AE, gives

$$AB \cdot EC \cdot OP = AP \cdot BE \cdot OC.$$

Multiplying together and cancelling common factors gives

$$PB \cdot EC \cdot FA = AP \cdot BE \cdot CF,$$

Fig. 242.

which was to be proved.

2. *The medials of a triangle meet in a point.*

For we have

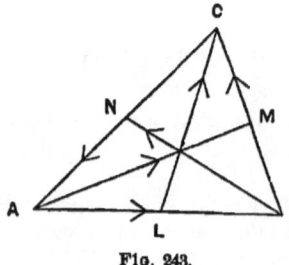

$$AL = LB,$$
$$BM = MC,$$
$$CN = NA;$$

and multiplying together gives

$$AL \cdot BM \cdot CN = LB \cdot MC \cdot NA,$$

Fig. 243.

which, according to Article 346, shows that the lines AM, BN, CL, meet in a point.

3. *The angle-bisectors of a triangle meet in a point.*

We have

$$\frac{CB}{CA} = \frac{LB}{AL}; \quad \frac{AC}{AB} = \frac{MC}{BM}; \quad \frac{AB}{BC} = \frac{NA}{CN};$$

and, multiplying together, we get

$$AL \cdot BM \cdot CN = LB \cdot MC \cdot NA;$$

which, according to Article 346, is the required result.

4. *The altitudes of a triangle meet in a point.*

From the similar right triangles CPB, AEB, we have

$$CP : PB :: AE : BE.$$

Similarly, $BF : FA :: CP : PA,$

also $AE : EC :: BF : FC;$

multiplying and dropping common factors, we find

$$PB \cdot EC \cdot FA = BE \cdot CF \cdot AP;$$

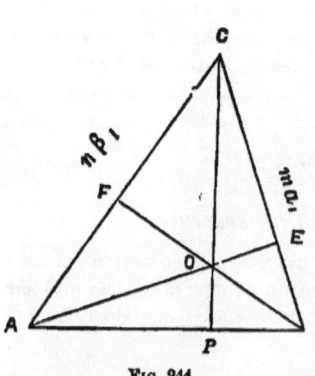

Fig. 244.

hence, (346), they have a common point O.

5. *If the corresponding vertices of two triangles are in lines radiating from a point, the corresponding sides prolonged will meet in a right line.*

Let ABC, $A'B'C'$ be two triangles having their vertices in lines radiating from O, then will the intersections of AB, $A'B'$; BC, $B'C'$; AC, $A'C'$, be in the line NL.

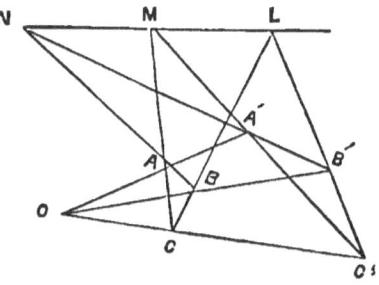

Fig. 245.

We have from

triangle OCA, secant $C'A'\ldots$ $OC'.CM.AA' = CC'.AM.OA'$,
" OCB, " $C'B'\ldots$ $CC'.BL.OB' = OC'.CL.BB'$,
" OBA, " $B'A'\ldots BB'.AN.OA' = OB'.BN.AA'$.

Multiplying together and cancelling common terms, gives

$$CM.BL.AN = AM.CL.BN;$$

hence, (344), N, M, L, are in a right line.

350. A complete quadrilateral is a four-sided figure in which each side cuts all the others, the sides being prolonged indefinitely. Thus, $ABCD$ is a complete quadrilateral, the side BC cutting the other sides in B, C, E; AB in the points B, A, F; AD in A, D, E; and CD in C, D, F. A line which is not a side, joining any two angles of a quadrilateral, is a diagonal. There are three diagonals, AC, BD, EF. There are also three quadrilaterals in the complete figure; the common convex $ABCD$, the uni-concave $EDFB$; and the bi-concave $EDFADC$, composed of two opposite triangles.

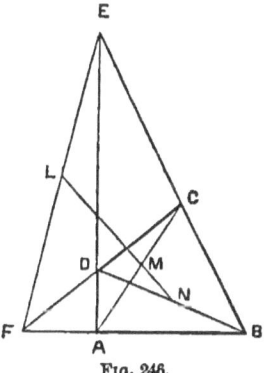

Fig. 246.

EXAMPLE.

The middle points L, M, N, of the diagonals of a complete quadrilateral are in a right line.

CHAPTER II.

MULTIPLICATION AND DIVISION OF VECTORS.

[REMARK.—The first time that a student reads the following chapter, he will naturally be led to ask, What right has one to make such assumptions, What led Hamilton to think of it, and, Of what use can it be? We cannot answer the first question more concisely, and, at the same time, comprehensively, than in Hamilton's words. He says: "*If the knowledge previously acquired, by any suitably performed analysis, be afterwards suitably applied, by the Synthesis answering to that Analysis, it will conduct to a suitable* RESULT. Or that whatever has been found by Analysis may afterwards be used by Synthesis; and that the thing which is reproduced by this synthetic process, will be the same with that which had been submitted to analysis previously."—(Hamilton's *Lectures on Quaternions*, p. 39.)

As a partial answer to the second, we will give, in an Appendix, a brief sketch of the history of the invention. It is unnecessary to answer the third, for the reader will soon see how the method can be used in the solution of problems. The problems here solved are simple, and are intended only to familiarize the reader with the use of the symbols, and not as a *test* of the merits of the system.]

Division of Vectors.

351. Notation.—Conceive three rectangular unit vectors, i, j, k, radiating from a common point O. Let i be positive to the right of O, j positive upward, and k positive in front of the plane of ij, and the opposite directions negative. Let a left-handed rotation be positive; thus if one is looking along the positive vector k toward the origin, then will the rotation of i upward toward j be

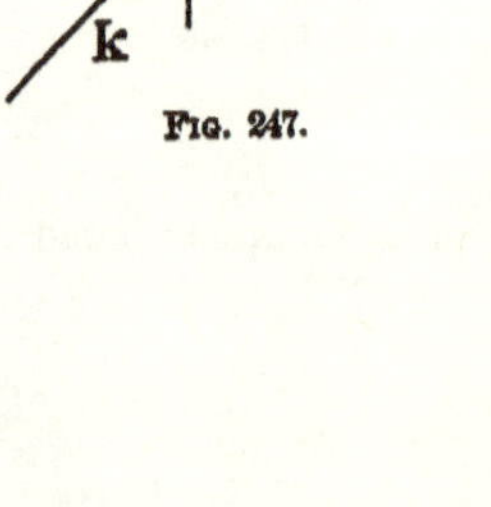

FIG. 247.

264

positive, and in the opposite direction, negative. Similarly, the positive rotation of j will be toward the front of the plane, and that of k will be downward.

352. Operation.—Conceive that j is attached to i so that as i is turned, j may be turned in either direction about i as an axis. Let j be turned through a quadrant in a positive direction; it will fall upon k, and the *line* which originally coincided with vector j will now coincide with vector k, and is thus said to produce k. The whole operation is thus expressed: i operating on j produces k, where vector i is spoken of as *the operator*. This statement is written

$$i = \frac{k}{j}, \qquad (1)$$

and is read, " vector i equals vector j divided into vector k," or, more simply, "i equals j divided into k," or, "i equals k divided by j." This is a vector equation.

In the same manner i operating upon k produces $-j$; and upon $-j$ produces $-k$, and so on, as shown by the figure.

If i operates twice successively upon j, thereby turning it positively through two quadrants, it will bring it into the position of $-j$. Successive operations of a *unit* vector upon a perpendicular one will be indicated by a repetition of the vector in the form of multiplication, or by means of an exponent. Thus the two successive quadrantal rotations above described will be written:

$$i.i = ii = i^2 = \frac{-j}{j} = -1, \qquad (2)$$

and, three successive quadrantal rotations will be written:

$$i^3 = \frac{-k}{j}.$$

Generally, if the rotation be through t successive quadrantal rotations producing some vector β, whose position can be definitely designated only when the *numerical* value of t is known, we have, according to the above plan:

$$i^t = \frac{\beta}{j}. \qquad (3)$$

266   QUATERNIONS.   [352.

The analysis indicated by Equation (3) may be so extended as to make it a GENERAL EQUATION. Thus, while the base i is a *unit* vector, it may be *positive* or *negative*; for if the rotation of j about $+i$ be *positive*, we have only to rotate *negatively* about $-i$ to produce the same result. Again, the exponent t may be not only *entire*, and *positive* or *negative*, but also fractional.

Thus, if vectors OB and OA are equal in length and perpendicular to vector i, the arc BA be represented by t where t is a fractional part of a quadrant (say one-third of a quadrant, or $t = \frac{1}{3}$), then if vector OB be revolved in a positive direction about $+i$ to coincide with vector OA, we have, according to the notation of Equation (3),

$$i^t = i^{\frac{1}{3}} = \frac{OA}{OB}.$$

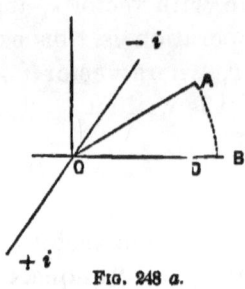

FIG. 248 a.

If $-i$ be the axis of rotation, then will the direction of rotation of OB into OA be negative, that is, t will be negative, and we have

$$(-i)^{-t} = \frac{OA}{OB}; \therefore i^t = (-i)^{-t}.$$

If OA be revolved about $+i$ to coincide with OB, t will be negative; and if about $-i$, t will be positive, hence

$$i^{-t} = \frac{OB}{OA} = (-i)^t.$$

The two preceding equations show that the value of an expression is not changed by changing the signs of both the base and the arc t in the exponent. Employing in this science the algebraic notation for the reciprocal, we have

$$i^{-t} = \frac{1}{i^t} = (-i)^t. \qquad (4)$$

In the expression $-i^t$, the *minus* sign applies to the entire expression, so that we have the identical equation

$$-i^t = -(i^t).$$

In this case the effect of the negative sign is to reverse the result produced by i^t. Thus, if $i^t = OA + OB$ as given above, then will $-i^t = -(OA + OB)$ be equivalent to revolving OB into the position of OA, and then reversing the result, producing $-OA$. But this is the same as revolving OB at once into the position of $-OA$, or of $-OB$ into OA, the angle of which is $2 + t$; or if the rotation be about $-i$, the positive angle will be $2 - t$; hence

$$-i^t = -(i^t) = -\frac{OA}{OB} = \frac{-OA}{OB} = \frac{OA}{-OB} = i^{2+t} = (-i)^{2-t}.$$

If OA were the divisor, we would have

$$-i^{-t} = -\frac{OB}{OA} = -\frac{1}{i^t} = i^{2-t}. \qquad (5)$$

353, 354.]   THE VERSOR.   267

If the angle AOB be 0 degrees, then $t = 0 \div \frac{1}{2}\pi$ (where $\frac{1}{2}\pi$ represents 90°) and (3) becomes

$$i^{\frac{2\theta}{\pi}} = \frac{\beta}{j}.$$

On the same plan, successive rotations equal respectively to x, y, z, etc., is the same as one rotation equal to the sum of $x + y + z +$ etc., or

$$i^x \cdot i^y \cdot i^z \cdot \text{etc.} = i^{(x+y+z+\text{etc.})},$$

where the exponents follow the algebraic law.

From the above it appears that in Equation (3) the vectors i, j, β, may be positive or negative, that j and β may be of any length, provided only that their lengths are equal.

353. If the vectors of the divisor and dividend are of unequal lengths, let α and β be unit vectors in a plane perpendicular to the axis i, a and b their lengths respectively, and $c = b \div a$; then if t be the angle between α and β, we have

$$c i^t = \frac{b\beta}{a\alpha} = \frac{b}{a} \cdot \frac{\beta}{\alpha};\qquad(6)$$

where the operation is conceived to be that of turning the vector $a\alpha$ about the axis i through an angle t, to coincide in direction with β; then comparing the length of $b\beta$ with that of $a\alpha$. The latter consists of an algebraic division of b by a.

If $c = d^t$, we have

$$(di)^t = \frac{b\beta}{a\alpha}.\qquad(7)$$

354. The Versor.—A versor, literally, implies that which turns about, and refers to the agent producing the rotation. In this system, the entire expression i^t, where i is a unit vector, and t an exponent, entire or fractional, positive or negative, is called *a versor*; and, *when operating as a quotient upon a perpendicular line, its effect is to turn that line through t quadrants*. If $t = 1$, the rotation will be through one quadrant; hence, every unit vector, *as a versor*, turns a perpendicular line through a quadrant, and is called a *quadrantal versor*. According to Equation (3) $\frac{\beta}{j}$ represents a versor, β and j being unit vectors perpendicular to i.

An operator is conceived to be an agent which changes the direction of a line, by rotating it ; or its length, by stretching it. A versor as an operator rotates a line, and a tensor elongates it. In division, we conceive that the versor operates upon the divisor line, turning it into the direction of the dividend line, after which the tensor of the quotient stretches the divisor line to the length of the dividend line.

Multiplication of Vectors.

355. Equation (3) may be written :

$$i'j = \beta, \qquad (8)$$

where the versor is the multiplier, the vector operated upon is the multiplicand, and is perpendicular to the axis of the versor, and the vector produced is the product, also perpendicular to the axis of the versor. This mode of transformation from division to multiplication corresponds to that of clearing an algebraic equation of fractions, but it will soon be shown that the *order* of the factors given above must be *particularly* observed.

Equation (8) is read "i' multiplied into j," or, "j multiplied by $i' = \beta$"; but in no case is it read "i' multiplied by j."

If, in the preceding discussion of the division of *unit* vectors, *multiplicand* be substituted for *divisor*, and *product* for *dividend*, it will be applicable to this form of multiplication. If the lengths of the vectors are unequal, the multiplicand is revolved about the axis of the versor to coincide in direction with that of the product, and the multiplicand is then *stretched* the amount required by the multiplier, thus producing the product. Thus, from Equation (6), we have

$$ci'. a\alpha = b\beta.$$

This is not an ordinary algebraic * multiplication ; it is a *vector multiplication*. We are not, therefore, at liberty to apply it to the rules of ordinary algebra, but must develop the rules which govern the operations in accordance with the principles upon which it is founded.

356. The non-commutative principle.—Suppose that j operates upon i, turning the latter left-handed so as to cause it to fall on $-k$; then we have (Fig. 247),

$$ji = -k.$$

---

* *Ordinary algebra* treats of operations upon literal *numbers*. The *analytical* part of quaternions is considered as *an algebra*. . Prof. Benjamin Pierce, of Harvard College, discovered 162 systems of algebra (see Johnson's *Encyclopædia*, Article QUALITATIVE ALGEBRA).

If i operates on j, we have

$$ij = k,$$

which shows that in the multiplication of rectangular vectors, reversing the order of the factors changes the sign of the result. This is one of the most peculiar principles of this system. In algebra the factors are interchangeable, thus $ab = ba$, $3 \times 4 = 4 \times 3$, and for this reason are said to be *commutative*, but in quaternions we have for rectangular vectors $ij = -ji$, and the factors are *non-commutative*.

The *plus* and *minus* signs are also commutative. Thus $i(-j) = -ij$. For if in Fig. 247, $-i$ operate upon j, it will produce $-k$, which is the same as $i(-j)$.*

357. The associative principle consists in grouping the factors in different ways, thus,

$$i \cdot jk = ij \cdot k = ijk.$$

If the *cyclical* order of the factors be preserved, the products will be equal; thus we have

$$ijk = jki = kij.$$

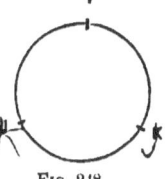

Fig. 248.

But if the cyclical order be deranged, the sign of the product will be changed, thus

$$ijk = -ikj = -kji = -jik,$$

a result which follows directly from the preceding Article.

EXAMPLES.

Show, from the figure, that the following equations are correct:

1. $ji = -k,$ $j(-k) = -i,$ $j(-i) = k,$ $jk = i.$
2. $ki = j,$ $kj = -i,$ $k(-i) = -j,$ $k(-j) = i.$
3. $-ij = -k,$ $i(-k) = j,$ $-i(-j) = k,$ $-ik = j.$
4. $-ji = k,$ $-jk = -i,$ $-j(-i) = -k,$ $-j(-k) = i.$
5. $-ki = -j,$ $-k(-j) = -i,$ $-k(-i) = j,$ $-kj = i.$
6. $ijk = i^2 = j^2 = k^2 = -1.$

---

* There are, however, many cases in algebraic analysis where the *elements* in the expression are non-commutative. Thus a^x, $\log^2 x$, $\log \sin x$, are *not* equal respectively to x^a, $\log^x 2$, $\sin \log x$.

358. The square of any vector *equals minus the square of a line of equal length.*

Let a be the length of the vector, then if i be a unit vector, the entire vector will be ai, and we have (since $i^2 = -1$, (352)),
$$(ai)(ai) = a^2 i^2 = -a^2,$$
which was to be proved. This furnishes a convenient method of changing vectors to lines.

359. The reciprocal of a *unit* vector equals minus the same vector.

Equation (2), (352), gives
$$-ii = 1 \; ; \; \therefore -i = \frac{1}{i} = i^{-1}. \qquad (9)$$

That is, a *positive* quadrantal rotation about $-i$ as an axis, is equivalent to a *negative* quadrantal rotation about $+i$; but the latter is called the reciprocal of the former.

It was observed (354) that $\beta \div j$ represents a versor, and (355) that the versor as a factor is the multiplier. These principles lead to another mode of changing division into multiplication; thus, multiplying both members of (8) *into* j^2 gives
$$i' j^2 = \frac{\beta}{j} j^2 = \beta j.$$

But, since $j^2 = -1$, this becomes
$$i'(-1) = -i' = \beta j = -\beta(-j) = -\beta \cdot \frac{1}{j} \text{ (by (9))} = -\frac{\beta}{j} \; ;$$
$$\therefore \frac{\beta}{j} = -\beta j = i'. \qquad (10)$$

Here we have another peculiar result, that while i' operating on j as a divisor produces β, by turning the former through an angle t, by operating upon the multiplier $-\beta$ it produces j, by turning the former through an angle equal to $2 - t$; that is, the supplement of the angle t of the versor.

We further observe that $\beta \div j$ as a versor, operating as a factor upon j, produces β; hence, its equal, $-\beta j$, must produce the same result. We have $-\beta j \cdot j$, which, by the associative principle, becomes $-\beta \cdot jj = -\beta \cdot j^2 = -\beta(-1) = \beta$, as it should.

Again, if β be rotated about $-i$, the positive angle will be t, and its supplement will be $2 - t$, hence, according to the preceding principle, we write at once, $\beta j = (-i^{2-t})$, and this compared with Equation (10) gives
$$-i' = (-i)^{2-t}.$$

The fact that the versor, when operating upon the multiplicand, is made the multiplier, or is the first one of the two factors, determines the proper order of the factors, so that cancellation may be performed as if it were algebraic. Thus we have (Equation (3)

$$ij = \frac{\beta}{j} j = \beta,$$

which is the same as if j were cancelled in the second member by a stroke inclined thus /. If written $j\frac{\beta}{j}$, the result will not, in general, be β, and the j's cannot be cancelled.

360. From Equation (2), (352), we have

$$i^2 = -1 = (-1)^1;$$

and, taking one-half the exponents of both terms, we have

$$i = (-1)^{\frac{1}{2}},$$

or, if we employ the radical sign as in ordinary algebra, this may be written

$$i = \sqrt{-1},$$

and the same result may be found for every vector which operates to turn a line through a quadrant. The expression $\sqrt{-1}$, separated from i, j, or any other axis, may be considered as an INDETERMINATE VECTOR-UNIT, or an unit-vector with *indeterminate direction*.*

The symbol $\sqrt{-1}$ in this system is, therefore, an analytical expression for the turning of a line through a quadrant about an axis perpendicular to the line.

This is of the same form as the even root of negative unity in algebra, but the significations of the two are very different.† The $\sqrt{-1}$ in algebra is impossible. It results from an attempt to find a number whose square is minus one; but the above symbol results from expressing by an exponent a *fact* which has no representation in ordinary algebra, combined with the laws already assigned to the multiplication of vectors.‡ The result was reached by first causing two

---

* Hamilton's *Lectures*, p. 178.
† Hamilton's *Lectures*, p. 635.
‡ There are several instances in which the same notation is employed, having very different meanings. Thus, $\sin^{-1} x$ does not mean the reciprocal of $\sin x$; in the expression Fx, read "function of x," F is not a factor.

successive operations upon j, producing $i^2j = -j$, then *going backward* one operation, producing $i = \sqrt{-1}$ (j having been dropped). The latter operation is *a division* in this system, and not an algebraic root.

Another Method.

361. Multiplication of Oblique Vectors.—Let $OA = \beta$, $OB = \alpha$, each being a unit-vector, $AOD = \theta$ be the angle between them; it is required to find the product of $\alpha\beta$ and $\beta\alpha$.*

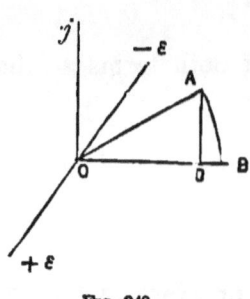

Fig. 249.

Take the unit-vector ε perpendicular to the plane of α and β, and positive in front of the plane; and vector j perpendicular to α and ε. The length of OA being unity, we have, as lines

$$OD = \cos\theta, \qquad DA = \sin\theta,$$

and as vectors, (328),

$$OD = \cos\theta.\alpha, \qquad DA = \sin\theta.j.$$

The triangle ODA gives, (330),

$$OA = OD + DA,$$

or
$$\beta = \cos\theta.\alpha + \sin\theta.j. \qquad (1)$$

Let α be the multiplier, then writing α *before* each term, we have

$$\alpha\beta = \cos\theta.\alpha^2 + \sin\theta.\alpha j. \qquad (2)$$

But (352), $\qquad \alpha^2 = -1,$

---

* The subscripts to the unit-vectors are dropped, and the Greek letters are generally restored for designating the vectors. Hamilton, for some reason, used the English letters *i, j, k*, to designate mutually *perpendicular* vectors.

and
$$\alpha j = \varepsilon;$$
$$\therefore \alpha\beta = -\cos\theta + \varepsilon\sin\theta, \quad (3)$$

which is the result sought.

If β be the multiplier, we have, writing α *after* each term,

$$\beta\alpha = \cos\theta.\alpha^2 + \sin\theta.j\alpha, \quad (4)$$
$$= -\cos\theta - \varepsilon\sin\theta, \quad (5)$$

which is the same as making ε negative in (3).

362. Division of Oblique Vectors.—Let α be the divisor, then from (1) we have

$$\frac{\beta}{\alpha} = \cos\theta + \sin\theta\frac{j}{\alpha},$$
$$= \cos\theta + \varepsilon\sin\theta. \quad (6)$$

The same result is found by dividing the left member of (5) by α^2, and the right member by its equal -1.

Similarly, dividing the left member of (3) by β^2, and the right by its equal -1, gives

$$\frac{\alpha}{\beta} = \cos\theta - \varepsilon\sin\theta, \quad (7)$$

the right member of which differs from the right member of the preceding equation by the sign of ε.

EXAMPLES.

1. Deduce the value of $\dfrac{\beta}{\alpha}$, equation (7), by taking the reciprocal of the second member of Equation (6).

2. Show that the product of the right members of Equations (3) and (5) is unity, and therefore, as unit-vectors $\alpha\beta = \dfrac{1}{\beta\alpha}$, (and not $\alpha\beta = \dfrac{1}{\alpha\beta}$).

3. Find the value of $\dfrac{\alpha}{\beta}$ for $\theta = 0°$, $45°$, $90°$, and $180°$.

4. Show that $(\alpha\beta)(\beta\gamma) = -\alpha\gamma$.

363. Generally, if α and β are not unit-vectors, we have

$$\alpha\beta = T\alpha T\beta \left(-\cos\theta + \varepsilon\sin\theta\right); \tag{9}$$

$$\beta\alpha = T\beta T\alpha \left(-\cos\theta - \varepsilon\sin\theta\right); \tag{10}$$

$$\frac{\alpha}{\beta} = \frac{T\alpha}{T\beta} \left(\cos\theta - \varepsilon\sin\theta\right); \tag{11}$$

$$\frac{\beta}{\alpha} = \frac{T\beta}{T\alpha} \left(\cos\theta + \varepsilon\sin\theta\right). \tag{12}$$

Since j has disappeared from the result, or, more generally, since α and β are simply confined to a plane perpendicular to the axis ε, it is evident that they may have *any* position in that plane. The axis ε fixes the position of the *plane* of the two vectors, but not their position in that plane. The parenthetical parts are versors.

It also appears that multiplying or dividing one vector by another involves the three rectangular dimensions of space.

364. Scalar.*—Since tensor α and tensor β are each numerical, their product or quotient will be numerical, and when multiplied by $\cos\theta$, the final product will also be numerical, and may be positive or negative. This product Hamilton called the SCALAR *part*. It represents the *reals* of algebra. It is represented by the letter S placed before an undeveloped expression; thus, $S\alpha\beta$ implies the scalar part of the product of α into β, and equals $T\alpha T\beta \cos\theta$. Similarly for $S\dfrac{\beta}{\alpha}$.

365. The Vector Part.—The other part of (9), or $T\alpha T\beta \sin\theta \cdot \varepsilon$, is a vector either longer or shorter than the unit-vector ε, and similarly in regard to the other three equations. The VECTOR *part* is indicated by the letter V placed before an undeveloped expression; thus, $V\alpha\beta$, $V\dfrac{\beta}{\alpha}$, etc.

366. A Quaternion is the result of multiplying or divid-

---

* Latin, *scala*, series of steps; so called because it represents *discontinuous number*.

ing one directed line by another in space. This name was first applied to the result, because the result, when found by means of rectangular vectors, gave an expression of *four* terms, of the form $w + ix + jy + kz$, where w is numerical, i, j, k, mutually perpendicular unit-vectors, and x, y, z, distances along the vectors.

A Quaternion is the product of a *tensor* and a *versor* (Eqs. (9) and (10·); or the product of a tensor into a unit-vector with a scalar exponent (Eq. (6), Art. 353); or the power of a vector (Eq. (7), Art. 353); or the power of a versor when the tensor is unity (Eq. (10), Art. 359, and Eq. (7), Art. 353); or the sum of a SCALAR and VECTOR (Eqs. (1) (2) (3) (4) of Article 367).

In each of these forms of the quaternion, *four* elements are involved : the tensor (which may be unity), the angle between the vectors, and two angles for fixing the direction of the unit axis.

367. General Expressions.—We may now write equations (9), (10), (11), (12), of Article 363, as follows, disregarding the algebraic signs,

$$\alpha\beta = S\alpha\beta + V\alpha\beta, \qquad (1)$$
$$\beta\alpha = S\beta\alpha + V\beta\alpha, \qquad (2)$$
$$\frac{\alpha}{\beta} = S\frac{\alpha}{\beta} + V\frac{\alpha}{\beta}, \qquad (3)$$
$$\frac{\beta}{\alpha} = S\frac{\beta}{\alpha} + V\frac{\beta}{\alpha}; \qquad (4)$$

and, by comparing these expressions with (9), (10), (11), (12), before referred to, we find

$$S\alpha\beta = - T\alpha T\beta \cos\theta. \qquad (5)$$
$$S\beta\alpha = - T\alpha T\beta \cos\theta. \qquad (6)$$
$$S\frac{\alpha}{\beta} = \frac{T\alpha}{T\beta} \cos\theta. \qquad (7)$$
$$S\frac{\beta}{\alpha} = \frac{T\beta}{T\alpha} \cos\theta. \qquad (8)$$
$$V\alpha\beta = + T\alpha T\beta \sin\theta . \varepsilon. \qquad (9)$$
$$V\beta\alpha = - T\alpha T\beta \sin\theta . \varepsilon. \qquad (10)$$
$$V\frac{\alpha}{\beta} = - \frac{T\alpha}{T\beta} \sin\theta . \varepsilon. \qquad (11)$$
$$V\frac{\beta}{\alpha} = + \frac{T\beta}{T\alpha} \sin\theta . \varepsilon. \qquad (12)$$

Comparing these expressions, we have

from (5) and (6) $\quad S\alpha\beta = S\beta\alpha,$ (13)

" (7) and (8) $\quad S\dfrac{\alpha}{\beta} = \dfrac{(T\alpha)^2}{(T\beta)^2} S\dfrac{\beta}{\alpha},$ (14)

" (9) and (10) $\quad V\alpha\beta = - V\beta\alpha,$ (15)

" (11) and (12) $\quad V\dfrac{\alpha}{\beta} = -\dfrac{(T\alpha)^2}{(T\beta)^2}\cdot V\dfrac{\beta}{\alpha},$ (16)

" (1), (2), (13), (15), $\alpha\beta + \beta\alpha = 2S\alpha\beta,$ (17)

" (1), (2), (13), (15), $\alpha\beta - \beta\alpha = 2V\alpha\beta.$ (18)

368. Discussion.—1°. Let the vectors be mutually perpendicular. Then $\theta = 90°$, and (5), (6), (7), (8) all reduce to zero; hence

$$Sij = 0. \tag{1}$$

Equations (9), (10), (11), (12), will all reduce to the forms given in Article 354, as they should.

2°. Let the vectors coincide; then $\theta = 0$ and the vector part reduces to zero, and (1) becomes

$$\alpha\beta = S\alpha\beta = -T\alpha T\beta \cos 0 = -ab, \tag{2}$$

where a and b are the lengths of the lines.

3°. Let $\theta = 270°$, then

$$\alpha\beta = V\alpha\beta = \sin 270°\cdot \varepsilon = -\varepsilon, \tag{3}$$

which agrees with the expression $i(-j) = -k$.

4°. The coefficient of the vector part of $\alpha\beta$ may be represented by a parallelogram.

From equation (9), Article 367, we have

$$V\alpha\beta = T\alpha T\beta \sin\theta\cdot\varepsilon.$$

Let $T\alpha = a$, $T\beta = b$, then will the coefficient of ε be

$ab \sin \theta.$

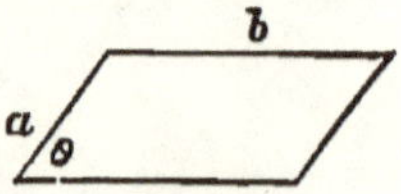

Let a and b be the adjacent sides of a parallelogram; then $a \sin \theta$ will be its altitude and $ab \sin \theta$, its area.

Fig. 250.

5°. The scalar of $\alpha\beta$ is *numerically* the area of a parallelo-

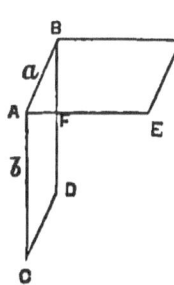

Fig. 251.

gram whose acute angle is the complement of θ.

From (5) of Article 367, we have

$$S\alpha\beta = -T\alpha T\beta \cos \theta$$
$$= -ab \cos \theta.$$

Draw AC perpendicular to AE and make it equal to b, and complete the parallelogram on a and b as sides. Then will $a \cos \theta = a \cos BAF = a \sin ABF = AF =$ the altitude of the parallelogram AD; hence the area of $ABDC$ will be

$$ab \cos \theta,$$

which is *numerically* the same as $-ab \cos \theta$.

EXAMPLES.

1. Show that $S\alpha^2 = -a^2$.

This is true because $\alpha^2 = -a^2$, and the latter being numerical is the scalar part.

2. Show that $V\alpha^2 = 0$.
3. Show that $S\alpha = 0$.
4. Show that $V\alpha = \alpha$.
5. Show that $(\alpha + \beta)^2 = \alpha^2 - 2T\alpha T\beta \cos \theta + \beta^2$.

SOLUTION.—We have

$$(\alpha + \beta)^2 = (\alpha + \beta)(\alpha + \beta),$$

Art. 367, Eq. (17), $\qquad = \alpha^2 + 2S\alpha\beta + \beta^2,$

Art. 367, Eq. (5), $\qquad = \alpha^2 - 2T\alpha T\beta \cos \theta + \beta^2,$

which was to be proved.

6. Find what the preceding example becomes when $\theta = 90°$.
7. Show that $(\alpha - \beta)^2 = \alpha^2 - 2S\alpha\beta + \beta^2$.
8. If $\dfrac{\alpha}{\gamma} = \cos(\theta-\varphi) - \varepsilon \sin(\theta-\varphi)$, write the scalar and vector parts.
9. If $\dfrac{\beta}{\gamma} = (\cos \theta + \varepsilon \sin \theta)(\cos \varphi + \varepsilon' \sin \varphi)$, find the scalar and vector parts.

SOLUTION.—Expanding the second member, we have

$$\cos \theta \cos \varphi + \varepsilon \sin \theta \cos \varphi + \varepsilon' \cos \theta \sin \varphi + \varepsilon\varepsilon' \sin \theta \sin \varphi.$$

But, (361, 3),

$$\varepsilon\varepsilon' = -\cos(\varepsilon,\varepsilon') + \varepsilon'' \sin(\varepsilon,\varepsilon'),$$

in which ε'' is a unit-vector perpendicular to the plane of $\varepsilon\varepsilon'$. Hence we have

$$S\frac{\beta}{\gamma} = \cos\theta\cos\varphi - \sin\theta\sin\varphi\cos(\varepsilon,\varepsilon'),$$

$$V\frac{\beta}{\gamma} = \varepsilon\sin\theta\cos\varphi + \varepsilon'\cos\theta\sin\varphi + \varepsilon''\sin\theta\sin\varphi\sin(\varepsilon,\varepsilon').$$

10. $S\alpha\beta\gamma$ represents the volume of a parallelopiped.

Let α, β, γ, be unit-vectors along AB, AC, AE.

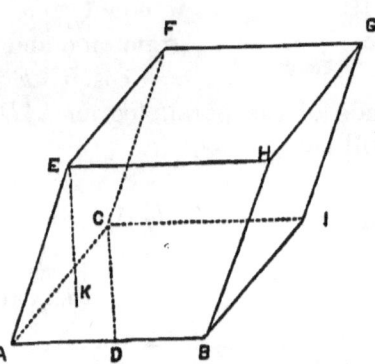

Fig. 252.

Then, (Art. 361, Eq. (3)),

$$\alpha\beta = -\cos(\alpha,\beta) + \varepsilon\sin(\alpha,\beta)$$

$$S\alpha\beta\gamma = S[-\cos(\alpha,\beta)\gamma + \varepsilon\gamma\sin(\alpha,\beta)]$$

$$= S\varepsilon\gamma(\sin\alpha,\beta),$$

because $S[-\cos(\alpha,\beta)]\gamma = 0$ being a vector.

But

$$\varepsilon\gamma = -\cos(\varepsilon,\gamma) + \varepsilon'\sin(\varepsilon,\gamma)$$

ε' being a vector perpendicular to ε, γ.

$$\therefore S\alpha\beta\gamma = S[-\cos(\varepsilon,\gamma)\sin(\alpha,\beta) + \varepsilon'\sin(\varepsilon,\gamma)\sin(\alpha,\beta)]$$

$$= -\cos(\varepsilon,\gamma)\sin(\alpha,\beta),$$

$$= -\cos(90° - EAK)\sin CAB \text{ (EK being perpendicular to the base)},$$

$$= -\sin EAK \sin CAB.$$

Let the length of the vectors be a, b, c, then

$$-S\alpha\beta\gamma = abc\sin CAB \sin EAK \qquad (1)$$

$$= a\cdot b\sin CAB\cdot c\sin EAK$$

$$= AB\cdot CD\cdot EK \text{ (CD being perpendicular to AB)},$$

$$= \text{Volume } BACI\text{-}E.$$

Since the volume will be the same whatever be the order of the vectors, we have

$$S\alpha\beta\gamma = \pm S\alpha\gamma\beta = \pm S\gamma\beta\alpha, \text{ etc.} \qquad (2)$$

If $EAK = 0$, the three vectors will be in one plane, and we have

$$S\alpha\beta\gamma = 0. \qquad (3)$$

Conversely, if $S\alpha\beta\gamma = 0$, neither α, β, nor γ being zero, the three vectors will be in one plane.

11. $-\frac{1}{6} S\alpha\beta\gamma$ represents the volume of a triangular pyramid.

12. Let α be a vector perpendicular to $\beta - \alpha$, find the value of $S\alpha\beta$.

We have from Article 368, 1°,

$$S\alpha(\beta - \alpha) = 0,$$

and expanding, $S(\alpha\beta - \alpha^2) = 0,$

or $\quad S\alpha\beta - S\alpha^2 = 0,$

or, (359), $\quad S\alpha\beta + a^2 = 0$;

$$\therefore S\alpha\beta = -a^2,$$

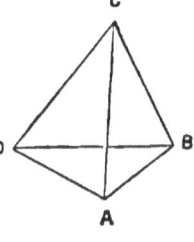

Fig. 253.

which is the result sought. The last equation may also be considered an *equation of condition* of the mutual perpendicularity of α and $\beta - \alpha$.

369. We have previously found (359), that

$$\epsilon^2 = -1, \quad (1)$$

which shows that the vector operated upon has been reversed, and hence is the measure of two right angles. Similarly,

$$\epsilon^4 = +1, \quad (2)$$

indicates a revolution through four right angles, and generally, if n be an integer, we have

$$\epsilon^{4n-2} = -1, \quad (3)$$

$$\epsilon^{4n} = +1 ; \quad (4)$$

in both of which n is one of the series of natural numbers, 1, 2, 3, etc., but the exponent of the former is an *odd* multiple of 2, and indicates n complete revolutions *less* two quadrants, and the exponent of the latter is an even multiple of 2, and indicates n complete revolutions.

EXAMPLES.

1. The sum of the angles of a triangle equals two right angles.

SOLUTION.—Let $CB = \alpha$, $CA = \beta$, $BA = \gamma$. Let CA be turned left-handed about ϵ (the axis of rotation being perpendicular to CAB) to a parallelism with CB, or α ; CB to a parallelism with AB, or γ ; and AB, or $-\gamma$, to a parallelism with AC, or $-\beta$. Or the last

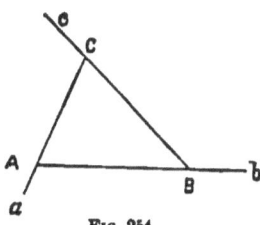

Fig. 254.

rotation may be right-handed, coinciding with CA prolonged. In either case we have

$$\varepsilon^{\frac{2C}{\pi}} = \frac{\alpha}{\beta}; \quad \varepsilon^{\frac{2A}{\pi}} = \frac{\beta}{\gamma}; \quad \varepsilon^{\frac{2B}{\pi}} = -\frac{\gamma}{\alpha}.$$

Multiplying together and cancelling common factors, we have

$$\varepsilon^{\frac{2}{\pi}(C+A+B)} = -1;$$

hence we have, (Eq. (3)),

$$\frac{2}{\pi}(C + A + B) = 4n - 2,$$

or
$$A + B + C = (2n - 1)\pi;$$

hence, *analytically*, the sum of the angles of a triangle may be any odd number of times π; that is, π, 3π, 5π, etc. But this result is obtained by supposing that we pass around the triangle repeatedly. *Arithmetically*, we pass around but once, and the corresponding value for the sum of the angles is found by making $n = 1$, and hence we have

$$A + B + C = \pi = 2 \text{ right angles.}$$

2. If the sides of a triangle be prolonged in the same direction, the sum of the external angles will equal four right angles.

(Obs.—In this case the sum of the exponents of ε will equal $4n$, and $n = 1$.)

3. The sum of all the angles on one side of a line formed by lines drawn from any point of it equals two right angles.

4. If the sides of a convex polygon be prolonged in the same direction, the sum of the external angles will equal four right angles.

370. Proposition.—*If an equation contains scalars and also vectors, the sum of the scalars on one side of the sign of equality will equal the sum of those on the other side; and also the sum of the vector parts on one side will equal the sum of those on the other.*

(Obs.—It should be understood that the vectors are of the first degree, and that each term contains only one vector, for otherwise the terms containing vectors may have scalar parts.)

Thus, if we have

$$x + x\alpha + (x + z)\beta = ax + y - b\beta,$$

then
$$x = ax + y,$$

and
$$x\alpha + (x + z)\beta = -b\beta.$$

For the scalars being numerical will exist independently of the parts which involve rotation.

EXAMPLE.

We found, in Article 369, that

$$\cos(\theta + \varphi) + \varepsilon(\sin\theta + \varphi) = (\cos\theta + \varepsilon\sin\theta)(\cos\varphi + \varepsilon\sin\varphi).$$

Expanding the second member, observing that $\varepsilon^2 = -1$, we have

$$\cos(\theta + \varphi) + \varepsilon\sin(\theta + \varphi) = \cos\theta\cos\varphi + \varepsilon\sin\theta\cos\varphi + \varepsilon\sin\varphi\cos\theta$$
$$- \sin\theta\sin\varphi.$$

Therefore, taking the scalars and vectors we have, according to the proposition (observing that ε cancels out),

$$\cos(\theta + \varphi) = \cos\theta\cos\varphi - \sin\theta\sin\varphi,$$
$$\sin(\theta + \varphi) = \sin\theta\cos\varphi + \cos\theta\sin\varphi;$$

which are well-known trigonometrical formulas.

APPLICATIONS.

371. Solutions of Plane Triangles.—We have already found, for a plane triangle, (330), that

$$\gamma = \alpha + \beta. \qquad (1)$$

By operating upon this equation by the multiplication (or division) of vectors, all the cases of plane triangles may be solved.

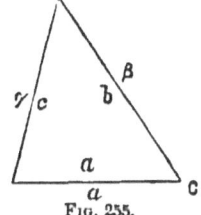
Fig. 255.

372. *The sines of the angles are proportional to the opposite sides.*

In solving this problem it is necessary to introduce the sines of two angles. Multiplying both sides of (1) into α, gives

$$\gamma\alpha = \alpha^2 + \beta\alpha, \qquad (2)$$

and taking the vectors of both sides, (see Ex. 2, Art. 368, and Eq. (9), Art. 367), we have

$$V\gamma\alpha = 0 + V\beta\alpha,$$

introducing tensors, we have

$$ca\varepsilon\sin(\gamma, \alpha) = ba\varepsilon\sin(\beta, \alpha),$$

dropping a and ε, $\quad c\sin B = b\sin C;$

$$\therefore c : b :: \sin B : \sin C, \qquad (3)$$

which was to be proved.

373. *To find a side in terms of the other two sides and the angles opposite those sides.*

Taking the scalars of equation (2), we have

$$S\gamma\alpha = -a^2 + S\beta\alpha,$$

with tensors, $\quad -ca\cos(\gamma, \alpha) = -a^2 - ba\cos(\beta, \alpha),$

cancelling a $\quad\quad a = c\cos(\gamma, \alpha) - b\cos(\beta, \alpha),$

or $\quad\quad\quad\quad a = c\cos B + b\cos C, \quad\quad (4)$

the sign of the last term being changed because C, the internal angle of the triangle, is the supplement of the angle between β and α. The angles between γ and α, and β and α are both at the right of the lines b and c and above a.

374. *To find the cosine of an angle in terms of the sides.*

Squaring (1), gives

$$\gamma^2 = \alpha^2 + 2S\alpha\beta + \beta^2, \quad\quad (5)$$

or, (359), $\quad -c^2 = -a^2 - 2ab\cos(\alpha, \beta) - b^2$

or $\quad\quad\quad c^2 = a^2 + 2ab\cos(\alpha, \beta) + b^2,$

or $\quad\quad\quad c^2 = a^2 - 2ab\cos C + b^2. \quad\quad (6)$

375. *The square of the hypothenuse equals the sum of the squares of the sides about the right angle.*

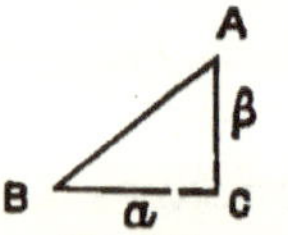

(This may be solved by making $S\alpha\beta = 0$ in equation (5)).

Fig. 256.

376. *The side adjacent either acute angle of a right-angled triangle equals the hypothenuse into the cosine of the angle.*

Let γ = the hypothenuse, then from (1)

$$\beta = \gamma - \alpha.$$

But β and α are mutually perpendicular, hence (see Ex. 12, Art. 368)

$$S\alpha(\gamma - \alpha) = 0;$$
$$\therefore S\alpha\gamma = -a^2,$$

or, (367, Eq. (5)), $\quad -ac\cos B = -a^2;$

$$\therefore a = c\cos B. \quad\quad (7)$$

377. Division may be used in the solution of the same examples. Thus, dividing both members of (1) by α, gives

$$\frac{\gamma}{\alpha} = 1 + \frac{\beta}{\alpha}. \qquad (8)$$

Taking the scalars, gives

$$\frac{c}{a} \cos\left(\frac{\gamma}{\alpha}\right) = 1 + \frac{b}{a} \cos\left(\frac{\beta}{\alpha}\right),$$

or

$$\frac{c}{a} \cos B = 1 - \frac{b}{a} \cos C,$$

which reduces at once, to equation (4).

Taking the vectors of (8) gives

$$V\frac{\gamma}{\alpha} = V\frac{\beta}{\alpha},$$

or

$$\frac{c}{a} \epsilon \sin B = \frac{b}{a} \epsilon \sin C,$$

which reduces to (3).

Squaring (8) and reducing, gives (5).

Three Co-initial Vectors.

378. Three vectors drawn from a point in any direction, will, generally, be the edges of a triedral. Let α, β, γ, be those vectors. We observe that a general relation between them cannot be found by addition. The process of turning a line from α so as to coincide in direction with β is, as we have seen, expressed by multiplication, or by division. Passing from α to β, β to γ, γ to α, whether they be unit-vectors or not, will be expressed by

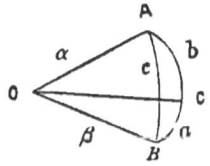

Fig. 257.

$$\frac{\alpha}{\beta} \cdot \frac{\beta}{\gamma} \cdot \frac{\gamma}{\alpha},$$

and as these, by cancellation, reduce to unity, we have the equation

$$\frac{\alpha}{\beta} \cdot \frac{\beta}{\gamma} \cdot \frac{\gamma}{\alpha} = 1, \qquad (1)$$

This equation is as general for triedrals, as is equation (1), Article 330, for triangles; and by resolving this equation, all the relations existing between the facial and diedral angles of a triedral may be determined; and further, it is applicable to the special case in which the three vectors are co-planar. We will consider the vectors as *unit-vectors*. The following form of the equation is more convenient for use.

$$\frac{\alpha}{\gamma} = \frac{\alpha}{\beta} \cdot \frac{\beta}{\gamma}. \qquad (2)$$

This, in the abridged notation, (369), becomes

$$\varepsilon^{\frac{2\mu}{\pi}} = \varepsilon_1^{\frac{2\theta}{\pi}} \varepsilon_2^{\frac{2\phi}{\pi}}. \qquad (3)$$

379. Three co-initial co-planar vectors.—Let the angle between α and β be θ, and between β and γ, φ; then will the angle between α and γ be $\theta + \varphi$.

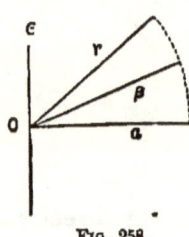

Fig. 258.

The vectors being in one plane and unity in length, we have

$$\varepsilon = \varepsilon_1 = \varepsilon_2,$$

and equation (3) gives

$$\varepsilon^{\frac{2}{\pi}(\theta+\phi)} = \varepsilon^{\frac{2\theta}{\pi}} \varepsilon^{\frac{2\phi}{\pi}},$$

which shows that a rotation from α to β followed by a rotation from β to γ equals a rotation from α to γ.

By restoring the values of these expressions we have

$$\cos(\theta+\varphi) + \varepsilon \sin(\theta+\varphi) = (\cos\theta + \varepsilon \sin\theta)(\cos\varphi + \varepsilon \sin\varphi), \quad (4)$$

which is the same as the example in Article 370; hence, expanding and reducing as in that example, we have

$$\cos(\theta+\varphi) = \cos\theta \cos\varphi - \sin\theta \sin\varphi,$$
$$\sin(\theta+\varphi) = \sin\theta \cos\varphi + \cos\theta \sin\varphi.$$

If γ were between α and β, we would have $\theta - \varphi$ for the angle between α and γ, and proceeding as before we would find

$$\cos(\theta-\varphi) = \cos\theta \cos\varphi + \sin\theta \sin\varphi,$$
$$\sin(\theta-\varphi) = \sin\theta \cos\varphi - \cos\theta \sin\varphi;$$

which are the well-known trigonometrical formulas for the cosine and sine of the sum and of the difference of two arcs. Next, let $\theta = \varphi$, then equation (4) gives

$$\cos 2\varphi + \varepsilon \sin 2\varphi = (\cos \varphi + \varepsilon \sin \varphi)^2,$$

and continuing this operation n times, we have

$$\cos n\varphi + \varepsilon \sin n\varphi = (\cos \varphi + \varepsilon \sin \varphi)^n, \qquad (5)$$

which follows the same law as De Moivre's formula. Substituting the algebraic expression for the indeterminate vector-unit, (354), we have

$$\cos n\varphi + \sqrt{-1} \sin n\varphi = (\cos \varphi + \sqrt{-1} \sin \varphi)^n, \qquad (6)$$

which is De Moivre's formula. It shows that one rotation through an angle $n\varphi$ is the same as n successive rotations through an angle φ.

EXAMPLE.

Find the three roots of unity.
Putting equation (6) under the more general form

$$(\cos \varphi + \sqrt{-1} \sin \varphi)^n = \cos n(2r\pi + \varphi) + \sqrt{-1} \sin n(2r\pi + \varphi),$$

in which $r = 0, 1, 2, 3$, etc., and making $\varphi = 0$, and $n = \tfrac{1}{3}$, we have

$$\sqrt[3]{1} = \cos \tfrac{2}{3} r\pi + \sqrt{-1} \sin \tfrac{2}{3} r\pi,$$

$$= 1, \text{ if } r = 0,$$

$$= -\tfrac{1}{2} + \tfrac{1}{2}\sqrt{-3}, \text{ if } r = 1,$$

$$= -\tfrac{1}{2} - \tfrac{1}{2}\sqrt{-3}, \text{ if } r = 2;$$

therefore the roots are 1, $\tfrac{1}{2}(-1 + \sqrt{-3})$, $\tfrac{1}{2}(-1 - \sqrt{-3})$.

[REMARK.—The abridged notation of Hamilton bears a close analogy to the corresponding expressions deduced from Euler's formulas. These formulas are

$$\cos \theta = \tfrac{1}{2}(e^{\theta\sqrt{-1}} + e^{-\theta\sqrt{-1}}), \qquad (a)$$

$$\sqrt{-1} \sin \theta = \tfrac{1}{2}(e^{\theta\sqrt{-1}} - e^{-\theta\sqrt{-1}}), \qquad (b)$$

in which e is the base of the Napierian system of logarithms. Adding gives

$$\cos \theta + \sqrt{-1} \sin \theta = e^{\theta\sqrt{-1}}. \qquad (c)$$

Hamilton's notation is

$$\cos \theta + \sqrt{-1} \sin \theta = e^{\frac{2\theta}{\pi}}. \qquad (d)$$

Equation (c) is an equality of values, while (d) is a notation.

In (c) let $\theta = (4n + \frac{1}{2})\pi$, in which n is one of the series 0, 1, 2, 3, etc., and it becomes

$$0 + \sqrt{-1} = e^{(4n+\frac{1}{2})\pi\sqrt{-1}};$$

$$\therefore \log \sqrt{-1} = (4n + \tfrac{1}{2}) \pi \sqrt{-1},$$

hence the log of $\sqrt{-1}$ has, analytically, an infinite number of imaginary values. If $n = 0$, we have

$$\log \sqrt{-1} = \tfrac{1}{2}\pi \sqrt{-1}.$$

This $\sqrt{-1}$ is not a unit-vector, but the *imaginary* of algebra. (See Art. 355.)]

APPLICATIONS.

380. Triedrals.—Let the diedral angle at α be A, at β, B, at γ, C, and the facial angles opposite these be a, b, c, respectively. Also refer to the plane AOB as c, and similarly for the others. At the initial point O erect unit-vectors perpendicularly to the respective planes

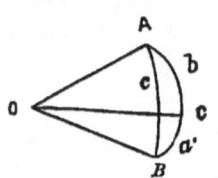

Fig. 259.

$$\varepsilon \perp b, \qquad \varepsilon_1 \perp c, \qquad \varepsilon_2 \perp a,$$

and they will be the operators for the respective vectors, and also form the edges of another triedral, the diedral angles of which will be the supplements of the diedrals of the given triedral. If an operator be drawn from any other point of the plane than at the vertex, it will either pass through the triedral, or else lie entirely without it. Consider those as positive which lie without, and the opposite as negative, and that the rotation from the vector of the denominator to the vector of the numerator is positive. Then we have

$$\frac{\alpha}{\gamma} = \cos b - \varepsilon \sin b,$$

$$\frac{\alpha}{\beta} = \cos c + \varepsilon_1 \sin c,$$

$$\frac{\beta}{\gamma} = \cos a + \varepsilon_2 \sin a,$$

and equation (3), Article 378, becomes

$$\cos b - \varepsilon \sin b = (\cos c + \varepsilon_1 \sin c)(\cos a + \varepsilon_2 \sin a), \quad (1)$$

the second member of which has already been developed in Example 9, Article 368. Taking the scalar part, observing that the angle between ε_1 and ε_2 is $180° - B$, we have

$$\cos b = \cos a \cos c + \sin a \sin b \cos B, \qquad (2)$$

which is *one of the fundamental equations of spherical trigonometry*.

If $B = 90°$, we find one of the formulas for right-angled spherical triangles.

Taking the vector parts and dividing by ε_1, we have

$$-\frac{\varepsilon}{\varepsilon_1}\sin b = \cos a \sin c + \frac{\varepsilon_2}{\varepsilon_1}\sin a \cos c + \frac{\beta}{\varepsilon_1}\sin a \sin c \sin B. \qquad (3)$$

Applying the preceding rules to the triedral $\varepsilon\,\varepsilon_1\,\varepsilon_2$, we have

$$\frac{\varepsilon}{\varepsilon_1} = \cos(180°-A) - \alpha\sin(180°-A)$$

$$= -\cos A - \alpha\sin A,$$

$$\frac{\varepsilon_2}{\varepsilon_1} = \cos(180°-B) + \beta\sin(180°-B)$$

$$= -\cos B + \beta\sin B,$$

$$\frac{\beta}{\varepsilon_1} = \cos 90° + \varepsilon_3\sin 90° = \varepsilon_3,$$

$$\frac{\varepsilon_3}{\beta} = \varepsilon_1.$$

Substituting in (3) and taking the scalars, gives

$$\sin b \cos A = \cos a \sin c - \sin a \sin c \cos B, \qquad (4)$$

which is *another important formula in spherical trigonometry*.

Taking the vectors gives, after dividing by β,

$$\frac{\alpha}{\beta}\sin b \sin A = \sin a \cos c \sin B + \frac{\varepsilon_3}{\beta}\sin a \sin c \sin B,$$

in which substitute the values of $\frac{\alpha}{\beta}$ and $\frac{\varepsilon_3}{\beta}$ given above, and it becomes

$$\sin b \cos c \sin A + \varepsilon_1 \sin b \sin c \sin A = \sin a \cos c \sin B +$$
$$\varepsilon_1 \sin a \sin c \sin B.$$

The scalars and the vectors of this equation give two identical equations, either of which is

$$\sin b \sin A = \sin a \sin B,$$

or $\quad\quad \sin b : \sin a :: \sin B : \sin A,$

that is,—*The sines of the angles of a spherical triangle are proportional to the sines of the opposite sides.*

Other relations may be found by making other combinations of the vector axes in equation (3).*

381. *Relation between the vector to a point and the Cartesian coördinates to the same point.*

Let α, β, γ, be unit-vectors along the axes x, y, z, respectively, and ρ the vector from the origin to the point; then

$$\rho = x\alpha + y\beta + z\gamma.$$

If the axes are rectangular, then

$$\rho = ix + jy + kz.$$

Multiply through by i, and we have

$$Si\rho = -x,$$

and similarly, $\quad Sj\rho = -y, \quad Sk\rho = -z,$

$$\therefore \rho^2 = x^2 + y^2 + z^2.$$
$$= (Si\rho)^2 + (Sj\rho)^2 + (Sk\rho)^2.$$

Let r be the scalar of ρ, then $Si\rho = -r\cos X$, etc., and we have

$$r^2 = r^2 \cos^2 X + r^2 \cos^2 Y + r^2 \cos^2 Z,$$

or $\quad\quad 1 = \cos^2 X + \cos^2 Y + \cos^2 Z,$

which is equation (3), Article 198.

382. Conjugate quaternions in reference to each other are such as are equal in all respects, *except* that their angles have contrary signs. Thus, in unit-vectors, if q be the given quaternion, and Kq is conjugate, we have

---

* Other relations are given in Hamilton's *Lectures*, p. 537.

$$q = -\cos\theta + \varepsilon\sin\theta,$$
$$Kq = -\cos\theta - \varepsilon\sin\theta.$$

Generally, if $\quad q = Sq + Vq,$

then $\quad Kq = Sq - Vq,$

and $\quad qKq = (Sq)^2 + (TVq)^2,$

since, (359), $\quad (Vq)^2 = -(TVq)^2.$

CHAPTER III.

OF THE LINE, PLANE, SPHERE, AND CONIC SECTIONS.

Right Line.

383. Vector Equation to a Right Line.—The position of a right line may be determined in several different ways.

1. A right line is determined when two points of it are known. Let A and B be two points of the line, O the origin of vectors, α and β the known vectors, C any point of the line, and ρ the vector to C.

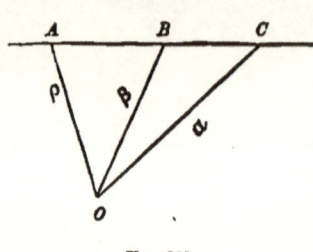

Fig. 260.

Then since AC is parallel to AB, (in this case coinciding with it), we have (**327**) $AC = xAB$. The triangles OAB and OAC give

$$OA + AB = OB,$$
$$OA + AC = OC;$$

or

$$\alpha + AB = \beta,$$
$$\alpha + xAB = \rho. \qquad (1)$$

Eliminating AB gives

$$\rho = \alpha + x(\beta - \alpha), \qquad (2)$$

which is the required equation.

2. A right line is also known when the length and position of a perpendicular to the line are known.

Let AP be the line, O the initial point of vectors, $OA = \delta$ the vector perpendicular to AP, P any point of the line, $OP = \rho$, and γ a unit-vector to which the line is parallel, and $AB = \gamma$; then
$$AP = x\gamma,$$
and
$$\rho = OA + AP$$
$$= \delta + x\gamma. \qquad (3)$$

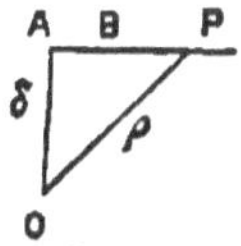

Fig. 201.

This may be put in another form. Multiplying both sides into δ gives
$$\rho\delta = \delta^2 + x\gamma\delta. \qquad (4)$$

But $x\gamma$ and δ being mutually perpendicular, we have $xS\gamma\delta = 0$, (Art. 368, Ex. 12); and if the tensor of δ be d and of ρ be p, (4) becomes
$$S\rho\delta = -d^2, \textbf{(359)}, \qquad (5)$$
or
$$-\rho d \cos(AOP) = -d^2,$$
or
$$\rho \cos \varphi = d, \qquad (6)$$
which is a polar equation to a line, (**37**, 7°). In (5) γ has disappeared, hence the line may be positive in either direction. The line is in the plane of δ and ρ, which fact may be expressed by the equation
$$S\varepsilon\rho = 0. \qquad (7)$$

(Obs.—Dividing (3) by δ, and taking the scalars, (**367**, (7)) gives
$$\frac{\rho}{d} \cos \varphi = 1,$$
or
$$\rho \cos \varphi = d,$$
as before.)

3. A right line is given when its direction and any point of it are known.

Let B be the given point, P any point, γ a unit vector parallel to the line;

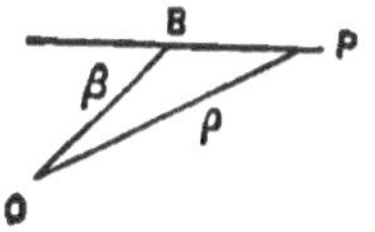

then
$$BP = x\gamma,$$
and
$$\rho = OB + BP$$
$$= \beta + x\gamma, \qquad (8)$$

Fig. 202.

which is the required equation.

[Remark.—Of the equations (2), (5), and (8), sometimes one is better adapted to the solution of a particular example than either of the others.]

EXAMPLES.

1. To find the equation to a line which shall pass through a given point.

Let β be the vector to the point; then equation (8) will be the required equation, in which ρ and γ are both unknown; hence an infinite number of lines may be drawn through a point.

2. To find the equation of a line which shall pass through a point and be perpendicular to a given line.

Let β be the vector to the given point, and δ the vector to which the line is perpendicular.

Draw a vector γ through the given point perpendicular to δ, then

$$S\gamma\delta = 0.$$

Let ρ be a vector to any point of γ, then (Eq. (8)),

$$\rho = \beta + x\gamma,$$

will be the required equation. We also have

$$\rho\delta = \beta\delta + x\gamma\delta,$$

or
$$S\rho\delta = S\beta\delta,$$

since $S\gamma\delta = 0$. Since γ has disappeared from this equation it appears that *an infinite number of lines may pass through a point perpendicular to a line*. But if the lines δ and γ are embraced by one plane, we have the additional condition

$$S\varepsilon\rho = 0,$$

in which case only one line can be drawn perpendicular to the given line.

3. Find the equation to a line which shall pass through two points. (This is equation (2).)

4. Find the distance between two given points.

Let ρ and ρ' be vectors to the given points, and AB the distance between them; then

$$AB = \rho - \rho'.$$

For the remainder of the solution, consult Article 374.

5. To find the length of a perpendicular from a given point to a given line.

Let the line be given by equation (8), the initial point being at the given point. When ρ is perpendicular to BP, let its value be δ, and we have

$$\delta = \beta + x\gamma,$$

or
$$\delta^2 = \delta\beta + x\delta\gamma,$$

or
$$S\delta^2 = S\delta\beta + 0,$$

or
$$-d^2 = S\delta\beta;$$

$$\therefore d = -\frac{S\delta\beta}{d} = -S\beta U\delta,$$

$U\delta$ being a unit-vector, d the length of the perpendicular, and therefore $d . U\delta = \delta$.

6. Find the equation to a line perpendicular to each of two given lines.

The Plane.

384. Equation to the Plane.—If the plane be given by the condition that it shall pass through a point and be perpendicular to a line, its essential equation will be the same as equation (5) of the preceding Article, subjected to the condition that ρ shall not be confined to a plane; hence the equations are

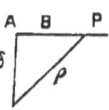

Fig. 263.

$$S\rho\delta = -d^2 = C \text{ (a constant)}, \qquad (1)$$

and $\qquad S\varepsilon\rho = $ an indeterminate quantity.

Equation (1) may be written

$$S\frac{\rho}{\delta} = 1. \qquad (2)$$

If the origin be in the plane, $\cos\theta$ of equation (5), Article 367, will be zero, and the equation becomes

$$S\rho\delta = 0;$$

in which case ρ will be independent of θ.

EXAMPLES.

1. To find the equation of a plane which shall pass through three given points.

Let α, β, γ, be the given vectors, and ρ the variable vector. Then $\rho - \alpha$ is a vector in the given plane, and similarly, $\alpha - \beta$, $\beta - \gamma$; hence, [Art. 368, Ex. 10, Eq. (3)],

$$S(\rho - \alpha)(\alpha - \beta)(\beta - \gamma) = 0, \qquad (1)$$

which may be reduced to

$$S\rho(V\alpha\beta + V\beta\gamma + V\gamma\alpha) - S\alpha\beta\gamma = 0, \qquad (2)$$

which is the equation sought.

[For expanding we have

$$S[\rho\alpha\beta - \rho\beta^2 - \alpha^2\beta + \alpha\beta^2 - \rho\alpha\gamma + \rho\beta\gamma + \alpha^2\gamma - \alpha\beta\gamma] = 0.$$

But, $S\rho\beta^2 = 0$, for $S\rho\beta^2 = S\rho(-1) = 0$, since it has no scalar part,

(Art. 368, Ex. 3), and, similarly, for other terms containing squares. Hence we have
$$S[\rho\alpha\beta - \rho\alpha\gamma + \rho\beta\gamma - \alpha\beta\gamma] = 0,$$
or, (357), $\quad S[\rho\alpha\beta + \rho\gamma\alpha + \rho\beta\gamma] - S\alpha\beta\gamma = 0.$

But $S\rho\alpha\beta = -rab \sin(a, b) \sin(r, \overline{ab})$, (Art. 368, Ex. 10, Eq. (1)).

We have $\quad V\alpha\beta = ab \sin(a, b).\varepsilon$, (Art. 367, Eq. 9)),

and $\quad S\rho V\alpha\beta = Sab \sin(a, b) \rho\varepsilon,$

also $\quad \rho\varepsilon = r(-\cos(r, \varepsilon) + \varepsilon' \sin(r, \varepsilon)\,;$

$\therefore S\rho V\alpha\beta = Srab[-\sin(a, b)\cos(r, \varepsilon) + \varepsilon' \sin(a, b) \sin(r, \varepsilon)]$

$= -rab \sin(a, b) \cos(r, \varepsilon)$

$= -rab \sin(a, b) \sin(r, \overline{ab})\,;$

$\therefore S\rho V\alpha\beta = S\rho\alpha\beta,$

and similarly for the others.]

2. A plane cuts off a pyramid from the three rectangular coördinate planes; required the locus of the foot of the perpendicular from the origin upon the plane when the volume is constant.

Let the three rectangular unit-vectors be i, j, k; and the vectors ai, bj, ck; and δ the perpendicular from the origin upon the plane; then, as in the preceding example, we have
$$\delta - ai, \quad \delta - bj, \quad \delta - ck,$$
three vectors in the plane.

But since δ and these vectors are perpendicular to each other, we also have, (Art. 368, Ex. 12),

$S\delta(\delta - ai) = 0, \quad S\delta(\delta - bj) = 0, \quad S\delta(\delta - ck) = 0\,;$

or $\quad -d^2 = aS\delta i, \quad -d^2 = bS\delta j, \quad -d^2 = cS\delta k.$

Multiplying together gives
$$-d^6 = abc\, S\delta i\, S\delta j\, S\delta k,$$
which will give the perpendicular upon the plane for any given position of δ. The product abc is, according to the problem, constant. Let $-d^2 = \delta^2$ be variable, and we have, from Cartesian coördinates,
$$\delta^2 = x^2 + y^2 + z^2,$$
$$S\delta i = -\delta \cos(\delta, i) = -\delta \cos(\delta, x),$$
$$\cos(\delta, x) = \frac{x}{\delta}\,;$$
$$\therefore S\delta i = -x,$$
and we have
$$(x^2 + y^2 + z^2)^3 = abcxyz,$$
a surface of the sixth order.

A solution might be effected by operating upon equation (2) of the preceding example, but it is not so direct as the above.

The Circle.

385. Equation to the Circle.—Let C be the centre of the circle, O the origin of vectors, $CP = r =$ the radius, $OC = \gamma =$ the vector to the centre, P any point whose vector is $OP = \rho$, and $\alpha =$ vector CP.

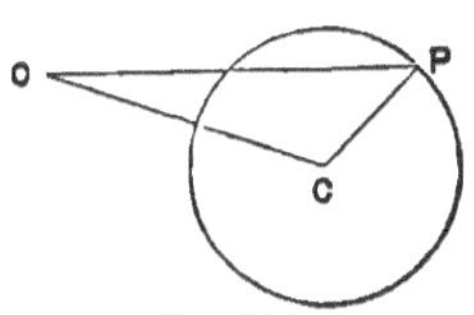

FIG. 264.

The triangle OCP gives

$$\rho = \gamma + \alpha, \tag{1}$$

which must be subjected to the condition that $CP = r =$ *a constant.* We find

$$(\rho - \gamma)^2 = \alpha^2 = -r^2, \tag{2}$$

which is one form of the required equation. Squaring the first member, making $OC = c$, this becomes, (Art. 368, Ex. 6),

$$\rho^2 - 2S\rho\gamma = c^2 - r^2. \tag{3}$$

If the origin O be upon the circumference, we have $OC = r$ and (3) becomes

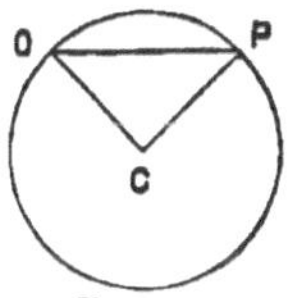

FIG. 265.

$$\rho^2 - 2S\rho\gamma = 0. \tag{4}$$

Substituting the value of $S\rho\gamma = -\rho' r \cos\theta$, (4) becomes

$$\rho' - 2r\cos\theta = 0,$$

which is the polar equation to the circle, the origin being on the circumference, and the diameter the initial line. If $OC = \frac{1}{2}\gamma$, (4) becomes

$$\rho^2 - S\rho\gamma = 0. \tag{5}$$

If the origin be at the centre, the length of γ will be zero, $c = 0$ and (2) becomes

$$\rho^2 = \alpha^2 = -r^2, \tag{6}$$

that is the radius vector is constant. If $-\rho'^2$ be the scalar of ρ^2 the equation becomes

$$\rho'^2 = r^2.$$

EXAMPLES.

1. An angle inscribed in a semicircumference is a right angle.

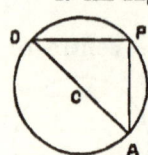

Fig. 266.

Equation (4) may be written

$$S\rho(\rho - 2\gamma) = 0;$$

hence the vectors ρ and $\rho - 2\gamma$ make a right angle, (Art. 368, Ex. 12). But $2\gamma = 2OC = OA$, a diameter, and the vector equation

$$OA + AP = OP,$$

gives
$$AP = \rho - 2\gamma;$$

hence OP and PA make a right angle.

2. A line is drawn at random through a fixed point A and a perpendicular is let fall upon it from another fixed point P; to find the locus of the intersection.

Let α be the vector joining A and P, β a unit-vector through P and ρ the variable vector through A, then will the equation of the line be, (382, Eq. (8)),

$$\rho = \alpha + x\beta.$$

Let δ be the perpendicular from A, then we have

$$\delta = \alpha + y\beta,$$

and
$$\delta^2 = \delta\alpha + y\delta\beta;$$

$$\therefore \delta^2 = S\delta\alpha,$$

for $S\delta\beta = 0$. Hence, (Eq. (5)), the locus is a circle whose diameter is $\alpha = r$.

386. Equation of a Tangent Line to a Circle.—Take the origin at the centre of the circle, α a radius vector to T the point of tangency, τ a vector from the centre to any point A of the tangent, then we have

$$AT = \tau - \alpha.$$

But α and AT make a right angle, hence

$$S\alpha(\tau - \alpha) = 0, \qquad (1)$$

or
$$-r^2 = \alpha^2 = S\alpha\tau, \qquad (2)$$

or
$$1 = S\frac{\tau}{\alpha}, \qquad (3)$$

is the required equation.

EXAMPLE.

The square of a tangent to a circle equals the product of the whole secant into its external segment.

Let TA be a tangent, AD the secant, vector $CT = \alpha$, $CA = r$, $TA = \gamma$, and lines $CT = CB = CD = r$, $CA = s$, $TA = t$.

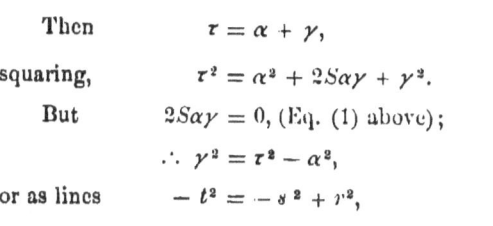

Fig. 267.

Then $\qquad r = \alpha + \gamma,$

squaring, $\qquad r^2 = \alpha^2 + 2S\alpha\gamma + \gamma^2.$

But $\qquad 2S\alpha\gamma = 0,$ (Eq. (1) above);

$\qquad \therefore \gamma^2 = r^2 - \alpha^2,$

or as lines $\qquad -t^2 = -s^2 + r^2,$

or $\qquad t^2 = (s + r)(s - r)$

$\qquad = AD \cdot AB.$

The Sphere.

387. Equations of the Sphere.—If the lines in Fig. 264, Article 385, are not restricted to one plane, equations (3), (4), (6), will be the equations of the sphere, and equations (1), (2), (3), of Article 386 will be the equation of a tangent plane.

EXAMPLES.

1. Every section of a sphere made by a plane is a circle.

Let δ be a vector perpendicular to the plane, d the scalar of δ, r the radius of the sphere, ρ the vector to the line of intersection of the plane and sphere, β a vector in the plane connecting δ and ρ. Then we have the vector equation

$$\rho = \delta + \beta,$$

squaring, $\qquad \rho^2 = \delta^2 + 2S\delta\beta + \beta^2;$

or as lines, $\qquad -r^2 = -d^2 - 2d\beta \cos 90° - b^2;$

$\qquad \therefore S\beta^2 = d^2 - r^2 = a\ constant;$

hence, (Eq. (6), Art. 385), the line of intersection is a circle. It is real for $d < r$. (Observe that $S\beta^2$ is negative.)

2. Find the curve of intersection of two spheres.

Take the origin anywhere; then equation (3), Article 385, gives

$$\rho^2 - 2S\gamma\rho = C,$$
$$\rho^2 - 2S\gamma'\rho = C',$$

subtracting, $\qquad 2S(\gamma' - \gamma)\rho = C - C' = a\ constant,$

which is the equation of a plane, hence, from example 1, we see that the intersection is a circle.

The Ellipse.

388. Equations to the Ellipse.—1. Let the origin be at the centre, ρ a vector to any point P in the curve, γ a vector to the focus F, the length of CF being c,

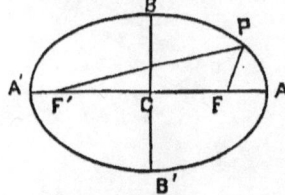

Fig. 268.

$$a = CA,\ b = CB,\ e = \frac{c}{a}.$$

We have

vector $FP = \rho - \gamma$,

$F'P = \rho + \gamma$,

and by the definition of an ellipse

$$T(\rho - \gamma) + T(\rho + \gamma) = 2a,$$

or $\quad \sqrt{-(\rho-\gamma)^2} + \sqrt{-(\rho+\gamma)^2} = 2a;$

$\therefore \sqrt{-(\rho-\gamma)^2} = 2a - \sqrt{-(\rho+\gamma)^2}.$

Squaring gives

$$-(\rho^2 - 2S\rho\gamma + \gamma^2) = 4a^2 - 4a\sqrt{-(\rho+\gamma)^2} - (\rho^2 + 2S\rho\gamma + \gamma^2);$$

or $\quad S\rho\gamma = a^2 - a\sqrt{-(\rho+\gamma)^2}.$

Transposing a^2, squaring and reducing gives

$$a^2\rho^2 + (S\rho\gamma)^2 = -a^4 - a^2\gamma^2$$
$$= -a^4 + a^2c^2$$
$$= -a^4 + a^4e^2$$
$$= -a^4(1 - e^2), \qquad (1)$$

which is the required equation.

Substituting $\quad (S\rho\gamma)^2 = (-\rho\gamma\cos\theta)^2$
$$= \rho^2 c^2 \cos^2\theta$$

(observing that ρ is used to represent both vector and scalar so that ρ^2 as vector $= -\rho^2$ as a line), gives

$$\rho = \frac{ab}{\sqrt{a^2 - c^2\cos^2\theta}}$$

$$= \frac{ab}{\sqrt{a^2\sin^2\theta + b^2\cos^2\theta}}, \qquad (2)$$

which is the polar equation to the ellipse, the pole being at the centre, Article 174.

2. Let i, j, be unit-vectors along the axes then will the equation be

$$\rho = xi + yj, \qquad (3)$$

in which x and y are related by the equation

$$y^2 = \frac{b^2}{a^2}(a^2 - x^2).$$

3. Let α, β, be unit-vectors along conjugate diameters. Then the equation becomes

$$\rho = x\alpha + y\beta, \qquad (4)$$

in which x and y are related by the equation

$$y^2 = \frac{b'^2}{a'^2}(a'^2 - x^2).$$

The Hyperbola.

389. The Equations to the Hyperbola are the same as for the ellipse excepting that e exceeds unity.

The Parabola.

390. Equation to the Parabola.—Let the origin be at the focus F, $FP = \rho$, $FA = \gamma$, $OG = \beta$, the directrix; then $FO = 2\gamma$, $PG = x\gamma$, and as lines $FP = PG$, $FA = AO = p$.
We have the vector equation

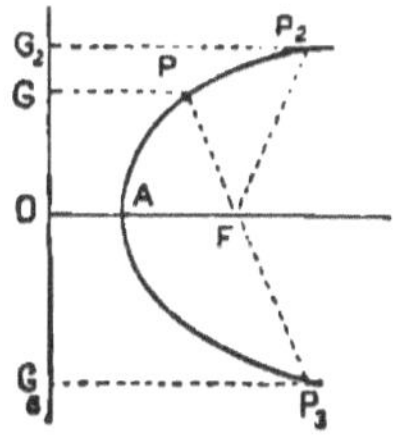

$$FP + PG = FO + OG,$$

or $\qquad \rho + x\gamma = 2\gamma + \beta;$

then $\qquad \rho\gamma + x\gamma^2 = 2\gamma^2 + \beta\gamma,$

and $\qquad S\rho\gamma + Sx\gamma^2 = 2S\gamma^2, \qquad (1)$

for $S\beta\gamma = 0$, since β and γ are mutually perpendicular;

then $\qquad S\rho\gamma - xp^2 = -2p^2;$

$$\therefore x = \frac{S\rho\gamma + 2p^2}{p^2}. \qquad (2)$$

Fig. 269.

From the property of the parabola, we have
$$FP = PG$$
or $$T\rho = xT\gamma;$$
hence as vectors,
$$-\rho^2 = -x^2\gamma^2, \qquad (3)$$
and as lines
$$r = xp, \qquad (4)$$
where r is a variable line. Eliminating x between (2) and (4) gives
$$pr = 2p^2 + S\rho\gamma \qquad (5)$$
which is the required equation. Substituting $S\rho\gamma = -rp\cos\theta$, gives
$$pr = 2p^2 - rp\cos\theta,$$
$$\therefore r = \frac{2p}{1+\cos\theta}, \qquad (6)$$
which is the polar equation to the parabola, the parameter here being $4p$, (Art. 168).

(Since γ^2 has no vector part, the value will be the same if S be omitted before γ^2 and equation (1) becomes
$$S\rho\gamma + x\gamma^2 = 2\gamma^2;$$
$$\therefore x = \frac{2\gamma^2 - S\rho\gamma}{\gamma^2},$$
which in (3) gives
$$\gamma^2\rho^2 = (2\gamma^2 - S\rho\gamma)^2, \qquad (7)$$
which is the required equation in terms of vectors only. Substituting in it the values
$$\gamma^2 = -p^2, \; \rho^2 = -r^2, \; S\rho\gamma = -rp\cos\theta,$$
we have
$$r = 2p - r\cos\theta;$$
$$\therefore r = \frac{2p}{1+\cos\theta},$$
which is the same as equation (6)).

By means of these equations all the properties of the conic sections may be discussed.

PART III.

MODERN GEOMETRY.

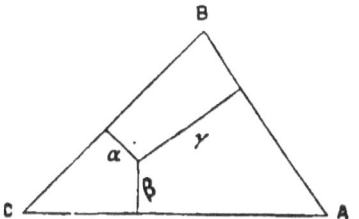

MODERN GEOMETRY.

CHAPTER I.

391. Modern Geometry includes all the systems of geometry which have arisen since the Cartesian system was recognized; but the term is generally restricted to the system called *Trilinear Coördinates*.

392. Definition of a Locus extended.—Any system of lines which make known a curve, may be called a system of coördinates; hence *the equation to a locus is an equation which expresses the relation between any system of lines which determine a locus.*

Tangential System.

393. Conceive that tangents are drawn at all the points of any curve, and that the curve is then removed. It will be easy to retrace the locus by drawing a curve which will be tangent to the consecutive lines. Such a locus is said to be an *envelope* of the system of right lines, and any equation which will determine the consecutive positions of the tangents will, in this way, determine the locus.

394. The four cusp Hypocycloid.—To illustrate, it is known that the tangent to the four cusp hypocycloid between the rectangular axes is constant. Therefore if a right line, equal in length to the radius of the directrix, be made to slide on the axes CB and CD keeping both ends in contact with these lines, they will in all positions be tangent to the hypocycloid DB. And if an indefinite number of lines be drawn upon the paper, coinciding with the several positions of the tangent line, the hypocycloid may be drawn by making it tangent to the successive lines.

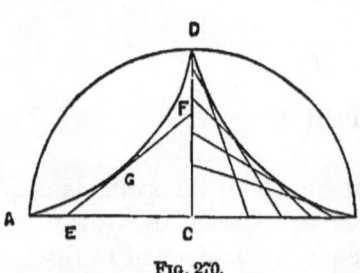

Fig. 270.

To find its equation in terms of the tangential intercepts.

Let EF be one of the tangents; then will CE be the intercept on the axis of x, and CF on the axis of y. Also let
$$CE = -x_t, \quad CF = y_t, \quad EF = l, \text{ a constant,}$$
then we have $\quad x_t^{\frac{2}{3}} + y_t^{\frac{2}{3}} = l^{\frac{2}{3}},$
for the required equation.

395. Equation to the Ellipse *in terms of tangential intercepts.* The equation of the tangent is, (Art. 100),

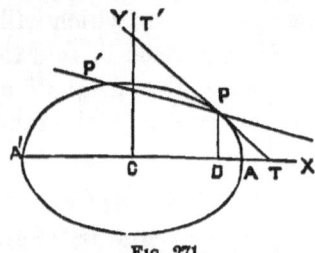

Fig. 271.

$$b^2 x' x + a^2 y' y = a^2 b^2,$$

and the intercepts are, (104),

$$CT = \frac{a^2}{x'}; \quad \therefore \quad x' = \frac{a^2}{x_t},$$

$$CT' = \frac{b^2}{y'}; \quad \therefore \quad y' = \frac{b^2}{y_t},$$

and since the point $x'y'$ is on the curve, we have
$$b^2 x'^2 + a^2 y'^2 = a^2 b^2,$$
in which substitute the values of x' and y', and we have
$$\frac{a^2}{x_t^2} + \frac{b^2}{y_t^2} = 1,$$
which is the required equation.

Let $\quad \dfrac{1}{x_t} = \xi, \text{ and } \dfrac{1}{y_t} = \eta,$

and the equation becomes
$$a^2 \xi^2 + b^2 \eta^2 = 1.$$

396. Equation to a point *in terms of tangential intercepts.*—Every line passing through a point is (by an extension of the definition) tangent to the point. Let P be the point and KH a line passing through it;

$$OA = a, \ AP = b = OB, \ \frac{1}{OH} = \xi, \ \frac{1}{OK} = \eta.$$

We have

$$HO : PB :: OK : KB,$$

or $\quad \dfrac{1}{\xi} : a :: \dfrac{1}{\eta} : \dfrac{1}{\eta} - b;$

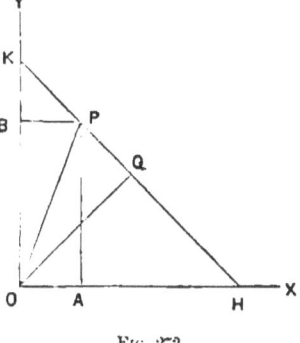

Fig. 272.

$$\therefore \ a\xi + b\eta = 1, \tag{1}$$

which is the required equation. By means of this equation any number of lines may be constructed, all of which will pass through the point P, and thus the point becomes determined.

If the line HK be given in terms of the perpendicular, $OQ = p$, and the angle $QOX = \varphi$, we have

$$\xi = \frac{\cos \varphi}{p}, \qquad \eta = \frac{\sin \varphi}{p},$$

and the equation becomes

$$a \cos \varphi + b \sin \varphi = p. \tag{2}$$

Still further, if $OP = c$, and $POA = \alpha$, equation (2) becomes

$$p = c \cos (\varphi - \alpha), \tag{3}$$

which latter equation is called the *tangential polar coördinates* of the locus. (See *Quarterly Journal of Pure and Applied Mathematics*, Vol. I., p. 210.)

397. Tangential equation to a Right Line.—Only one right line can be tangent to a right line, hence its intercepts will be constant, and we have

$$\xi = a, \qquad \eta = b,$$

for the required equations.

We see that, in this system, a point is given by *one* equation, while a line requires *two* equations.

EXAMPLES.

1. Find the equation to a circle in terms of tangential intercepts.
2. Find the equation to a hyperbola.
3. Find the equation to a parabola.
4. Prove that the distance between the points $a\xi + b\eta = 1$, and $a'\xi + b'\eta = 1$, is $\sqrt{(a'-a)^2 + (b'-b)^2}$.
5. Prove that $p = d + n \cos \varphi$ represents a circle, d and n being fixed numbers, and determine its radius.

398. A line may be determined by the perpendicular distances to it from three fixed points. The fixed points will be the vertices of a triangle. Let A, B, C, be the points, then will the perpendiculars α, β, γ, determine the line RP.

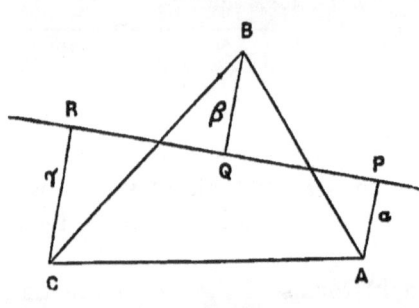

Fig. 273.

The perpendiculars will have the same sign, if the required line does not pass between the points from which the perpendiculars are drawn, otherwise they will have contrary signs. Thus, in the figure, α and γ will have the same signs, both *plus* or both *minus;* but α and β, and β and γ will have contrary signs. If the required line bisects AB, we have

$$\alpha = -\beta, \text{ or } \alpha + \beta = 0,$$

which may be considered as the equation of the point of bisection of AB. If the line RP were parallel to AC, we would have

$$\alpha = \gamma,$$

and if α and β have contrary signs it will cut the other two sides of the triangle, but otherwise it will lie either entirely above or entirely below the triangle.

Any curve may be determined by this system by making it the *envelope* of a system of lines determined according to a proper law. This system is generally considered as a modification of the trilinear system.

Trilinear System.

399. When a locus is determined by three perpendiculars from each point upon the three sides respectively of a given triangle, the system is called *trilinear*. The given triangle ABC is called *the triangle of reference*. Let P be any point; from it let fall the perpendiculars PE upon CA, PD upon CB, PF upon AB; and let the perpendicular upon the side opposite the angle A be α, opposite B be β, and opposite C, γ. Then, conversely, if α, β, γ, are given, the point P will be determined.

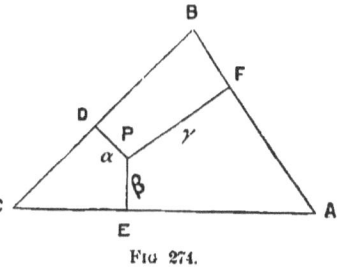

Fig 274.

We have previously seen, (Art. 8), that a point may be determined by its reference to *two* axes, and hence it appears that *three* are unnecessary; but it will soon appear that when three are used the expressions have a certain symmetry which they do not have when only two are used; and also that the angle between the two axes does not appear.

400. Signs of the Perpendiculars.—Those perpendiculars will be positive which lie on the same side of the line of reference as the corresponding angle. Thus, in Fig. 274, β and B are on the same side of AC, and hence β will be positive, but in Fig. 275 Pb, or β, and B are on opposite sides of AC, and hence β in this case will be negative.

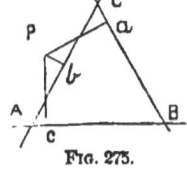

Fig. 275.

When the point is *within* the triangle of reference, *all* the perpendiculars will be positive, but when the point is without *one* or *two* will be negative, but in no case will all three be negative.

401. Relation of the Coördinates.—Let Δ be the area of the triangle of reference ABC, $a = CB$, the side of the triangle opposite A, $b = AC$, $c = AB$, and conceive lines to be drawn from P to the vertices of the triangle; then will

$a\alpha =$ twice the area of the triangle PBC,
$b\beta =$ "   "   "   "   "   " PCA,
$c\gamma =$ "   "   "   "   "   " PBA:
$\therefore a\alpha + b\beta + c\gamma = 2\Delta$. (1)

If the point is without the triangle, one or more of the coördinates α, β, γ, will be negative, and 2Δ will be the difference of the terms in equation (1).

Let ρ be the radius of the circle circumscribing the triangle ABC; then we have

$$\frac{a}{\sin A} = \frac{b}{\sin B} = \frac{c}{\sin C} = 2\rho, \qquad (2)$$

which combined with (1), gives

$$\alpha \sin A + \beta \sin B + \gamma \sin C = \frac{\Delta}{\rho} = S, \text{ (say).} \qquad (3)$$

Equations (1) and (3) are useful in transforming other equations so as to make them homogeneous. For from (1) we have

$$\frac{a\alpha + b\beta + c\gamma}{2\Delta} = 1,$$

and hence we may multiply any term of an equation by the fraction $\dfrac{a\alpha + b\beta + c\gamma}{2\Delta}$, thus raising by unity the *order* of the term without changing its value. Thus, to render homogeneous the equation

$$\alpha^3 + 3\beta^2 + 2\gamma = 1,$$

we raise each term to the third degree as follows:

$$\alpha^3 + 3\beta^2 \frac{a\alpha + b\beta + c\gamma}{2\Delta} + 2\gamma \left(\frac{a\alpha + b\beta + c\gamma}{2\Delta}\right)^2 = \left(\frac{a\alpha + b\beta + c\gamma}{2\Delta}\right)^3,$$

which may be expanded and reduced.

EXAMPLES.

1. Show that the coördinates of the middle point of BC are

$$\alpha = 0, \qquad \beta = \frac{\Delta}{b}, \qquad \gamma = \frac{\Delta}{c}.$$

2. Show that the coördinates of the angle A are

$$\frac{2\Delta}{a}, \qquad 0, \qquad 0.$$

3. The coördinates of the foot of the perpendicular from A upon BC are

$$0, \qquad \frac{2\Delta}{a} \cos C, \qquad \frac{2\Delta}{a} \cos B.$$

4. The coördinates of the centre of the circle inscribed in the triangle of reference are

$$\alpha = \beta = \gamma = \frac{2\varDelta}{a+b+c}.$$

5. The equations to the medials, (**337**, 2), are

$$a\alpha = b\beta, \quad b\beta = c\gamma, \quad c\gamma = a\alpha.$$

Since any one of these is the consequence of the other two, it follows that the medials of a triangle meet in a point, the coördinates of which point must satisfy the three equations simultaneously.

6. The equations to the perpendiculars of a triangle, (**340**, 3), are

$$\cos A = \beta \cos B, \quad \beta \cos B = \gamma \cos C, \quad \gamma \cos C = \alpha \cos A.$$

From which it appears that the altitudes meet in that point which satisfies the three equations simultaneously.

402. Bisectors.—Every point on the line bisecting the angle C, gives the equation

$$\alpha = \beta.$$

Hence the equations of the internal bisectors are

$$\alpha = \beta, \quad \beta = \gamma, \quad \gamma = \alpha;$$

and since each equation is the consequence of the other two, it follows that *the angle-bisectors of a triangle meet in a point*, which point is the centre of the inscribed circle, the coördinates for which must satisfy the three equations simultaneously.

The equations for the external bisectors will be

$$\alpha + \beta = 0, \quad \beta + \gamma = 0, \quad \alpha + \gamma = 0;$$

and hence they do not meet in a point.

The *external* bisectors of C and A, and the *internal* bisector of B give

$$\alpha + \beta = 0, \quad \beta + \gamma = 0, \quad \alpha - \gamma = 0,$$

the last of which is a consequence of the other two; hence, *the external bisector of two angles of a triangle, and the internal bisector of the other angle meet in a point*, which point is the centre of an enscribed circle.

403. Equation of a Line.—Let α_1, β_1, γ_1, be the trilinear coördinates of one point of the line, α_2, β_2, γ_2, the coördinates of another point, and α, β, γ, the coördinates of *any* point. Then by similar triangles (which the student can easily construct), we find the relations between these coördinates, and the resulting equation can be put under the form

$$l\alpha + m\beta + n\gamma = 0, \qquad (1)$$

in which $l = f(\beta_1, \beta_2, \gamma_1, \gamma_2)$, $m = f(\alpha_1, \alpha_2, \gamma_1, \gamma_2)$, and $n = f(\alpha_1, \alpha_2, \beta_1, \beta_2)$.

Equation (1) is the general equation of a right line in the trilinear system. We might now proceed to find the equation of a line passing through a point and parallel to a given line; determine the condition of parallelism; determine the distance between two points; the equation of a line perpendicular to a given one, etc.; but these must be left as exercises for the student.

404. Abridged Notation.—The equation of a line in terms of the perpendicular from the origin is, (30, (6)),

$$x \cos \alpha + y \sin \alpha - p = 0.$$

This is referred to as 'the line α,' since this notation at once calls to mind the original equation, and the equation to this line is written

$$\alpha = 0.$$

Similarly, $\qquad \beta = 0$

is the equation of the line β. The sides of the triangle of reference, are generally referred to by means of this abridged notation; thus

$$\alpha = 0, \qquad \beta = 0, \qquad \gamma = 0,$$

being the equations to the sides of the triangle of reference, then

$$l\alpha + m\beta + n\gamma = 0,$$

will be the equation to a right line.

As an application of this method, take Example 5, page 263. Let ABC be the triangle of reference, then

$$\alpha = 0, \quad \beta = 0, \quad \gamma = 0,$$

will be the equations of its sides.

The equations of the lines will be of the form

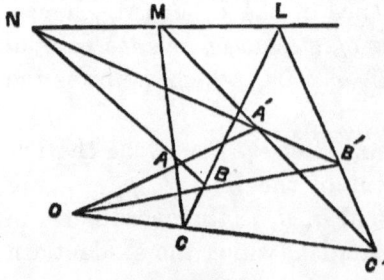

Fig. 276.

for $B'C'$ $l\alpha + m\beta + n\gamma = 0,$ (1)
for $C'A'$ $l\alpha + m'\beta + n\gamma = 0,$ (2)
for $A'B'$ $l\alpha + m\beta + n'\gamma = 0,$ (3)

in which each equation differs from each of the other two by two of its coefficients. By the conditions of the problem these lines are to intersect on lines radiating from O and passing through the vertices A, B, C. At the intersection C' of $B'C'$ and $A'C'$, γ will have a constant value for all lines passing through C'; hence eliminating γ between (1) and (2), we have

$$(l' - l)\, \alpha + (m - m')\beta = 0\,;\qquad (CC')$$

which is the equation of some line through C'. It is also the equation of a line through C, for every line through any angle of the triangle of reference may be reduced to the form

$$\alpha + k\beta = 0\,;$$

hence (CC') is the equation of the line CC'. Similarly, eliminating β between (1) and (3), we have

$$(l' - l)\, \alpha + (n - n')\, \gamma = 0,\qquad (BB')$$

which is the equation of a line BB'; and eliminating α between (2) and (3), gives

$$(m' - m)\, \beta + (n - n')\, \gamma = 0,\qquad (AA')$$

which is the line AA'. Since the last equation is a consequence of (CC') and (BB'), they will have a common point. Such values may therefore be assigned to the coefficients l', m, n, etc., in (1), (2), (3), as will cause them to be the equations of the lines $A'B'$, $B'C'$, $A'C'$.

At the point of intersection L of the lines $B'C'$ and BC, the perpendicular α will be zero. (Here it is the perpendicular that is zero, and not the equation of the line.) Then (1) gives

$$m\beta + n\gamma = 0.$$

Similarly, for the point M, $\beta = 0$, and (2) gives

$$l\alpha + n\gamma = 0\,;$$

and for N, $\gamma = 0$, and (3) gives

$$l\alpha + m\beta = 0.$$

Adding these gives

$$l\alpha + m\beta + n\gamma = 0,$$

which is the equation of a line, and it contains the three points L, M, N; hence the intersections of the corresponding sides of the triangles meet in a right line.

405. Triangular System.—Dividing the equation

$$a\alpha + b\beta + c\gamma = 2\Delta,$$

by 2Δ gives

$$\frac{a\alpha}{2\Delta} + \frac{b\beta}{2\Delta} + \frac{c\gamma}{2\Delta} = 1.$$

Let $\quad x = \dfrac{a\alpha}{2\Delta}, \quad y = \dfrac{b\beta}{2\Delta}, \quad z = \dfrac{c\gamma}{2\Delta},$

and the preceding equation will become

$$x + y + z = 1,$$

and since x, y, z, are connected with α, β, γ, by known relations, the last equation may be called the equation of a point P. It will be observed that x is the ratio of the area of the triangle $a\alpha$, to that of the triangle of reference. This system is called *triangular*.

406. The Quadrilinear System consists in the use of four lines of reference.

407. Equation of the Second Degree.—Every equation of the second degree (being made homogeneous) (**401**), may be written in the form

$$u\alpha^2 + v\beta^2 + w\gamma^2 + 2u'\beta\gamma + 2v'\gamma\alpha + 2w'\alpha\beta = 0.$$

Intersecting this by the straight line

$$l\alpha + m\beta + n\gamma = 0,$$

and we find that there may be two real points of intersection, two coincident points, or two imaginary points, which results correspond with those obtained by a similar operation in Cartesian geometry; hence *every equation of the second degree in trilinear coördinates represents a conic section.*

Other Systems.

408. Twenty-two Systems.—The Rev. Thomas Hill, D.D., LL.D., late President of Harvard College, in the year 1857, gave a list of twenty-two systems of coördinates, as follows.* Let x and y be the coördinates in the Cartesian system, r and φ the polar coördinates, s the length of the curve, ρ its radius of curvature, ε the angle between the radius vector r and the tangent, τ the angle between the tangent and an assumed axis, and ν the angle between the normal and the same axis. Then we have †

(1) $y = f(x)$. (2) $r = f(\varphi)$. (3) $\rho = f(\nu)$. (4) $\tau = f(s)$.
(5) $\rho = f(s)$. (6) $\varepsilon = f(\varphi)$. (7) $\tau = f(\varphi)$. (8) $r = f(x)$.
(9) $x = f(\varphi)$. (10) $x = f(\varepsilon)$. (11) $\tau = f(x)$. (12) $\rho = f(x)$.
(13) $x = f(s)$. (14) $\varepsilon = f(r)$. (15) $\tau = f(r)$. (16) $\rho = f(r)$.
(17) $r = f(s)$. (18) $\rho = f(\varphi)$. (19) $\varphi = f(s)$. (20) $r = f(\varepsilon)$.
(21) $\rho = f(\varepsilon)$. (22) $\varepsilon = f(s)$.

To transform a curve from one of these systems to another often involves the Calculus; their discussion, therefore, is unsuited to this work. The following remarks in regard to these systems is an abstract of those made by Dr. Hill, to which we have added some examples by way of illustration.

(1). $y = f(x)$. This is the ordinary system of bilinear coördinates, and the leading system discussed in the first 147 pages of this work.

(2). This is the polar system of coördinates.

(3). This is a system of circular coördinates, much used by Professor Peirce. Dr. Hill shows that the equation

$$\rho = A \sin^n \nu$$

includes a great variety of curves, such as the catenary, parabola, cycloid, etc.‡

---

* Proceedings of the American Association for the Advancement of Science, 11th meeting (1857), p. 43, 12th meeting (1858), p. 1.
† Mathematical Monthly, 1858-9, p. 363.
‡ Gould's Astro. Jour., Vol. II., p. 84.

(4). This is substantially the *Intrinsic equation of a curve*. The intrinsic equation of a curve is defined to be the relation between the length of arc of the curve and the change of direction of the tangent. If the arc begin at $s = 0$, and the tangent at that point be the line of reference, the equation will be

$$s = f(\tau),$$

which is the same as (4) reversed.

EXAMPLES.

1. The intrinsic equation of the circle is $s = R\tau$.
2. Of the involute of the circle, $s = \frac{1}{2}R\tau^2$.
3. Of the catenary, $s = c \tan \tau$.
4. Of the cycloid, $s = 4R \sin \tau$.
5. Of the parabola, $s = \frac{1}{2} p \log \frac{1 + \sin \tau}{1 - \sin \tau} + \frac{p \sin \tau}{1 - \sin^2 \tau}$.

(5). "An interesting point in the 5th system is the ease with which a curve expressed in it may be produced by the deformation or metamorphosis of a curve in rectangular coördinates. The curve, for example, $\rho = f(s)$ may be produced from the curve $y = f(x)$, by conceiving the ordinates of the latter curve to be fastened at right angles to the axis of x, and the axis of x to be afterwards curved until the ordinates become tangents to a new curve thus generated. The ordinates thus become the radii of curvature to the axis, and the axis becomes the curve s. This can be done rudely, by drawing $y = f(x)$ upon a card, bending it on the axis of x to a right angle, cutting it on the ordinates into strips down to the axis, and then laying these strips over each other so as to make the points of the curve become points of tangency to the involute of the new curve $\rho = f(s)$. Thus the straight line $y = Ax$ will give the logarithmic spiral $\rho = As$; the parabola $y^2 = 4Px$, will give the involute of a circle $\rho^2 = 4As$; the hyperbola $xy = A$ produces the beautiful volute or double finite spiral $\rho s = A$; the parabola $y = Ax^2$ produces the spiral $\rho = As^2$; the ellipse $Ax^2 + By^2 = 1$, produces the epicycloids $A\rho^2 + Bs^2 = 1$, or hypocycloids $As^2 + B\rho^2 = 1$, which are cycloids when $A = B$."

The equation of the cycloid will be $\rho = \sqrt{A^2 - s^2}$, from which it would appear that it has only one real branch, an example showing that what appears to be imaginary in one system may be real in another.

(6). These, in many cases, may readily be reduced to $r = f(\varphi)$ and investigated in this form. Thus $\varepsilon = -\varphi$ is a straight line; $\varepsilon = \varphi$, a circle; $\varepsilon = -\frac{1}{2}\varphi$, a parabola; $\varepsilon = -\frac{1}{2}(\varphi - \pi)$, a cardioid; $\varepsilon = -2\varphi$, an equilateral hyperbola.

(7). The equation $\tau = a\varphi$ is equivalent to $\varepsilon = (a-1)\varphi$; hence $\varepsilon = b$, or $\tau = \varphi + b$, is the equation of a logarithmic spiral; $\tau = \frac{1}{2}\varphi$ is the equation of a parabola.

(8). This may be changed to rectangular coördinates, thus $r = \sqrt{x^2 + y^2} = f(x)$. Changing $y = \sec x$, $y = \tan x$, $y = \log x$, into $r = \sec x$, $r = \tan x$, $r = \log x$, is singular and beautiful.

(9). This system may be changed to polar coördinates, thus, $x = r \cos \varphi = f(\varphi)$.

(10). $x = B \cot \varepsilon$ and $x = A \sqrt{\cot \frac{\varepsilon}{2}}$, are two equations of an equilateral hyperbola.

(11). In this we have for the equation of the parabola, $\tan \tau = Ax$, and for the circle $\sin \tau = Ax$.

(12). This may be constructed approximately to a scale. The straight line $y = Ax$ may be metamorphosed into $\rho = Ax$.

(13). The straight line will be $ax = s$, the cycloid $x = as^2$.

(16). The equation $\rho = Ar$ includes the logarithmic spirals; and $\rho = Ar^3$ is the equation of the equilateral hyperbola.

(17). $r = as$ includes logarithmic spirals.

(18). Form of curve not easily detected.

(19). The equation of the circle is $\varphi = as$.

(20). The equation $r = a\varepsilon$ is equivalent to $r = \dfrac{a}{a-1}\varphi$, or to $\varepsilon = \dfrac{1}{a-1}\varphi$, the latter of which is the 6th system. The equation of the circle is

$$\sin \varepsilon = a \cos (\log r + b)\sqrt{-1}.$$

409. Another system investigated by Dr. Hill consists in using as variables, the perpendicular from the origin upon the tangent and the variable angle between this perpendicular and a fixed axis.*

410. Other systems may be found by introducing as variables, the tangent, subtangent, normal, subnormal, perpendicular from the origin on the normal, of different lines having a fixed relation between them, etc.

---

* Proceedings of the American Association for the Advancement of Science, 1873 and 1875.

APPENDIX I.

THE system of Quaternions grew out of an effort to geometrize the *imaginaries* of Algebra. Hamilton sought to establish a system which would, at the outset, give a clear interpretation to the square roots of *negatives*, without introducing considerations so expressly geometrical, as those which involve the conception of an angle. This idea led him to consider Algebra as the SCIENCE OF PURE TIME,* and an essay containing his views upon the subject was published in 1835.† From this, as a starting point, he proceeded by a system of logical reasoning to make a new system of mathematics; and the invention of this system was made when he established the relation $i^2 = j^2 = k^2 = -1$;
$$ij = k, jk = i, ki = j;$$
$$ji = -k, kj = -i, ik = -j;$$
which he did in an article first published in 1843.‡ Out of this grew his system of Quaternions.§ In presenting the subject in the preceding pages we have completely ignored the abstract reasoning given by him, and have presented it entirely in its geometrical aspects.∥

---

* Hamilton's *Lectures on Quaternions*, Preface, p. (2).

† Theory of Conjugate Functions or Algebraic Couples; with a Preliminary Essay upon Algebra as the Science of Pure Time.—*Trans.* of the *Royal Irish Academy*, Vol. xvii., Part II., pp. 293–422.

‡ *Proc. Royal Irish Academy*, 1843.

§ Hamilton's *Lectures on Quaternions*, 1853. Also the following Articles by Hamilton in the *Philos. Mag.* and *Jour. of Science*, (Lond.):—Vol. xxv., 1844, pp. 10, 241, 489; Vol. xxvi., 1845, p. 220; Vol. xix., pp. 26, 113, 326; Vol. xxx., p. 458; Vol. xxi., pp. 214, 278, 511; Vol. xxxii., p. 367; Vol. xxxiii., p. 58 (Brit. Assoc. Rep., 1844, Pt. 2, p. 2); Vol. xxxiv., pp. 294, 340, 425; Vol. xxv., pp. 133, 200; Vol. xxxvi., p. 305; Vol. lii., 1852, p. 371; Vol. iv., 1852, p. 303; Vol. v., 1853, pp. 117, 236, 321; Vol. vii., 1854, p. 492; Vol. viii., 1854, pp. 125, 261; Vol. ix., 1855, pp. 46, 280. Vols. iii., iv., v., are on continued fractions in Quaternions. Vols. vii., viii., ix., on some extensions of Quaternions. See also Nichol's *Cyclopædia of Physical Science*, article QUATERNIONS, which article was prepared by Hamilton and is written in a remarkably free and easy style, pp. 706–720.

∥ Hamilton's *Lectures*, p. 72.

APPENDIX.

There were many attempts during the half century preceding Hamilton's work, to accomplish the same result. In order to understand more clearly the nature of the problem, it is necessary to go back and consider what had been accomplished by algebraic analysis.

The corner-stone of algebra, or, at least, that which made it a distinctive science, was the establishment of the law of *minus*.* According to algebra, we not only understand the meaning of the symbol minus when it indicates an operation, but we also know how to interpret it when it appears in the result of the solution of a problem. The law thus determined was universal, being applicable to all kinds of problems, geometrical, physical, and abstract. In the course of analysis, a new expression appeared, the $\sqrt{-1}$, which, in an *algebraic sense*, is an expression for an *impossible operation*. It is impossible to find a NUMBER *whose square is negative*, and the $\sqrt{-a}$ (a being always positive) is an *expression*, the operation indicated by which cannot be performed. Still, there are many expressions resulting from the application of algebraic analysis to geometrical problems containing the *imaginary*, the values of which are known to be real; as, for instance, Euler's formulas, which are

$$2\sqrt{-1}\sin x = e^{x\sqrt{-1}} - e^{-x\sqrt{-1}},$$

$$2\cos x = e^{x\sqrt{-1}} + e^{-x\sqrt{-1}};\dagger$$

and De Moivre's formula, which is

$$(\cos x + \sqrt{-1}\sin x)^m = \cos mx + \sqrt{-1}\sin mx;$$

all of which *appear* to be imaginary, but are actually real for all values of x, in which x is expressed as an arc of a circle, the radius being unity. These formulas (and many others) are the results of pure analysis, and it is very natural to seek for a geometrical interpretation of them. Hamilton's Quaternions furnishes an easy and natural mode of interpreting the $\sqrt{-1}$

---

* Algebra has finally come to be 'The Science of the Equation.'

† Let $x\sqrt{-1} = x'$, and the equations become

$$2\sqrt{-1}\sin x'\sqrt{-1} = e^{x'} - e^{-x'}$$

$$2\cos x'\sqrt{-1} = e^{x'} + e^{-x'},$$

the second members of which are in the *form of real* quantities. Some writers put Sin x for $\sqrt{-1}\sin x\sqrt{-1}$, and Cos x for $\cos x\sqrt{-1}$, and call the expressions Sin x and Cos x *potential* functions. Dropping the accents from the preceding expressions, we have

$$2\operatorname{Sin} x = e^x - e^{-x},$$

$$2\operatorname{Cos} x = e^x + e^{-x},$$

and $\qquad \operatorname{Cos} x + \operatorname{Sin} x = e^x.$

APPENDIX. 319

when it is the result of an operation in that system, but it does not cover the imaginaries of ordinary algebra (Art. 355). The interpretations of *negatives* and of *imaginaries* differ widely in this regard ; for while the former covers all cases which have been known to arise, the latter, as explained by Quaternions, is applicable only to the system which was invented expressly for the purpose of giving them a rational existence. A moment's consideration, however, will show that a different result could hardly be expected ; for it is found that the *negatives* in algebra always have, in a certain sense, a real existence, while the *imaginaries* have not. Thus, in algebra, if a negative result does not agree with the wording of the problem, we know that by stating the problem in an opposite sense the result becomes real ; but when an *imaginary* occurs in the solution of an *algebraic* problem, we cannot, by any transformation of the language make the result real ; it is necessary to change the data. To illustrate by means of some examples : take first, the following,

A cistern can be filled in 56 minutes by two faucets flowing together; if they flow separately, it will take one faucet 66 minutes longer to fill the cistern than the other; in what time will the cistern be filled by each ?

Let x be the time required to fill it from the larger one, then will $x + 66$ be the time required by the other, and we have the equation

$$\frac{56}{x} + \frac{56}{x + 66} = 1,$$

by solving which we find $x = 88$ or -42, and $x + 66 = 154$ or 24 ; hence the required times are 88 and 154 minutes respectively; and -42 and 24 minutes respectively. The last results are incompatible with the statement of the problem ; but if we consider that -42 is the time required for the first to *empty* the cistern if it were full and permitted to flow out through that faucet, and 24 the time required for the other to fill it ; the results become real. By using these numbers, it will be found that if both faucets be opened at the same time the cistern being empty and is being filled by the latter while, at the same time, it flows out through the former, it will require 56 minutes to fill it.

Next, take the following example,

The sum of the times required for two faucets to fill the cistern is 8 minutes, and the product of the times is 20 minutes ; required the times.

Here we have the equations

$$x + y = 8,$$

$$xy = 20,$$

a solution of which gives $x = 4 \pm 2 \sqrt{-1}$ and $y = 4 \mp 2 \sqrt{-1}$; hence the conditions are *impossible*, and no *interpretation* of the language will make it real. By changing the data by making the sum of the times 9, or the product 15, the problem becomes real.

The *negatives* and *imaginaries* of algebra are not considered as quanti ties;* they are, primarily, *symbols* of unexecuted operations, but when they appear in the *results* of analysis, their chief use is to aid in the interpretation of the problem from which they were deduced. In regard to the attempts to geometrize the *imaginaries* and the views held in regard to them we make a few extracts from Hamilton's writings, and from the references therein given.

Dr. Wallis of Oxford, in his "Treatise of Algebra," published in 1685, proposed to interpret the imaginary roots of a quadratic equation, by going *out of the line*, on which if real they should be measured. He says "so that whereas in the case of Negative roots, we are to say the point B cannot be found, so as is supposed in AC forward, but backward it may be in the same line; we must here say, in the case of a Negative Square the point B cannot be found so as was, in the line AC; but above that Line it may be in the same Plane. This I have the more largely insisted on, because the notion (I think) is new; and this, the plainest Declaration that at present I can think of, to explicate what we commonly call the *Imaginary Roots* of Quadratick Equations. For such are these."

In June, 1805, M. Abbé Buée read a paper entitled *Mémoires sur les Quantités Imaginaires*, which was printed in the *Philosophical Transactions* (London) for 1806. This writer has the credit of being the first to formally maintain that the $\sqrt{-1}$ as a symbol denoted *perpendicularity*, though this view had been *suggested* by others. This view has never led to the development of a system, and in explaining his methods, Buée *expressly excludes* the consideration of the position of the factor-lines in *multiplication*.

The preceding are simply *interpretations* of the *imaginary*. Hamilton gives full credit to M. Argand as being the first writer *to multiply together* (as well as add) *directed lines* in one plane; which he did in an "Essay on a manner of representing Imaginary Quantities," published in 1806. This method was reproduced independently by a Mr. Warren in 1828, and in the same year by M. Mourey in a work entitled : "The True Theory of Negative Quantities and of the so-called Imaginary Quantities" (Paris, 1828). As the method of Argand is of considerable interest historically, we here illustrate its method by the following extract taken from Mourey's work.

Take the expression $a + b \sqrt{-1}$, and give it, at first, the form

$$\sqrt{a^2 + b^2} \left[\frac{a}{\sqrt{a^2 + b^2}} + \frac{b}{\sqrt{a^2 + b^2}} \sqrt{-1} \right].$$

If we take the sum of the squares of the fractions, which are between the brackets, we find that this sum is equal to 1; and from thence we conclude that these two fractions can be regarded as being the sine and cosine of the same angle α. Designate also the modulus $\sqrt{a^2 + b^2}$ by A; the imaginary expression can be put under the form $A(\cos \alpha + \sqrt{-1} \sin \alpha)$. Considering that this expression contains really but two quantities, the *modulus A* and

---

* *Lectures*, p. (2).

the angle α, M. Mourey proposed to regard the modulus A as expressing the length of a right line OA, and α as being the angle AOB, which this line makes with a fixed axis OB. In other words, the modulus A represents a line of a certain length, which at first lay upon the axis OB, and which, by making a movement round the origin O upward, has departed from this axis by an angle α. M. Mourey gives the name *verser* to this angle, or, rather, to the arc which measures it; and then, instead of the imaginary expression, he writes simply $A\alpha$, a notation very suitable to recall at the same time the modulus A and the *verser* α. He proposes even to give the name *route*, or *way*, to the length OA, placed in its true position with regard to OB, so that A verser α, or $A\alpha$, is the route from O toward A.

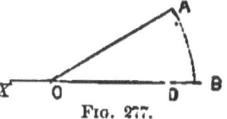

Fig. 277.

As a line can make around the origin O as many revolutions as we please, and that, also, as well by commencing its rotation below as well as above OB, it follows that the verser may pass through all states of magnitude, and be as well negative as positive. It will be positive when the movement of the line shall have commenced above; it will be negative when the movement commenced below. From this it follows that the same route can be represented with a verser which is positive, or one which is negative, provided that the sum of the versers, abstraction being made of the signs, is 360°.

From the preceding conventions it results that a *way* can be represented by giving to the length A an infinity of different versers. Suppose, to fix the ideas, that OA should be a determinate way, and that then the verser AOB, should be an acute angle α; it is evident that the position of OA will undergo no change if we add or subtract from α any number whatever of entire circumferences. Thus is established this important remark, that if we designate by 2π an entire circumference, or 360°, and by n any whole number whatever, positive or negative, the expression $A(2\pi n + \alpha)$ will represent the same route as $A\alpha$; this is expressed by the equality

$$A(2\pi n + \alpha) = A\alpha.$$

When we give to A a verser equal to zero, the length A lies upon the line OB. When the verser is equal to π or 180°, this length is found in the opposite direction, OX; then it is nothing else than the negative quantity $-A$. Thus we ought to regard as altogether equivalent the two expressions $-A$ and $A\pi$.

After these preliminaries, M. Mourey establishes the rules of algebraic calculus. Next determine the rule to be followed in the multiplication of any two quantities whatever, $A\alpha$ and $B\beta$. Here the two factors are the magnitudes A and B, measured upon two lines OA and OB, which make, with a fixed axis OX, angles AOX, BOX, represented by the versers α and β. (The reader can draw any line through O to represent OX.) It is necessary, then, first of all, to give to the definition of multiplication the extension suitable to render it applicable to the case in ques-

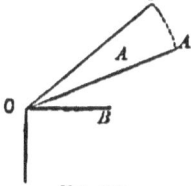

Fig. 278.

tion. But, considering that the multiplier $B\beta$ indicates a line B, which departs from the fixed line OX by an angle equal to β, M. Mourey regards multiplication as having for its object to take at first the length A in its actual direction as many times as there are units in B, giving the line OA', and to turn the new line OA' around the point O, to depart from this direction by an angle equal to β, and to give it the position OC. From this it follows that, in designating by AB the product of the two magnitudes, abstraction being made of all idea of position, the product sought will be $AB(\alpha+\beta)$. Thus we have

$$A\alpha \times B\beta = AB(\alpha+\beta);$$

that is to say, we *multiply the moduli according to the ordinary rules of arithmetic, and take the sum of the versers.*

If the versers are each equal to π or $180°$, we shall have $A\pi \times B\pi = AB(2\pi)$. But $A\pi$ and $B\pi$ are nothing else than $-A$ and $-B$, and $AB2\pi$ is the same thing as $+AB$; then $-A \times -B = +AB$. This is the known rule, $-$ by $-$ gives $+$.

According to this rule, the square of $A\alpha$ will be $A^2(2\alpha)$; that is to say, *we take the square of the modulus and double the verser.* Then, reciprocally, the square root is obtained by *extracting the square root of the modulus without regarding the verser; then take half the verser.*

Let us come now to the interpretation of the imaginary expression $\sqrt{-A^2}$. For this purpose, let us observe, first, that it is equivalent to $\sqrt{A^2(2n\pi+\pi)}$; then extracting the square root,

$$\sqrt{-A^2} = A(n\pi + \tfrac{1}{2}\pi).$$

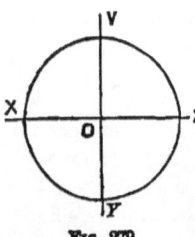

Fig. 279.

If n is even, the verser $n\pi + \tfrac{1}{2}\pi$ places the length A in the same position as $\tfrac{1}{2}\pi$; that is to say, in the position OY, perpendicular to OX. If n is uneven, the verser $n\pi + \tfrac{1}{2}\pi$ will place the length A in a position OY', perpendicular to OX, but below. Thus, in the system of M. Mourey, the expression $\sqrt{-A^2}$ offers no longer to the mind any idea of impossibility. It represents two routes, OY and OY', equal and opposite, both perpendicular to the fixed axis OX.

We see shadowed here some of the elements of Hamilton's Quaternions, still the systems have very little in common. The difference between them is great. In Mourey's system all the operations are in one plane, two dimensions only of space being necessary; and the multiplication of two lines produces a third *in the same plane.* In this system it is necessary to know the angle between the lines and another fixed line, and if the axis coincides with one of the lines, the multiplication of the two lines will be expressed by laying off on one of them the numerical product of the two. It is quite evident that Hamilton received little or no aid from these writings in the establishment of his system; though some may have served as hints, but

APPENDIX. 323

nothing more. Most of the other important investigations which have a bearing upon the subject, were contemporary with Hamilton's work. It is proper to note that M. Servois seems to have made the nearest approach to an anticipation of quaternions.* He inferred from analogy, that if α, β, γ, be the angles between a right line and the three rectangular axes, the following expression ought to be true :

$$(p \cos \alpha + q \cos \beta + r \cos \gamma)(p' \cos \alpha + q' \cos \beta + r' \cos \gamma) = \cos^2 \alpha + \cos^2 \beta + \cos^2 \gamma = 1 ;$$

but he could not determine the values of p, q, r, p', q', r', and asked "Will they be *imaginaries*, reducible to the general form $A + B \sqrt{-1}$?" It is now known that they are identical with the $+i$, $+j$, $+k$, $-i$, $-j$, $-k$, of quaternions.

M. Cauchy, in his *Cours d'Analyse* (Paris, 1821), remarks, "Every imaginary equation is only the symbolic representation of two equations between real quantities."

We would do great injustice to Hamilton's worthy friend Mr. John T. Graves, were we to make no mention of his labors. He seemed, if possible, to be the more enthusiastic of the two, in trying to overcome the difficulties which beset these men in their endeavors to construct the new system. If his labors were not crowned with being the successful inventor, he, at least, encouraged Hamilton in his labors,† and brought to his notice the results of his predecessors and of his contemporaries. Still, however much he may have profited by his suggestions, or by the suggestions of others, Hamilton appears justly to have the full credit of conceiving and first applying the fundamental principles of his system.

We add a few words more upon the invention. Hamilton supposed that he could retain the *commutative* principle,‡ or the interchangeability of the factors in regard to order; but after many trials, with expressions of various forms, he was obliged to abandon the principle, and make $ij = -ji$. He had previously assumed that $i^2 = j^2 = -1$; but the value of k, which appeared later in the discussion, had not been fixed. At first he tried $k^2 = +1$, because $i^2 j^2 = +1$, but as this failed to work, he tried $k^2 = -1$, and thus completed his fundamental *assumptions* § for the multiplication of two vectors. Vectors generally are represented by the letters of the Greek alphabet, but I presume that Hamilton used the Roman and italic letters for the mutually perpendicular vectors, because the letter i had long been used in analysis for the $\sqrt{-1}$, and also because his friend, Mr. Graves, was using it, and had in some of his investigations made $i^2 = -1$, and $j^2 = -1$.|

---

* *Lectures*, p. (57). † *Lectures*, Preface, p. (35).
‡ *Lectures*, Preface, p. (43). § *Lectures*, Preface, p. (46).

| Hamilton was not the only one to invent a *non-commutative* system. Professor H. Grassman, in 1844, the year following that of the first publication by Hamilton of his Quaternions, published a very original and remarkable work (Aus Dehnungslehre, or *Doctrine of Extension*), involving the

The development of this subject may be considered in another light. By examining the successive processes in mathematics we find that *it has been extended by the removal of restrictions*. This principle makes its appearance in the earliest stages of mathematics, and may be recognized in all departments of the subject; and may be applied to many of the principles of quaternions.* We will illustrate the principle by means of a few examples.

In arithmetic we are taught that multiplication is a short process of making *repeated additions*. According to this definition multiplication can be performed by a process of *counting*. It answers well so long as the numbers are positive integers, but fails when the multiplier is a proper fraction. In order, therefore, to multiply by a fraction it is necessary to make a new definition, or *extend* the meaning of the old one; and the latter alternative is the one chosen. In the extended sense, multiplication is the process of finding the result of taking one number several times, or part or parts of a time. The process of involution is a further extension of the principle, being a process of repeated multiplication, and hence is the repeating of the result of the result of repeated additions. When logarithms are employed, the original conception of addition is almost entirely lost sight of; indeed to a certain extent the process is reversed, multiplication being performed by addition by the aid of logarithms. When the exponent is literal, it is read as a power, thus 10^x is read 'the x power of 10,' but if the exponent is fractional its meaning is entirely changed. If $x = \frac{2}{3}$, we have $10^{\frac{2}{3}}$, which *arithmetically* is the square of the cube root of 10, but *by the removal of restrictions* the definition is *extended* so that a *power* is defined to be *a number, entire or fractional, positive or negative, which is placed at the right and above another number*. According to this definition we read the expression 'ten to the two-thirds power.' This reading indicates the *form* of the expression but not the arithmetical operations to be performed in reducing it. The same remark applies to the reciprocals, as $a^{-x}, 10^{-\frac{2}{3}}$, etc.

In algebra, negative qualities are considered, and multiplication includes the process of performing *repeated subtractions;* but instead of making

---

multiplication of inclined lines (äussere, *outer*, multiplikation), which had the *non-commutative* principle. Neither Grassman nor Hamilton owed anything, in their original papers, to the works of the other, and the methods of the two are quite distinct from each other. The work of Grassman has been admired by mathematicians since his day. Hamilton speaks of it in the highest terms as an original work, and Professor Clifford, in an article entitled, "Applications of Grassman's Extensive Algebra," in the *American Journal of Mathematics*, Vol. I., No. 4, p. 350, endeavors to determine the place of Quaternions in the more extended system; and a demonstration that the algebra obtained by a generalization of the laws of more extended systems of algebras is always a compound of quaternion algebras which do not interfere with one another.

* Hamilton's *Lectures*, Preface, p. (50).

APPENDIX. 325

a new definition to cover this idea, the old definition is *extended* by removing *the restrictions* which were previously given to it, and with the new idea it covers the case of *minus* by *minus*, or of subtracting the result of repeated subtractions. If the exponent is zero, we have

$$a^0 = 1,$$

whatever be the value of a, and if a also is zero we have the expression

$$0^0,$$

which, according to the *arithmetical* definition of number, has no meaning. Yet in the Differential Calculus we not only meet with this expression, but also such as

$$1^\infty, \; 0 \times \infty, \; \frac{\infty}{\infty}, \; \infty^0,$$

and they are interpreted as having rational meanings. *Arithmetically* quantity is something which can be measured, and *number* is employed to express the measure. Originally it implied 'how many,' as 5, 9, 12, etc., but by an *extension* it includes fractions; it includes all measurements which can be expressed by means of *figures*. But 0 expresses, *arithmetically*, the absence of all quantity, and hence, technically, would have no place in the science of mathematics were it not for the principle of *extension*. For the same reason *infinites* would find no place; but as defined, both form exceedingly important elements in this science. Zero (so called) in the Calculus is not an arithmetical zero; it is an infinitesimal. Shall we form a new symbol for the new meaning, or shall we *extend* the definition so as to give a new meaning to the old symbol? The former might have been done, but the latter was chosen. The *restrictions* which had been given to it *were removed*, and the definition thus *extended*. Similarly, the *infinites*, according to the original conception of number, do not exist; but, as defined, they are realities. According to this view, we might found the Differential Calculus (substantially as many writers have done) after this manner :—In arithmetic we have considered only one class of numbers, called *finite* and *discontinuous*, as 3, 12. $\frac{3}{4}$, etc., with which our daily experience makes us familiar. But we find that great power is given to mathematics by including other numbers which we will now proceed to define and to which we will assign certain laws. One of these we will call *Infinitesimal*. This we *define* to be so small (and if you please, of such a peculiar character), that it cannot be added to an arithmetical number; so that if it be represented by δ we will have

$$5 + \delta = 5, \; 7 - \delta = 7, \; a + \delta = a, \text{ etc.}$$

But it can be multiplied by an arithmetical number, thus

$$5 \times \delta = 5\delta, \; -7 \times \delta = -7\delta, \; a \times \delta = a\delta, \text{ etc.}$$

If divided by an arithmetical number the quotient will be called zero: thus,

$$\frac{\delta}{5} = 0, \; \frac{\delta}{a} = 0, \text{ etc}$$

but if an arithmetical number be divided by an infinitesimal the result will be an *infinite*, the laws for which will now be assigned. Briefly, we would define an *infinite* as one to which neither a finite nor an infinitesimal can be added, which may be multiplied by a *finite*, which cannot be divided by a *finite* (or if so divided will give an infinite of a higher order), and if divided into a *finite* will give a result called zero. In a similar manner we would define *infinites* and *infinitesimals* of different orders, the relations between the successive orders being similar to those just defined. Next we would proceed to determine how the ratios of the infinitesimals of the first order could be determined from given relations between the *finites;* the result would be the *first differential coefficient*.

This process will appear (as it really is) arbitrary to the student, and, at first, would doubtless be considered by many, as unworthy of being called a *system;* still as an application of the principle of *extension by the removal of restriction*, it is worthy of being presented to every student while studying the calculus. The main object is to arrive at truth, and hence we should not be confined to one system, or mode of presenting it. In the light of what has been said we will interpret one of the preceding expressions; thus

$$1^{\infty}$$

means (in the light of the Calculus) that 1 is not an arithmetical 1 but a thing which differs from 1 by an infinitesimal, and may be

$$1 + \delta,$$

the infinite power of which will be some *finite*. The other expressions may be interpreted in a similar manner. Those who have not studied Vanishing Fractions may not understand the conclusions here given.

But to proceed. In finite geometry the parameters are considered as *fixed* quantities, but in the Calculus these are sometimes made to vary; so that by the *removal of restrictions* we state that a geometrical constant is one which varies infinitely slow compared with the rate of change of the variables. Returning to multiplication, we observe that multiplication of a line by a number, as $3b$, is represented by a line whose length is three times as long as b; but we are not *restricted* to this representation, for the product of a into b, or ab, is represented by a rectangle; neither are we *restricted* to these, for in quaternions ij is represented by rotation. We observe that the removal of restrictions sometimes imposes *new restrictions*. Thus, in quaternions *restrictions* were removed and *extensions* made in regard to the algebraic symbols $+$, $-$, $=$, $\times$, $\div$, but they resulted in destroying the *commutative* principle of algebra.

But we must bring these remarks to a close, as they have been already extended far beyond what was originally intended. We did not intend to write a critical history of the subject, but simply to give an outline of the manner in which the subject was developed. As for the extent of the usefulness of quaternions we express no opinion; but we can safely assert that no principle so original, novel, and comprehensive as that given by Hamilton can be introduced into any science without yielding much good fruit.

APPENDIX. 327

It seems that Euler's trigonometrical formulas ought readily to be deduced from Hamilton's system, and such is the case ; but we could not introduce it into the body of the work without assuming the development of log $(1 + x)$, and the value of π as given by a certain series. But now, assuming the development, we proceed as follows :

$$\log (1 + x) = M[x - \tfrac{1}{2}x^2 + \tfrac{1}{3}x^3 - \tfrac{1}{4}x^4 + \text{etc.}]$$
$$\log (1 - x) = M[-x - \tfrac{1}{2}x^2 - \tfrac{1}{3}x^3 - \tfrac{1}{4}x^4 - \text{etc.}],$$

the latter being deduced from the former by changing x to $-x$. Subtracting we have

$$\log (1 + x) - \log (1 - x) = 2Mx [1 + \tfrac{1}{3}x^2 + \tfrac{1}{5}x^4 + \text{etc.}],$$

or
$$\log \left[\frac{1 + x}{1 - x}\right] = 2Mx [1 + \tfrac{1}{3}x^2 + \tfrac{1}{5}x^4 + \text{etc.}].$$

Now let $x = \sqrt{-1}$, and we have

$$(1 + \sqrt{-1}) \div (1 - \sqrt{-1}) = \sqrt{-1},$$

and substituting the same value of x in the second member, we have

$$\log \sqrt{-1} = 2M \sqrt{-1} [1 - \tfrac{1}{3} + \tfrac{1}{5} - \tfrac{1}{7} + \text{etc.}].$$

The quantity within the brackets is $\tfrac{1}{4}\pi$,* therefore

$$\log \sqrt{-1} = \tfrac{1}{2}M\pi \sqrt{-1}, \qquad (a)$$

or $$\log i = \tfrac{1}{2}M\pi i,$$

which is substantially the same as the equation given on page 286 ; but we could not *assume* the value there given in this place, for it was found by means of the formulas which we are not seeking. Now find the value of $\varepsilon^{\frac{2\theta}{\pi}}$ given on page 278, and to avoid confusion in regard to letters and to simplify the expression, we will use the equally general one

$$i^\phi,$$

i being an indeterminate unit-vector, $\sqrt{-1}$.

Take $\log i = \log i,$

passing to exponentials, $i = e^{\log i},$

raising to φ power, $i^\phi = e^{\phi \log i}$
$= e^{\phi \log \sqrt{-1}}$

and by Eq. (a), $= e^{\tfrac{1}{2}M\phi\pi i}.$ (b)

Making $\varphi = \dfrac{2n\theta}{\pi}$, we have

$$i^{\frac{2n\theta}{\pi}} = e^{Mn\theta i}, \qquad (c)$$

---

* See advanced works on the Calculus. (Courtenay's Calculus, p. 52.)

the left member of which is a general expression for n rotations through an angle θ; hence, (**379**, (5)),

$$i^{\frac{2n\theta}{\pi}} = \cos n\theta + i \sin n\theta = e^{Mn\theta i}, \qquad (d)$$

also
$$\cos n\theta - i \sin n\theta = e^{-Mn\theta i}; \qquad (e)$$

adding,
$$\cos n\theta = \tfrac{1}{2}\left(e^{Mn\theta i} + e^{-Mn\theta i}\right), \qquad (f)$$

subtracting,
$$\sin n\theta = \frac{1}{2\sqrt{-1}}\left(e^{Mn\theta i} - e^{-Mn\theta i}\right).$$

In the last two equations, let $M = 1$, $n = 1$, $i = \sqrt{-1}$, and we have

$$\cos \theta = \tfrac{1}{2}\left(e^{\theta\sqrt{-1}} + e^{-\theta\sqrt{-1}}\right), \qquad (g)$$

$$\sin \theta = \frac{1}{2\sqrt{-1}}\left(e^{\theta\sqrt{-1}} - e^{-\theta\sqrt{-1}}\right), \qquad (h)$$

which are known as *Euler's formulas*.

In equation (*b*) let $\varphi = i$, and observing that $i^2 = -1$, we have

$$i^i = e^{-\tfrac{1}{2}M\pi}, \qquad (l)$$

which in the Napierian system becomes (since M will be unity)

$$i^i = e^{-\tfrac{1}{2}\pi},$$

and restoring the value of $i = \sqrt{-1}$, we have

$$\sqrt{-1}^{\sqrt{-1}} = e^{-\tfrac{1}{2}\pi},$$

or squaring,
$$(-1)^{\sqrt{-1}} = e^{-\pi},$$

taking reciprocals,
$$(-1)^{-\sqrt{-1}} = e^{\pi}, \qquad (m)$$

or squaring again,
$$1^{-\sqrt{-1}} = e^{2\pi}, \qquad (n)$$

which equations express a peculiar relation between π and e. Equation (*n*) at first sight appears to be incorrect for *arithmetically* $\log 1 = 0$; but upon examination it will be found that this is a special and not a general value, and that it does give correctly one of the *imaginary* values of log 1. From it we find

$$-\sqrt{-1} \log 1 = 2\pi;$$
$$\therefore \log 1 = -\frac{2\pi}{\sqrt{-1}}$$
$$= 2\pi \sqrt{-1}. \qquad (o)$$

From equation (*d*) we find

$$\log(\cos n\theta + \sqrt{-1} \sin n\theta) = Mn\theta \sqrt{-1}.$$

APPENDIX. 329

Let $\theta = \pi$, then $\cos n\theta = 1$, and $\sin n\theta = 0$, for all integer values of n, and we have

$$\log 1 = Mn\pi \sqrt{-1},*$$

which gives the real as well as imaginary values of log 1. If $n = 0$, then

$$\log 1 = 0,$$

which is its only arithmetical value, and which also is independent of the system in which it is taken.

If $M = 1$ and $n = 1$

$$\log 1 = \pi \sqrt{-1},$$

which is one of the imaginary values of the logarithm.

If $M = 1$ and $n = 2$,

$$\log 1 = 2\pi \sqrt{-1},$$

which is the value found in equation (o).

We see then, that in Euler's formulas, De Moivre's formula, and in the imaginary exponentials involving π, the *imaginary* $\sqrt{-1}$ of algebra becomes the same as the *real* INDETERMINATE UNIT-VECTOR of quaternions.

---

* Professor Graves gave a more general equation, thus,

$$\log 1 = \frac{2\omega'\pi}{2\omega\pi - \sqrt{-1}},$$

in which ω and ω' are independent integers. (*Phil. Trans.*, 1829.)

APPENDIX II.

HYPER-SPACE.

From any point in space a line may be drawn in any direction; hence if every such line were considered *a dimension of space*, the latter would have an unlimited number of dimensions, and the expression "space of n dimensions" would be perfectly rational. But the dimensions of space are not so determined. A right line is conceived as determining one dimension, and space confined to one right line is called *space of one dimension*. If from any point of this line a perpendicular be drawn, then will the space limited to the plane of these lines be called *space of two dimensions*. Similarly, if through the common point of the two lines, a third be drawn perpendicular to the former ones, any point will be determined by its distance from the planes of the lines, taken two and two, and such space is called *space of three dimensions*. No more than three mutually perpendicular lines can pass through a point; hence higher orders of space, as the *fourth, fifth*, nth, etc., are imaginary; in other words, have no real existence. Tri-dimensional space is the highest order of which we have any knowledge—it is natural space. As this mode of determining the dimensions of space dates at least from Euclid, it is sometimes called *Euclidean space* to distinguish it from the lower orders of real space, and the higher orders of imaginary space, and all the higher orders may be classed as *hyper-space*.

The solution of problems involving hyper-space depends upon the laws and conditions assigned. It generally, however, falls under the following principle, viz.: *The laws and formulas applicable to the solution of problems having real conditions may be extended to those involving* UNREAL *conditions*. This principle we will illustrate by a few examples from different branches of mathematics.

Beginning with arithmetic, we take the simple problem—If one and one are three, what will two and two be? Now one and one are two and nothing else, and in the nature of things can be nothing else so long as language retains its present meaning. We are aware that it has been asserted that if we were in some part of the universe where in the process of putting one thing by (or with) another, the result was always three, we would admit that one and one make three. But this is a misstatement of the case. In arithmetic, one and one do not *make* anything—*they are two*. It is a statement of a fact. Everywhere in the universe *one and one are two*, and two and two are four. Hence our problem may read—if two be three, what will four be? The problem involves *unreal* conditions; but by *assuming* that the same ratio exists between four and the required result as between two and three, the problem may be solved by proportion, giving six as the result.

By introducing different units into this problem, the unreal conditions may be removed; thus, if two apples cost three cents, what will four apples cost?

Again, if a melon be worth 20 cents, and it be divided into two equal parts so that one part shall be twice as large as the other, what will each part be worth? Here the conditions are self-contradictory, and there is no rational solution; still, by assuming that $\frac{1}{2}$ is $\frac{1}{3}$, then by proportion 10 cents will be $6\frac{2}{3}$ cents; then the other half will be two-thirds; and the cost $13\frac{1}{3}$ cents. But by a slight change in the wording the contradictory character is removed; thus, if one half is worth twice as much as the other, what will each half be worth?

In trigonometry the cosine of a circular function cannot exceed radius or unity; but if it be assumed that $\cos x = 2$, a value, or rather an expression, may be found for each of the other trigonometrical functions by assuming that the same law holds for the unreal as for the real conditions. Thus we would have for the sine, $\sin x = \sqrt{(1 - \cos^2 x)} = \sqrt{(1 - 4)} = \sqrt{-3} = \sqrt{3}\sqrt{-1}$; a result which shows that $\sin x$ is impossible when $\cos x = 2$.

In geometry the square of the hypothenuse of a right-angled triangle equals the sum of the squares of the two sides. In a rectangular parallelopiped the square of the diagonal joining the opposite angles equals the sum of the squares of the three adjacent edges. If, therefore, a solid be constructed upon four (imaginary) mutually perpendicular lines as edges, the square of the diagonal through the extreme angles will equal the sum of the squares of the four adjacent edges, and hence, if the length of each edge be unity, the length of the diagonal would be 2.

Questions pertaining to space of *n-dimensions* is, as has been intimated, purely an extension of analysis from certain real to certain imaginary cases; and to illustrate this still further, we make use of the analysis of coördinate geometry.

Tabulating the well-known results for space of one, of two, and of three dimensions in terms of rectangular coördinates, we have

No. of Dimensions of Space.	Equations to a Point.	Equations to the Right Line.	Equations to the Plane.
1	$x - a = 0$	x indeterminate	None
2	$x - a = 0$ $y - b = 0$	$x + my - a = 0$	x and y indeterminate
3	$x - a = 0$ $y - b = 0$ $z - c = 0$	$x + mz - a = 0$ $y + nz - b = 0$	$x + Ay + Bz + D = 0$

APPENDIX II.

In this table, we observe that the equations to a point are of the first degree of two terms each, and the same in number as the corresponding coördinates in space. The equations to a right line are also of the first degree between two variables, and one less in number than the corresponding ones for a point, and the equation for a plane is of the first degree between three variables. We also observe that the *form* of the equations to the right line may be found by adding the last of the equations to a point to each of the preceding ones separately; thus $z - c$ added to $x - a$ gives $x + z - a - c = 0$, which is generalized in the form $x + mz - a = 0$; and similarly for the others. Proceeding in this way the following table is formed:

Number of rectangular axes or so-called dimensions of space.	Equations to a Point.	Equations to the Right Line.	Equations to the Plane.	Equations to the Hyper-plane of the first order.	Equations to the Hyper-plane of the second order.
1	$x - a = 0$	x indeterminate	None	None	None
2	$x - a = 0$ $y - b = 0$	$x + mz - a = 0$	x and y indeterminate	None	None
3	$x - a = 0$ $y - b = 0$ $z - c = 0$	$x + mz - a = 0$ $y + nz - b = 0$	$x + Ay + Bz + C = 0$	$x, y,$ and z indeterminate	None
4	$w - a = 0$ $x - b = 0$ $y - c = 0$ $z - d = 0$	$w + lz - a = 0$ $x + mz - b = 0$ $y + nz - c = 0$	$w + Ay + Bz + C = 0$ $x + Dy + Ez + F = 0$	$w + Ax + By$ $+ Cz + D = 0$	$w, x, y,$ and z indeterminate
5	$v - a = 0$ $w - b = 0$ $x - c = 0$ $y - d = 0$ $z - e = 0$	$v + kz - a = 0$ $w + lz - b = 0$ $x + mz - c = 0$ $y + nz - d = 0$	$v + Ay + Bz + C = 0$ $w + Dy + Ez + F = 0$ $x + Gy + Hz + I = 0$	$v + Ax + By$ $+ Cz + D = 0$ $w + Ex + Fy$ $+ Gz + H = 0$	$v + Aw + Bx$ $+ Cy + Dz + E$ $= 0$

In a similar manner, the equations to the hyper-solids may be written. Thus, since all spheres will have a constant radius, we have for the hyper-sphere of the first order, or the sphere of four-dimension space, the equation

$$w^2 + x^2 + y^2 + z^2 = r^2.$$

Similarly, the equation of the hyper-ellipse of the first order, the origin being at the centre, will be

$$\frac{w^2}{a^2} + \frac{x^2}{b^2} + \frac{y^2}{c^2} + \frac{z^2}{d^2} + 1 = 0,$$

a^2, b^2, c^2, d^2 being positive. For, the section of each of the coördinate planes with this surface will be an ellipse, to determine which eliminate two of the variables, as w and x for instance, by means of the two equations of the plane of four-dimension space. If the equation of the hyper-plane, $w + Ax + By + Cz + D = 0$, be combined with the preceding equation, only one of the variables can be eliminated, and the intersection will be given in terms of three variables, and in order to discuss the properties of the curve in this case, it will be necessary to determine the general properties of hyper-curves referred to hyper-coördinates. It is not our purpose to develop the subject further.

Another view—a moving point will generate a line; a moving line may generate a surface; a moving surface, a solid; hence, a moving solid might generate a hyper-solid, and so on.

Some amusing problems have already been stated involving four rectangular coördinates. In the American Journal of Mathematics vol. i., p. 1 (1878), it is shown that a spherical shell in four-dimension space may be turned inside out without tearing or stretching; and in the same Journal, vol. iii., p. 1, several regular solids in four-dimension space are determined and named—thus affording another instance that the mathematician may create (or imagine) that which is impossible!

APPENDIX III.

To find the distance between two lines in space. Let the lines be

$$x = mz + a \atop y = nz + b \Big\}, \quad (1) \qquad x = m'z + a' \atop y = n'z + b' \Big\}; \qquad (2)$$

and pass a plane through (1) parallel to (2); then will the required distance be the length of the perpendicular between (2) and the plane thus passed. To determine the plane, draw a line through any point of (1) parallel to (2); and for convenience take the point where (1) pierces the plane xy, which point will be

$$x = a, \qquad y = b, \qquad z = 0; \qquad (3)$$

and the line through this point parallel to (2) will be (Art. 197, Eq. (2)),

$$x - a = m'z \atop y - b = n'z \Big\}. \qquad (4)$$

This line pierces the plane xz in the point

$$y = 0, \qquad z = -\frac{b}{n'}, \qquad x = a - \frac{m'b}{n'}; \qquad (5)$$

and (1) pierces the same plane in the point

$$y = 0, \qquad z = -\frac{b}{n}, \qquad x = a - \frac{mb}{n}. \qquad (6)$$

The three points (3), (5), (6), will determine the plane. The equation of the plane being of the form (Art. 202),

$$Ax + By + Cz + D = 0, \qquad (7)$$

gives, with (3), (5), (6), the equations of condition,

APPENDIX.

$$\left.\begin{array}{c} aA + bB + 0 + D = 0 \\ \left(a - \dfrac{m'b}{n'}\right) A - \dfrac{b}{n'} C + D = 0 \\ \left(a - \dfrac{mb}{n}\right) A - \dfrac{b}{n} C + D = 0 \end{array}\right\} ; \qquad (8)$$

from which we find

$$\left.\begin{array}{c} A = \dfrac{n - n'}{a(n' - n) - b(m' - m)} D \\ B = \dfrac{m' - m}{a(n' - n) - b(m' - m)} D \\ C = \dfrac{mn' - m'n}{a(n' - n) - b(m' - m)} D \end{array}\right\} ; \qquad (9)$$

which in (7) gives for the equation of the plane

$$(n - n')x + (m' - m)y + (mn' - m'n)z + a(n' - n) - b'(m' - m) = 0. \quad (10)$$

The equation to the line perpendicular to this plane from the point

$$x = a', \qquad y = b', \qquad z = 0, \qquad (11)$$

where (2) pierces the plane xy will be (Art. 215),

$$x - a' = \dfrac{A}{C} z, \qquad y - b' = \dfrac{B}{C} z; \qquad (12)$$

and the point where this line pierces the plane (7) will be (Art. 212),

$$\left.\begin{array}{c} x_1 = -A \dfrac{a'A + b'B + D}{A^2 + B^2 + C^2} + a' \\ y_1 = -B \dfrac{a'A + b'B + D}{A^2 + B^2 + C^2} + b' \\ z_1 = -C \dfrac{a'A + b'B + D}{A^2 + B^2 + C^2} \end{array}\right\} . \qquad (13)$$

Hence, the required distance will be, Equations (11) and (13),

$$l = \sqrt{(x_1 - a')^2 + (y_1 - b')^2 + z_1^2},$$

which reduced by the aid of (13) gives

$$l = \frac{a'A + b'B + D}{\sqrt{A^2 + B^2 + C^2}} ; \qquad (14)$$

and this reduced by the aid of (9) gives

$$l = \frac{(a + a')(n - n') + (b - b')(m - m')}{\sqrt{(m'n - mn')^2 + (m' - m)^2 + (n' - n)^2}} . \qquad (15)$$

If θ be the angle between (1) and (2), we have, Art. 199, Eq. (6),

$$\cos \theta = \frac{mm' + nn' + 1}{\sqrt{m^2 + n^2 + 1} . \sqrt{m'^2 + n'^2 + 1}} ; \qquad (16)$$

$$\therefore \sin \theta = \frac{\sqrt{(m'n - mn')^2 + (m'-m)^2 + (n'-n)^2}}{\sqrt{m^2 + n^2 + 1} . \sqrt{m'^2 + n'^2 + 1}} ; \qquad (17)$$

and hence (15) may be written

$$l = \frac{(a + a')(n - n') + (b - b')(m - m')}{\sin \theta \sqrt{m^2 + n^2 + 1} . \sqrt{m'^2 + n'^2 + 1}} . \qquad (18)$$

If one of the lines, as (2) for instance, passes through the origin of co-ordinates, we have $a' = 0$, and $b' = 0$, and

$$l = \frac{a(n - n') + b(m - m')}{\sin \theta \sqrt{m^2 + n^2 + 1} . \sqrt{m'^2 + n'^2 + 1}} ; \qquad (19)$$

and further, if (2) is in the plane xy, we have $m' = 0$, and

$$l = \frac{a(n - n') + bm}{\sin \theta \sqrt{m^2 + n^2 + 1} . \sqrt{n'^2 + 1}} ; \qquad (20)$$

and still further, if the line (2) coincides with the axis of (z), we also have $n' = 0$, and (15) becomes

$$l = \frac{an + lm}{\sqrt{m^2 + n^2}} . \qquad (21)$$

If lines (1) and (2) are parallel, equation (15) reduces to $\frac{0}{0}$, as indeed we

APPENDIX. 337

might have anticipated, since by our mode of constructing a line through any point of (1) parallel to (2) would, in this case, coincide with (1), an unlimited number of planes could be passed parallel to (2). It will, therefore, be necessary to make an independent solution of this case.

DISTANCE BETWEEN TWO PARALLEL LINES.

Let the lines be

$$\left. \begin{array}{l} x = mz + a \\ y = nz + b \end{array} \right\} ; \quad (22) \qquad \left. \begin{array}{l} x = mz + a' \\ y = nz + b' \end{array} \right\} . \quad (23)$$

Line (23) pierces plane xy in the point $(a', b', 0)$, from which pass a perpendicular to (22); its equation will be of the form, Art. 197,

$$\left. \begin{array}{l} x - a' = m''z \\ y - b' = n''z \end{array} \right\} ; \quad (24)$$

and the equation of condition of perpendicularity will be, Art. 199,

$$mm'' + nn'' + 1 = 0, \quad (25)$$

and the condition that (24) and (22) intersect will be, Art. 196,

$$\frac{a' - a}{m - m''} = \frac{b' - b}{n - n''}. \quad (26)$$

From (25) and (26) we find

$$\left. \begin{array}{l} m'' = \dfrac{(b' - b)\, mn - (a' - a)(n^2 + 1)}{(b' - b)\, n + (a' - a)\, m} \\[4pt] n'' = \dfrac{-(b' - b)(m^2 + 1) + (a' - a)\, mn}{(b' - b)\, n + (a' - a)\, m} \end{array} \right\} ; \quad (27)$$

which in (24) will give the equations of the perpendicular, and eliminating between that and (22), the point of intersection of (22) and (24) will be found to be

$$\left. \begin{array}{l} x_1 = \dfrac{(b' - b)\, n + (a' - a)\, m}{m^2 + n^2 + 1}\, m + a \\[4pt] y_1 = \dfrac{(b' - b)\, n + (a' - a)\, m}{m^2 + n^2 + 1}\, n + b \\[4pt] z_1 = \dfrac{(b' - b)\, n + (a' - a)\, m}{m^2 + n^2 + 1} \end{array} \right\} . \quad (28)$$

APPENDIX.

The required distance will be

$$d = \sqrt{(x_1 - a')^2 + (y_1 - b')^2 + z^2}$$

$$= \sqrt{\frac{(b'-b)^2(m^2+1) - 2(b'-b)(a'-a)mn + (a'-a)^2(n^2+1)}{m^2+n^2+1}}. \quad (29)$$

If one of the lines, as (23), passes through the origin, a' and b' will be zero, in which case we have

$$d = \sqrt{\frac{b^2(m^2+1) - 2ab \cdot mn + a^2(n^2+1)}{m^2+n^2+1}}, \quad (30)$$

and also if the lines be parallel to the plane xz, we have $n = 0$, and the equation becomes

$$d = \sqrt{\frac{b^2(m^2+1) + a^2}{m^2+1}} \quad (31)$$

Still further, if the lines be in the plane xz, b will also be zero, and

$$d = \frac{a}{\sqrt{m^2+1}}. \quad (32)$$

If the lines be in a plane parallel to xz, and neither passes through the origin, we have $n = 0$, and $b = b'$, and (29) reduces to

$$d = \frac{a'-a}{\sqrt{m^2+1}}. \quad (33)$$

MATHEMATICS.

CALCULUS—GEOMETRY—TRIGONOMETRY, ETC.

CALCULUS OF VARIATIONS.
A complete work for Mathematicians, Instructors, and Post-Graduate Students. By Louis B. Carll, of Columbia College. Has received the endorsement of many distinguished Mathematicians..8vo, cloth, **$5 00**

"Undoubtedly one of the most important Mathematical Books yet issued by the American Press."

AN ELEMENTARY TREATISE ON THE DIFFERENTIAL CALCULUS.
Founded on the Method of Rates or Fluxions, with Numerous Illustrative Examples. By Prof. J. M. Rice, Head of Depart. of Applied Math. at U. S. Naval Acad., and Prof. W. W. Johnson, St. John's Col., Annapolis, Md. Fifth edition, revised...8vo, cloth, 3 50

"We heartily commend the book to all who want a good text-book on the subject it so fully treats."—*Mathematical Visitor.*

DIFFERENTIAL CALCULUS.
By Profs. Rice and Johnson. Abridged edition...12mo, cloth, 1 50

DIFFERENTIAL AND INTEGRAL CALCULUS.
By Profs. Rice and Johnson. (Abridged edition.)
2 vols. in one, 12mo, cloth, 2 50

AN ELEMENTARY TREATISE ON THE INTEGRAL CALCULUS.
Founded on the Method of Rates or Fluxions. By W. W. Johnson, Prof. U. S. Naval Academy. With the co-operation of Prof. J. M. Rice, of U. S. Naval Academy, and companion book to Rice and Johnson's Abridged Differential Calculus...12mo, cloth, 1 50

"I am delighted with your Integral Calculus."—Prof. J. D. RUNKLE, *Mass. Institute of Technology.*

THE THEORY OF ERRORS AND THE METHOD OF LEAST SQUARES.
By Prof. Wm. Woolsey Johnson, U. S. Naval Academy, Annapolis, Ind ....12mo, cloth, 1 50

CONTENTS: Introductory, Independent Observations of a Single Quantity, Principles of Probability, The Law of Probability of Accidental Errors, The Combination of Observations and Probable Accuracy of Results, The Facility of Error in a Function of One or More Observed Quantities, The Combination of Independent Determinations of the Same Quantity, Indirect Observations, Gauss's Method of Substitution. Values of the Probability Integral, Tables, etc.

CURVE TRACING IN CARTESIAN CO-ORDINATES.

By Prof. W. W. Johnson, U. S. Naval Academy. Fully illustrated...................................12mo, cloth, $1 00

"This little book will prove a source of delight to students who have just completed the ordinary academic course of analytical geometry."—*Van Nostrand's Magazine.*

DIFFERENTIAL EQUATIONS.

A Treatise on Ordinary and Partial Differential Equations. By Prof. W. W. Johnson, Professor of Mathematics at the U. S. Naval Academy, Annapolis, Md. Third edi. 8vo, cloth, 3 50

A TREATISE ON LINEAR DIFFERENTIAL EQUATIONS.

Equations with Uniform Coefficients. Vol. I. By Prof. Thomas Craig, of Johns Hopkins University................. 5 00

"This book should mark, if not make, an epoch in the history of mathematical study in America. If we except Pierce's "Analytic Mechanics," published thirty years ago, it is the first special treatise on any of the subjects now recognized as belonging to advanced mathematics that has appeared in this country We earnestly hope that the Instructors in our leading colleges will find it helpful in redeeming our country from the reproach of being, among enlightened ones, that in which there are fewest students of modern mathematics."—*Nation.*

TEXT-BOOK ON THE METHOD OF LEAST SQUARES.

Sixth edition. Revised and enlarged by Prof. Mansfield Merriman, of Lehigh University, Pa...................... 2 00

"This is a very useful and much needed text-book."—*Science.*
"Even the casual reader cannot fail to be struck with the value which such a book must possess to the working engineer."—*Engineering News.*

MANUAL OF LOGARITHMIC COMPUTATIONS.

With numerous examples, being introductory to the Study of Logarithms. By Alfred G. Compton. For High Schools, Academies, and Scientific Institutions. Third edition.
12mo, cloth, 1 50

"If the simplicity of calculation by means of Logarithms were taught at the time or thereabouts when we learned the Multiplication Table, it would be proved in after life a great saving of time and patience."—*Engineering News.*

AN INTRODUCTION TO THE LOGIC OF ALGEBRA.

By Prof. Ellery W. Davis........ ....8vo, cloth, 1 50

"It is very satisfactory to have a book of this character, short and easy of comprehension, fill up a gap in our mathematical literature. I shall advise every student of Algebra that exhibits to me any taste for mathematical studies to obtain the book."—THOMAS S. FISKE, A.M., PH.D., *Instructor in Mathematics at Columbia College, and at Barnard College.*

A PRIMARY GEOMETRY.

With simple and practical examples in Plane and Projective Drawing, and suited to all beginners. By S. Edward Warren, C.E., late Professor in the Rensselaer Polytechnic Institute, Troy, N. Y., etc., etc............................12mo, cloth, 0 75

"Prepared with great care . . . more elementary than most geometries . . . . Might be used in the grammar grade with good effect, etc. . . . ."—*New England Journal of Education.*

PLANE PROBLEMS IN ELEMENTARY GEOMETRY;

Or, Problems in one Elementary Conic Section—the Point, Straight Line, and Circle. In two divisions—I. Preliminary or Instrumental Problems. II. Geometrical Problems. By Prof. S. E. Warren. With an introduction, plates, and woodcuts..12mo, cloth, $1 25

ELEMENTS OF GEOMETRY.

By C. M. Searle. With an Appendix containing Problems and Additional Propositions......................8vo, cloth, 1 50

ELEMENTS OF GEOMETRY.

By Geo. Bruce Halsted, Professor of Pure and Applied Mathematics, University of Texas, etc. Sixth edition. 8vo, cloth, 1 75

"His Metrical Geometry (mensuration) is the best book of its kind that has been published in this country."—Prof. FLORIAN CAJORI, Colorado College, in *Teaching and History of Mathematics in U. S.*

"Taken as a whole, Dr. Halsted's Geometry is in advance of most textbooks on the subject."—*Academy News.*

SYNTHETIC GEOMETRY.

For Schools and High Schools. By George Bruce Halsted, Professor of Pure and Applied Mathematics, University of Texas. 8vo, cloth, 1 50

CONTENTS: Symmetry, Symcentry, Congruence, Pure Spherics, Equivalence, Proportion, Similarity, Mensuration, Modern Geometry, Recent Geometry.

ELEMENTS OF PLANE ANALYTIC GEOMETRY.

A Text-book, including numerous Examples and Applications, and especially designed for Beginners. By Geo. R. Briggs, Instructor in Mathematics, Harvard College. Adapted to Colleges, High Schools, and Scientific Institutions. Sixth edition..................................12mo, cloth, 1 00

THE ELEMENTS OF CO-ORDINATE GEOMETRY.

In three parts—I. Cartesian Geometry. II. Quaternions. III. Modern Geometry and an Appendix. By De Volson Wood, Professor of Mathematics and Mechanics in the Stevens Institute of Technology. Second edition, with additions. 8vo, cloth, 2 00

It was designed to make this a thoroughly practical book for class use. The more abstruse parts of the subject are omitted. The properties of the conic sections are, so far as practicable, treated under common heads, thereby enabling the author to condense the work. The most elementary principles only of quaternions and modern geometry are presented; but in these parts, as well as in the first part, are numerous examples.

THE ELEMENTS OF DESCRIPTIVE GEOMETRY, SHADOWS, AND PERSPECTIVE.

With a brief treatment of Trihedrals, Transversals, and Spherical, Axonometric, and Oblique Projections. For Colleges and Scientific Schools. By Prof. S. Edward Warren. With 24 folding plates. Revised edition. 1882...8vo, cloth, 3 50

PROBLEMS, THEOREMS, AND EXAMPLES IN DESCRIPTIVE GEOMETRY.

For Colleges and Mathematical Students, and Engineering and Architectural Schools. By Prof. S. Edward Warren. Numerous plates...........................8vo, cloth, 2 50

MATHEMATICS.

GENERAL PROBLEMS IN THE LINEAR PERSPECTIVE OF FORM, SHADOW, AND REFLECTION; OR, THE SCENOGRAPHIC PROJECTIONS OF DESCRIPTIVE GEOMETRY.
By Prof. S. Edward Warren, C.E. New edition, revised. Plates..8vo, cloth, $3 50
"It seems to me that your works only need a thorough examination to be introduced and permanently used in all the Scientific and Engineering Schools."—*Prof. J. G. Fox, Collegiate and Engineering Institute, N.Y. City.*

ELEMENTS OF PLANE AND SOLID FREE HAND GEOMETRICAL DRAWING.
With LETTERING and some elements of Geometrical Ornamental Design, including the Principles of Harmonic Angular Ratios, etc. In three parts. Part I—Plane Drawing, or from the Flats. Part II. Solid Drawing, or from the Round. Part III. Elements of Geometric Beauty. For Draughtsmen and Artisans, and Teachers and Students of Industrial and Mechanical Drawing. By S. Edward Warren, C.E., with 12 folding plates and many woodcuts...............12mo, cloth, 1 00

DESCRIPTIVE GEOMETRY,
As applied to the Drawing of Fortifications and Stereotomy (Stone Cutting). For the use of the Cadets of the U. S. Military Academy. By D. H. Mahan. Plates.........8vo, cloth, 1 50

AN ELEMENTARY COURSE IN DESCRIPTIVE GEOMETRY.
By Solomon Woolf, A.M., Professor of Geometry and Drawing in the College of the City of New York. Third Edition.
Royal 8vo, cloth, 8 00
"A most scholarly and conscientious work, both precise and concise (yet ample) in treatment. The language is clear and thoughtful, the arrangement methodical and natural. The work is very thoroughly and beautifully illustrated. The cuts number 289, and most of them are models of clearness. What is of prime importance, they are well *lettered*, apparently being done by the 'wax process,' which admits of lettering in *type*.
"The book is admirably printed on excellent paper, and in the clearest of type. In short, it is, in all respects, a pattern of what a text-book ought to be."—*Journal of Franklin Institute.*

AN ELEMENTARY COURSE IN THE THEORY OF EQUATIONS.
For the Junior Year in Colleges. By C. H. Chapman, Johns Hopkins University, Baltimore. Treating of Determinants, Algebraic Equations and the Computation of the Real Roots of Numerical Equations.........................12mo, cloth, 1 50

DRAFTING INSTRUMENTS AND OPERATIONS.
In four Divisions.
    Div. I.—INSTRUMENTS AND MATERIALS.
    Div. II.—FUNDAMENTAL OPERATIONS.
    Div. III.—PLANE PROBLEMS AND PRACTICAL OPERATIONS.
    Div. IV.—ELEMENTS OF TASTE IN GEOMETRICAL DRAWING.
A Text-book for Schools and Artisan Classes, and for self-instruction. Thoroughly revised, with additions, by S. Edward Warren, C.E............................12mo, cloth, 1 25

MATHEMATICS.

A MANUAL OF ELEMENTARY PROJECTION DRAWING.
Involving three Dimensions, designed for use in High Schools, Academies, Engineering Schools, etc., and for the self-instruction of Inventors, Artisans, etc. In five Divisions—I. Elementary Projections. II. Details of Construction in Masonry, Wood, and Metal. III. Rudimentary Exercises in Shades and Shadows. IV. Isometrical and Oblique or Pictorial Projections. V. Elementary Structural Drawing. By S. Edward Warren, late Professor in the Rensselaer Polytechnic Institute. Fourth edition, revised and enlarged. Plates. 12mo, cloth, $1 50

A MANUAL OF ELEMENTARY PROBLEMS IN THE LINEAR PERSPECTIVE OF FORM AND SHADOW;
Or, The Representation of Objects as they appear, made from the Representation of Objects as they are. In two parts—1. Primitive Methods, with an Introduction. II. Derivative Methods, with some Notes on Aërial Perspective. By Prof. S. Edward Warren. Wood engravings.......... 12mo, cloth, 1 00

GENERAL PROBLEMS OF SHADES AND SHADOWS,
Formed both by Parallel and Radial Rays; and shown both in Common and in Isometrical Projections, together with the Theory of Shading. By Prof. S. Edward Warren. With folding plates.................................8vo, cloth, 3 00

TRIGONOMETRY, ANALYTICAL, PLANE, AND SPHERICAL.
With Logarithmic Tables. By Prof. De Volson Wood. Fourth edition.............................12mo, cloth, 1 00
"The author need make no apology for the publication of this textbook."—*School Journal.*

ELEMENTS OF TRIGONOMETRY.
With Logarithmic and other Tables. By Lt. H. H. Ludlow and Prof. Edgar W. Bass, U.S. Military Academy. 8vo, cloth, 3 00

LOGARITHMIC, TRIGONOMETRIC AND OTHER MATHEMATICAL TABLES.
By Lt. Henry H. Ludlow, U.S.A., with the co-operation of Prof. Edgar W. Bass, U. S. Military Academy.....8vo, cloth, 2 00

ELLIPTIC FUNCTIONS.
By Prof. Arthur L. Baker, Stevens School. 8vo, cloth....... 1 50
"I cannot speak too highly of this work, not only as an admirable presentation of the subject, but also as suitable for a college text-book, from the concise, simple, and systematic way in which this high subject is treated."
—Prof. J. E. DAVIES, *Univ. of Wisconsin.*

QUADRATURE OF THE CIRCLE.
Containing demonstrations of the errors of Geometers in finding the Approximations in use. With an Appendix and Practical Questions on the Quadrature applied to the Astronomical Circles; to which are added Lectures on Polar Magnetism and Non-existence of Projectile Forces in Nature. By John A. Parker.... .... 8vo, cloth, 2 50

MATHEMATICS.

CONVERSION TABLES
Of the Metric and British or United States WEIGHTS AND MEASURES. With an Introduction by Robt. H. Thurston, A.M., C.E............ ....8vo, cloth, $1 00

METROLOGY.
Based upon recent and original discoveries. A challenge to the Metric System, and an earnest word with the English speaking peoples on their Ancient Weights and Measures. By Chas. A. L. Totten, M.A.................................... 2 50

THE SOLUTION OF THE PYRAMID PROBLEM.
Or, Pyramid Discoveries, with a New Theory of their Ancient Use. By Robt. Ballard, C.E. With plates. 8vo, cloth, 1 50

THE METROLOGICAL SYSTEM OF THE GREAT PYRAMID.
An Answer to Prof. Piazzi Smythe and other Pyramid students. By F. A. P. Barnard, President of Columbia College, N. Y.
8vo, cloth, 1 50

We make No Introductory Rates. A single copy will be sent to a Teacher or Professor for examination with view to Introduction at one-third discount. Students and Colleges supplied at twenty per cent. discount, carriage extra.

JOHN WILEY & SONS,
53 E. 10th St., New York.

www.ingramcontent.com/pod-product-compliance
Lightning Source LLC
Chambersburg PA
CBHW030254240426
43673CB00040B/967